Agricultural Statistics

AGRICULTURAL STATISTICS

(A Guide for Competitive Examinations)

by :

Kushwaha K.S.

M.Sc.,P.S.C.C., Ph.D. (Statistics)
Recipient of Dr. Radha Krishnan Award (1992)
Associate Professor (Statistics)
Department of Mathematics & Statistics
College of Agriculture Engineering, J.N.K.V.V, Jabalpur-482 004, M.P. (India)

nipa® books for everyone
PAPERBACKS

New Delhi – 110 034

nipa® books for everyone
PAPERBACKS

NEW INDIA PUBLISHING AGENCY
101, Vikas Surya Plaza, CU Block, LSC Market
Pitam Pura, New Delhi – 110 034, India
Email: info@nipabooks.com
Web: www.nipabooks.com

For customer assistance, please contact
Phone: + 91-11-27 34 17 17 Fax: + 91-11- 27 34 16 16
E-Mail: feedbacks@nipabooks.com

ISBN: 978-93-81450-31-4

Composed and Designed by NIPA.

Dedicated to Beloved Parents

Late Sh. B.S. Kushwaha
&
Smt J. Kushwaha

PREFACE

The book entitled "Agricultural Statistics" has been designed for all U.G. and P.G. Students of "Pure Statistics, Agricultural Statistics, Biological & Social Sciences" and those who have to appear in competitive examinations of I.S.S., S.S.S., State's P.S.C.' and I.A.S. This book is also useful for faculties of "Department of Statistics" of Indian Universities. The book is the outcome of 28 years of teaching experience of U.G., P.G. and Ph. D. students of different disciplines of Agriculture, Agil. Engg. and Agril. Statistics. in J.N.K.V.V. Jabalpur. The content of the book covers the syllabus on the topic "Statistical Methods."

The book contains 15 chapters, out of which the chapters (1-4) deal with Discriptive Statistics, (5) Probability Theory, (6-10) Random variables and Distribution Functions, (11-14) Theory of Tests of Significance based on small and large samples and (15) Simple & Multiple Correction - Regression Analysis. In each chapter, three types of question" True/ False, Fill in the Blanks and Multiple Choice Questions & Answers" alongwith the "Key Answers" have been provided.

During the preparation of this manuscript, the author has incorporated the fruitful academic advices provided by Dr. H.P. Singh, Prof & Head, School of Studies in Statistics, Vikram University, Ujjain and Dr. B.B. Khara, Prof. (Statistics) Department of Statistics, B.H.U., Varanasi. It is expected to have a good popularity due to its usefulness amongst students and its users.

The author is very much thankful and extends his honourable honour of higher order to Prof. Kalloo Gautam, The Vice -chancellor

(J.N.K.V.V Jabalpur) for his kind permission to hand the manuscript to "New India Publishing Agency" New Delhi to publish it in the present shape of a book.

The author is thankful to Sh. Harishchand Maurya; Smt. Sangeeta Maurya and their lovely children (Mr. Rohit & Ku. Richa) Maurya for providing moral support during the preparation of the manuscript.

The author acknowledges very strongly to his wife Smt. Veena Kushwaha, daughter Miss Akansha Kushwaha and son Mr. Utkarsh Kushwaha who cared a lot during the preparation of this manuscript.

Last but not the least, the author is very much thankful to the entire staff of "New India Publishing Agency," New Delhi for taking keen interest to publish the manuscript in its present shape.

Although the author has tried his best in designing the subject matter very nicely but that can be further improved based on the suggestions received from its readers & users. Therefore, the fruitful academic suggestions received from all sides will be welcomed and will be highly appreciated by the author.

20th January, 2012

K.S. KUSHWAHA
Jabalpur

CONTENTS

Chapter - 1

Introduction to Statistics

Preliminaries

Section-A : Write True / False

Q.1 Statistics in singular is used to denote a subject (branch) of science.

Q.2 Statistics in plular is used to denote data or methods.

Q.3 The data first time collected is known as primary data.

Q.4 Primary data is collected from published report.

Q.5 Editing of data is not necessary.

Q.6 Consistency must be maintained in data.

Q.7 Baised and unbiased errors are known as statistical error.

Q.8 Arrangement of data in different, rows & columns is known as classification.

Q.9 Inclusive class interval is also known as discontinuous class.

Q.10 Exclusive class interval is same as continuous class interval.

Q.11 Class interval is used for numerical classification.

Q.12 Number of repetition of an item is known as frequency.

Q.13 There are two type of cumulative frequency.

Q.14 Ordinary frequency and simple frequency are the same thing.

Q.15 Cumulative frequency is prepared from simple frequency.

Q.16 Cumulative frequency can be changed into simple frequency.

Q.17 Putting data in tabler form is known as tabulation.

Q.18 Tally mark is used to prepare the frequency distribution.

Q.19 Statistical data are quantitative.

Q.20 Secondary data are more reliable than the primary data.

Q.21 Less than cumulative frequency has decreasing trend.

Q.22 Statistics is liable to be misued.

Q.23 Direct personal investigation is the method of collection of primary data.

Q.24 The editing of data is not essential before its use for analysis work.

Q.25 Completeness, consistency, accuracy and uniformity are the points w.r. to which the editing of data is done.

Q.26 The process of arranging the data into groups or classes according to some characteristics is known as classification.

Q.27 Classification is the preliminary to its tabulation.

Q.28 The classes 0-4, 5-9, 10-14, 15-19, 20-24, etc. are called exclusive classes.

Q.29 The classes 0-5, 5-10, 10-15, 15-20, 20-25, etc. are called inclusive classes.

Q.30 Tabulation of data is preliminary to its analysis and interpretation.

Q.31 The classification of data according to time or sequence of time is known as chronological classification.

Q.32 A table showing the distribution of frequencies over the classes is known as the frequency distribution.

Section-B : Fill in the Blanks

Q.1 The word statistics is used in singular to denote a .... of science.

Q.2 The statistics in plular is used to denote the .... or

Q.3 The data collected first time by an investigator is known as .... data

Q.4 The primary data is collected from the ....

Q.5 Biased and unbiased errors are known as .... errors.

Q.6 Arrangement of data in different rows and columns is known as ....

Q.7 Inclusive class intervals are also known as .... class intervals.

Q.8 Exclusive class intervals are also known as .... class intervals.

Q.9 Number of repetition of an item is known as .... of that item.

Q.10 Cumulative frequencies are of .... types.

Q.11 Ordinary frequency is also known as .... frequency.

Q.12 Cumulative frequency is prepared from .... frequency.

Q.13 Cumulative frequencies can be transformed back to .... frequencies.

Q.14 Putting classified data in tabular form is known as ....

Q.15 Tally mark is used to prepare a .... distribution.

Q.16 Less than cumulative frequencies has .... trend.

Q.17 More than cumulative frequencies has .... trend.

Q.18 Direct personal investigation is the method of collection of .... data.

Q.19 Classification is preliminary to its ....

Q.20 The .... of data is essential before its use for analytical work.

Q.21 Completeness, consistency, accuracy and uniformity are the characteristics of the data w.r. to which the .... of data is done.

Q.22 Classification of data is preliminary to its ....

Q.23 The classes 0-4, 5-9, 10-14, 15-19, 20-24 etc. are known as .... classes.

Q.24 The classes 0-5, 5-10, 10-15, 15-20, 20-25 etc. are known as .... classes.

Q.25 Tabulation of data is preliminary to its .... and ....

Q.26 Classification of data according to time or sequence of time is known as .... classification.

Q.27 A table showing the distribution of frequencies over the different classes is known as .... distribution.

Section -C : Multiple Choice Questions

Q.1 Data taken from the journal will be considered as

(a) Primary data (b) Secondary data

(c) Primary and Secondary data (d) None of the above

Q.2 For the middle values given as below 7, 10, 13, 16, 19, 22, etc. the first class of distribution is

(a) 5.5 - 8.5 (b) 4-10

(c) 5.5 - 9.5 (d) 5.5 - 7.5

Q.3 Frequency of a value or class interval is always considered as

(a) In fraction (b) An integer

(c) In percentage (d) None

Q.4 The class interval (length) of the class 5-9.5, 10-14.5, 20-24.5 etc. is equal to

(a) 5.0 (b) 4.5

(c) 5.5 (d) 4.0

Q.5 Ogive curves occur for
(a) Less than type cumulative frequency distribution
(b) More than type cumulative frequency distribution
(c) Both (a) & (b)
(d) None of the above

Q.6 Ogive of less than type and more than type intersects at
(a) Origin (b) Mode
(c) Median (d) Mean

Q.7 With the help of ogive curve, one can estimate the
(a) Median (b) Mode
(c) G.M. (d) A.M.

Q.8 Histogram can always be drawn for
(a) An inclusive class interval
(b) An exclusive class interval
(c) An equal and unequal class intervals
(d) Both (b) & (c)

Q.9 Less than c.f. should always correspond to
(a) Lower limits of class intervals of the distribution
(b) Upper limits of class intervals of the distribution
(c) Mid values of the class intervals of the distribution
(d) None of the above

Q.10 More than c.f. in the preperation of ogive must always correspond to the
(a) Lower limits of the class intervals
(b) Upper limits of the class intervals
(c) Mid values of the class intervals
(d) None of the above

Q.11 If the mid points of the tops of rectanges in a histogram are joined by straight lines then the palygone obtained is known as
(a) Frequency curve (b) Frequency polygone
(c) Histogram (d) None

Q.12 If the mid points of the tops of rectanges in a histogram are joined by a smooth hand curve then this curve is known as :
(a) Frequency polygone (b) Frequency curve
(c) Mid curve (d) None

Q.13 Frequency polygone can be drawn for continuous frequency distribution

(a) With the help of histogram
(b) Without the help of histogram
(c) With and without the help of histogram
(d) None

Q.14 In case of frequency distribution in the form of class intervals, the frequency density is defined as

(a) Frequency of the class/Length of the class
(b) Length of the class/Frequency of the class
(c) Frequency of the class/Average of class limits
(d) Frequency of the class/Mid value of the class

Q.15 In case of unequal class intervals of frequency distribution, the histogram can be plotted with the help of

(a) Relative frequencies (b) Frequency densities
(b) Frequency (d) Cumulative frequencies

Q.16 The word "statistics" can be considered as in

(a) Singular (b) Plular
(c) In singular & plular both (d) Undecided

Q.17 Editing of data before analysis is

(a) Necessary (c) Not necessary
(b) Essential (d) Both (a) & (c)

Q.18 For the valid conclusion to be drawn about the information contained in data, the statistical data must be

(a) Inconsistent (b) Consistent
(c) Relatively uniform (d) Both (b) & (c)

Q.19 Inclusive class intervals are of

(a) Continuous nature (b) Discontinous nature
(c) Discreate nature (d) Both (b) & (c)

Q.20 In case of frequency distribution, the number of repetition of an item is known as

(a) Replication (b) Frequency
(c) Relative frequency (d) Frequency density

Q.21 The process of putting data in tabler form is known as

(a) Classification (b) Tabulation

(c) Edition (d) None

Q.22 Less than cumulative frequencies are of

(a) Increasing trend (b) Decreasing trend

(c) Both of (a) & (b) type (d) None

Q.23 More than cumulative frequencies are of nature

(a) Decreasing (b) Increasing

(c) Both (a) & (c) (d) None

Q.24 Direct personal investivation is the method of collection of

(a) Secondary data (b) Primary data

(c) Experimental data (d) Both (b) & (c)

Q.25 The editing of statistical data is done to maintain the following characteristic of data

(a) Completeness (b) Consistency

(c) Uniformity (d) All the three

Q.26 Classification of data is preliminary to its

(a) Tabulation (b) Analysis

(c) Interpretation of results (d) Collection

Q.27 Tabulation of data is preliminary to its

(a) Analysis (b) Interpretation

(c) Collection (d) Compilation

Q.28 Classification of data according to time or sequence of time is known as

(a) Periodical classification

(b) Chronological classification

(c) Qualitative classification

(d) Manifold classification

Key Answers

Section-A : True False

1. True	2. True	3. True
4. Flase	5. False	6. True
7. True	8. True	9. True
10. True	11. True	12. True
13. True	14. True	15. True
16. True	17. True	18. True
19. True	20. False	21. False
22. True	23. True	24. False
25. True	26. True	27. True
28. False	29. False	30. True
31. True	32. True	

Section-B : Fill in the Blanks

1. Branch	2. Data, Methods
3. Primary	4. Field
5. Statistical	6. Classification
7. Discontinuous	8. Continuous
9. Frequency	10. Two
11. Simple	12. Simple
13. Simple	14. Tabulation
15. Frequency	16. Increasing
17. Decreasing	18. Primary
19. Tabulation	20. Editng
21. Editing	22. Tabulation
23. Inclusive	24. Exclusive
25. Analysis, Interpretation	26. Chronological
27. Frequency	

Section-C : Multiple Choice Questions and Answers

1. (b)	2. (a)	3. (b)
4. (b)	5. (c)	6. (c)
7. (a)	8. (d)	9. (b)

Section-C : Multiple Choice Questions and Answers

10.	(a)	11.	(b)	12.	(b)
13.	(c)	14.	(a)	15.	(b)
16.	(c)	17.	(d)	18.	(d)
19.	(d)	20.	(b)	21.	(b)
22.	(a)	23.	(a)	24.	(b)
25.	(d)	26.	(a)	27.	(a)
28.	(b)				

❑ ❑ ❑

Chapter - 2

Diagrammatic and Graphic Presentation of Data

Section-A : Write True / False

Q.1 A better and more beautiful way to present the data in a more visual form can be done by means of diagram.

Q.2 Description of data through diagrams and graphs is better for layman.

Q.3 Diagram does not register a meaningful impression almost before we start to think about it.

Q.4 Diagrams are never used by the business man and government agencies for advertisement and educating people.

Q.5 Diagrams do not give visual presentation of data.

Q.6 The diagrams put a rememberance for a longer time in mind of human being.

Q.7 Diagrams facilitate comparison and reveal pattern of data.

Q.8 Diagrams show approximate values of items in data.

Q.9 Diagrams are more attractive, appealing and impressive than tables.

Q.10 Diagrams can be more easily compared and interpreted than the tables.

Q.11 Tables contain precise figures while diagrams indicate approximate figures.

Q.12 Tables give more precise idea of the situation than the diagrams.

Q.13 Tables are useful for a statistician for further analysis.

Q.14 Bar diagram is an example of one dimensional diagrams.

Q.15 Rectangles, squares, circles and pie diagrams are the examples of two dimensional diagrams.

Q.16 Simple bar diagrams are suitable for discrete variable (x).

Q.17 For birth rates, death rates and production of a commodity etc. a simple bar diagram is highly useful.

Q.18 Data consisting of production of different commodities of a firm over a period of time can not be represented by a simple bar diagram.

Q.19 If a phenomenon is consists of different components, then "subdivided bar diagrams" or "component bar diagrams" can be used to represent such data.

Q.20 The classwise strength of students of a college or the crop wise production of food grains over a period of time can be suitably represented by "subdivided bar diagrams".

Q.21 The different parts of the subdivided bar diagrams are shown by the same shade or same colour.

Q.22 If the components of a phenomenon are expressed in percentage to the total then the corresponding bar diagram is known as "percentage chart".

Q.23 All the bars of percentage bar chart will not be of the same height.

Q.24 Data which are built up of two or more components, are represented by "Multiple bar diagrams".

Q25 In multiple bar diagrams, the component figures are shown as separate bars adjoining each others.

Q.26 In multiple bar diagrams, the different components are shown in different shades or different colours.

Q.27 Multiple bar diagrams are used where "only changes in actual values" of the component figures are required to be shown but the over all total value is "not of importance"

Q.28 Multiple and component bar charts can only be used when there are not more than three or four components.

Q.29 In one dimensional diagrams, only length is of significance.

Q.30 In two dimensional diagramms both of the length and width are of equal importance.

Q.31 In two dimensional diagrams, the area of each of them is proportional to the size what it represents.

Q.32 When there are wide variation among the values of the data, the bar diagrams are quite suitable.

Q.33 In case of there being wide variation among values of data, the square diagrams can be used to depict that data.

Q.34 Squar diagrams must be drawn in such a way that their areas are proportional to the values what they represent.

Q.35 In case of drawing the square diagrams, the sides of the squares are decided (taken) by taking the square roots of all values of the data.

Q.36 In case of drawing square diagrams, the sides of the squares are proportional to the values of the data.

Q.37 The square diagrams are not very much suitable when there is a wide variation among the values of data.

Q.38 To overcome the difficulties of squared diagrams, the circles can be conveniently drawn.

Q.39 The area of a circle is proportional to the square of its redious.

Q.40 In case of circle diagrams, the square roots of values of the data are calculated and the circles with the radious proportional to these calculated values are constructed.

Q.41 When a phenamenon is built of more than three or four components then subdivided bar diagrams or multiple bar diagrams are not suitable to draw for such data.

Q.42 In case of more than three or four components of the phenamenon, the pie diagrams are more suitable to carry the proper impression of such data.

Q.43 Pie diagram is also known as angular diagram.

Q.44 An angular diagram is a circle divided by radial lines into sections called sectors.

Q.45 If the total of values in data is equal to v and a component value is equal to vi then the angle require to draw the corresponding sector $= \left(\frac{Vi}{V} \text{ x } 360\right)^{o}$

Q.46 A graph is a visual representation of data by a continuous curves on a squared paper.

Q.47 In case of plating a graph, an independent variable should always be drawn along horizental axis.

Q.48 Histogram is a graph of one variable.

Q.49 Birth rates and death rates drawn on the same graph paper, are the examples of graphs of "two variables".

Q.50 Cubical, cylindrical and spherical diagrams are the examples of three dimensional diagrams.

Q.51 Pictograms are shown by pictures.

Q.52 Cartograms are the statistical maps in which different symbals are used to represent the magnitude of values in data.

Q.53 Ogive is a curve drawn between variable values and cumulative frequencies.

Section - B : Fill in the Blanks

Q.1 A better and more beautiful way to present the data in a more visual form can be done by means of ....

Q.2 Description of data by means of diagrams and graphs is better for ....

Q.3 A meaningful impression about data is registered by .... almost before we start to think about it.

Q.4 Diagrams are used by the .... man and .... agencies for advertisement and educating people.

Q.5 Diagrams give .... impression (presentation) of data.

Q.6 Diagrams put a rememberance for a .... time in mind of human being.

Q.7 Diagrams facilitate .... and reveal pattern of data.

Q.8 A more attractive, appealing and impressive way than table to present data can be done by means of ....

Q.9 Data can be compared and intepreted more easily than tables by means of ....

Q.10 Diagrams show .... values of items in data.

Q.11 Tables indicate more .... feagures while .... indicate approximate feagures of data.

Q.12 Tables containg the values of data is useful for a .... for further analysis.

Q.13 Bar diagram is an example of .... dimensional diagram to present data.

Q.14 Rectangles, squares, circles and pie diagrams are the examples of two dimensional diagrams.

Q.15 Simple bar diagrams are suitable for .... variables.

Q.16 To present data on birth rates, death rates, and production of a commodity etc., a simple .... is highly useful.

Q.17 Data on production of different commodities of a firm over a period of time can be represented by " ...." .

Q.18 If a phenomenon is consists of different components, then " ...." bar diagrams or " ...." bar diagrams can be used to present such data.

Q.19 The different parts of subdivided bar diagrams are shown by " ...." shades or " ...." colours.

Q.20 If the components of a phenomenon are expressed in percentage to the total then the corresponding bar diagrams is known as " .... " chart.

Q.21 All the bars of the percentage bar charts will be of the .... height.

Q.22 Data which are built of two or more components, are presented by " .... " diagrams.

Q.23 In multiple bar diagrams, the component figures are shown as " ...." adjoining each others.

Q.24 In multiple bar diagrams, the different components are shown in " .... " shades or " .... " colours.

Q.25 Multiple and component bar charts can only be used where there are not more than " .... " or " ...." components.

Q.26 In one dimensional bar diagrams, only " .... " is of importance.

Q.27 In two dimensional diagrams, both of the " ...." and " .... " are of equal importence.

Q.28 In " .... " dimensional diagrams, the area of each of them is proportional to the size what they represent.

Q.29 It the wide variation exists among the values of data then to depict such data, the suitable diagram is the " ...." diagram.

Q.30 The square diagrams must be drawn in such a way that their " ...." are proportional to the "..............................." of data what they represent.

Q.31 In case of drawing the " .... " diagrams, the length of sides of the squares are decided by taking the " ...." routs of all values of data.

Q.32 To overcome the dificulties of " .... " diagrams, the circle can be drawn conveniently.

Q.33 The area of a " .... " diagram is proportional to the square of its redious.

Q.34 In drawing circular diagrams, the " .... " of the values of data are calculated and the circles with the

" .... proportional to these calculated values are constructed.

Q.35 When phenomenon is built of more than three or four components, them " .... " bar diagrams or " ...." bar diagrams are not suitable to depict such data.

Q.36 Pie diagram is also known as " .... " diagram.

Q.37 An angular diagram is a " .... " divided by radial lines into sections called " ...."

Q.38 If the total of the values in data is equal to "V" and a component value is equal to "Vi" then angle required to draw the sector corresponding to that value vi is equal to " ...."

Q.39 A graph is a visual presentation of data by a continuous " ...." on a " ...." paper.

Q.40 In case of plating a graph, an independent variable should always been plotted along " .... " axis.

Q.41 Histogram is a graph of .... variable.

Q.42 Birth rates and death rates drawn on the same graph paper, are the graphs of " .... " variables.

Q.43 Cubical, cylindrical and spherical diagrams, are the examples of " .... " dimensional diagrams.

Q.44 Pictograms are shown by " ...." .

Q.45 The statistical maps in which different symbals are used to represent the magnitude of values in data, are called " ...." .

Q.46 Ogive is a curve drawn between variable values and " ...." frequencies.

Section -C : Multiple Choice Questions

Q.1 A better and more attractive way to present the data in a more visual form can be done by means of

(a) Diagram (b) Graph

(c) Curve (d) Polygon

Q.2 Description of data through diagrams and graphs is better for
(a) Statistician (b) Layman
(c) Scientist (d) Chemist

Q.3 Diagrams are used by business man for
(a) Saling purpose (b) Advertisement
(c) Forecasting (d) Showing profit

Q.4 Diagrams give the visual presentation of
(a) Data (b) Numerical values
(c) Both (a) & (b) (d) None

Q.5 Diagram facilitates the
(a) Comparison (b) Pattern of data
(c) Analysis (d) Both (a) & (b)

Q.6 Diagrams show
(a) Approximate value of items in data
(b) Exact value of items in data
(c) Analysis of data
(d) None of the above

Q.7 Diagrams are more attractive, appealing, and impressive than
(a) Table (b) Numerical values
(c) Freanency curve (d) All the above

Q.8 Diagrams can be more easily compared and interpreted than
(a) Tables (b) Numerical values
(c) Mathematical expressions (d) All the three

Q.9 Tables contain precise figures while diagrams indicate
(a) Approximate figures (b) Exact figures
(c) Frequency curve (d) Frequency polygon

Q.10 Tables are useful for a statistician for
(a) Further analysis (b) Ploting the curves
(c) Fitting the statistical model (d) All the three

Q.11 Bar diagram is an example of
(a) One dimensional diagrams
(b) Two diamensional diagrams
(c) Particular case of histograms
(d) Both (a) & (b)

Q.12 Rectangles, squares, circles and pie diagrams are the examples of
(a) Two diamensional diagrams
(b) The same dimensional diagrams
(c) Different diamensional diagrams
(d) Both (a) & (c)

Q.13 Simple bar diagrams are suitable for
(a) Discreate variables (b) Continuous variables
(c) Discreate distributions (d) Both (a) & (c)

Q.14 Most suitable diagram for birth rate, death rate & production of a commodity etc. is
(a) Simple bar diagram
(b) One dimensional bar diagram
(c) Histogram
(d) Both (a) & (b)

Q.15 Production of different commodities of a firm over different periods can be represented by
(a) Simple bar diagrams (b) Multiple bar diagrams
(b) Subdivided bar diagrams (d) Percentage bar diagrams

Q.16 A phenomenon consisting of different components can be represented by
(a) Subdivided bar diagrams (b) Multiple bar diagrams
(c) Component bar diagrams (d) Both (a) & (c)

Q.17 The number of classwise students of a college or cropwise production of food grains over a period of time, can be represented suitably by
(a) Sub-divided bar diagrams (b) Component bar diagrams
(c) Histograms (d) Both (a) & (b)

Q.18 The different components of the subdivided bar diagrams are shown by the
(a) Different shade (b) Different colours
(c) Same shade or colour (d) Both (a) & (b)

Q.19 All the bars of percentage bar charts will be of the
(a) Same height (b) Different height
(c) Different colours (d) Same colour

Q.20 Data built up of two or more components are represented by
(a) Simple bar diagrams (b) Subdivided bar diagrams
(c) Multiple bar diagrams (d) None

Q.2 In multiple bar diagrams, the different components figures are shown by
(a) Separate bars adjoining each other (b) Histograms
(c) Subdivided bar diagrams (d) Pie charts

Q.22 In one dimensional diagrams, the importent measure of the diagram is the
(a) Length (b) Area
(c) Rdious of the diagram (d) Diameter

Q.23 In two dimensional diagrams, the length and width
(a) Are of equal importance
(b) Are of unequal importance
(c) Have no importance
(d) Do not contribute to area

Q.24 If there is wide variation among values of data, then suitable diagram to depict them is the
(a) Bar diagram (b) Rectangle
(c) Square (d) Multiple bars

Q.25 In case of square diagrams, their areas must be
(a) Inversely proportional to the values
(b) Directly proportional to the values
(c) Proportional to the square root of the valves
(d) Both (b) & (c)

Q.26 In square diagrams, the sides of the squares are proportional to
(a) Square root of values (b) Values
(c) Squares of the values (d) Both (a) & (c)

Q.27 To overcome the difficulties of squared diagrams, the convenient diagram are the
(a) Rectangles (b) Circles
(c) Multiple bars (d) Subdivided bars

Q.28 The area of circle diagrams are proportional to the
(a) Square of the redious
(b) Radious
(c) Diameter
(d) Perimeter

Q.29 In circle diagrams, the radious of the circles will be proportional to the
(a) Values of data
(b) Square roots of values
(c) Squares of values
(d) Both (b) & (c)

Q.30 If the data contain more than three or four components of phenomenon then the suitable diagram for that is
(a) Pie diagram
(b) Sub divided bars
(c) Multiple bars
(d) Both (b) & (c)

Q.31 Pie diagram is also known as
(a) Angular diagram
(b) Circular diagram
(c) Areal diagram
(d) Both (a) & (b)

Q.32 Angular diagram is a circle divided by
(a) Radious
(b) Radial lines
(c) Diameters
(d) Both (a) & (c)

Q.33 Let V be the total of values of data, then the angle for a particular value vi to draw pie diagram is equal to
(a) $\left(\frac{V}{vi} \times 360\right)^{o}$
(b) $\left(\frac{vi}{V} \times 360\right)^{o}$
(c) $\left(\frac{vi}{V} \times 180\right)^{o}$
(d) None

Q.34 A graph is a visual representation of data by a
(a) Continuous curve on a squared sheet
(b) Polynomial on a squared sheet
(c) Histogram on a squared sheet
(d) All the above

Q.35 Histogram is a graph of
(a) One variable
(b) Two variables
(c) Simple bars attached with each
(d) All the above

Q.36 Birth rates and death rates drawn on the same graph paper are the examples of graphs of

(a) Two variables (b) One variable

(c) Three variables (d) None of above

Q.37 Cubical, cylindrical and spherical diagrams are the examples of

(a) Three dimensions (b) Two dimensions

(c) One dimension (b) Simple bars

Q.38 Pictograms are shown by

(a) Pictures (b) Simple bars

(c) Histograms (d) Pie diagrams

Q.39 Cartograms are the statistical maps in which the magnitude of values are represented by

(a) Different symbals (b) Same symbal

(c) Same colour (d) Different colours

Q.40 Ogive is a curve drawn between

(a) Variable and cumulative frequencies

(b) Variable and simple frequencies

(c) Variable and frequency densities

(d) Variable and frequency functions

Key Answers

Section-A : True False

1.	True	2.	True	3.	False
4.	Flase	5.	False	6.	True
7.	True	8.	True	9.	True
10.	True	11.	True	12.	True
13.	True	14.	True	15.	True
16.	True	17.	True	18.	True
19.	True	20.	True	21.	False
22.	True	23.	False	24.	True
25.	True	26.	True	27.	True
28.	True	29.	True	30.	True
31.	True	32.	False	33.	True

34. True
35. True
36. True
37. False
38. True
39. True
40. True
41. True
42. True
43. True
44. True
45. True
46. True
47. True
48. True
49. True
50. True
51. True
52. True
53. True

Section-B : Fill in the Blanks

1. Diagrams
2. Layman
3. Diagram
4. Business, Government
5. Visual
6. Larger
7. Comparison
8. Diagrams
9. Diagrams
10. Approximate
11. Precise or acurate
12. Statistician
13. One
14. Two
15. Discrete
16. Bar diagram
17. Multiple bar diagrams
18. Multiple, Subdivided
19. Diferent, Different
20. Percentage
21. Same
22. Multiple
23. Simple bars
24. Different, Different
25. Three, Four
26. Height
27. Height, Width
28. Two
29. Square
30. Areas, Values
31. Square, Square
32. Square
33. Pie
34. Square root, radious
35. Multiple, Subdivided
36 Angular
37. Circle, Sectors
38 $\left(\frac{V_i}{V}\right) \times 360$
39. Curve, square
40. Horezental
41. One
42. Two
43. Three
44. Pictures
45. Cartograms
46. Cumulative

Section-C : Multiple Choice Questions

1.	(a)	2.	(b)	3.	(b)
4.	(c)	5.	(d)	6.	(a)
7.	(d)	8.	(d)	9.	(a)
10.	(d)	11.	(d)	12.	(d)
13.	(d)	14.	(d)	15.	(b)
16.	(d)	17.	(d)	18.	(d)
19.	(a)	20.	(c)	21.	(a)
22.	(a)	23.	(a)	24.	(c)
25.	(d)	26.	(a)	27.	(b)
28.	(a)	29.	(b)	30.	(a)
31.	(a)	32.	(b)	33.	(b)
34.	(d)	35.	(d)	36.	(a)
37.	(a)	38.	(a)	39.	(a)
40.	(a)				

❑❑❑

Chapter - 3

Measures of Central Tendency

Section-A : Write True / False

Q.1 Measures of central tendency do not give an idea about the concentration of values in the central part of the distribution.

Q.2 Measures of central tendency should be based on all the values.

Q.3 Measures of central tendency are not suitable for further mathematical treatment.

Q.4 Measures of central tendency should be affected as little as possible by fluctuation of sampling.

Q.5 Measures of central tendency should not be affected much by extreme observations.

Q.6 The arithmetic mean of a set of observations say $x_1, x_2, ..., x_n$ is defined as $\bar{x}_n = \frac{1}{n} \sum_{i=1} xi$

Q.7 In case of frequency distribution (fi, xi, i = 1, 2, ..., n), the A.M. is defined as $\bar{x} = \frac{1}{\sum fi} \sum_{i=1}^{n} fixi$

Q.8 In case of grouped or continuous frequency distribution, the value of x is taken as the middle value of the corresponding class.

Q.9 If di = (xi - A) then $\bar{x} = (A - \bar{d})$.

Q.10 In question (9), A taken as the middle part of the distribution will not be much more convenient in finding the value of $\bar{x}$.

Q.11 If di is defined as di = $\frac{xi - A}{h}$, then following result hold good $\bar{x} = (A - h\bar{d})$ in case of continuous frequency distribution having h as the common class length.

Q.12 The algebraic sum of the deviations of a set of values from their A.M. is not zero.

Q.13 The S.S. of the deviations of a set of values is minimum when taken about A.M.

Q.14 The A.M. of K composite series (xi, ni, i = 1, 2, ..., k) can be worked from the expression

$$\bar{x} = \left(\sum_{i}^{k} ni\,\bar{xi} \right) \Big/ \sum_{i}^{k} ni$$

Q.15 Of all the averages, the A.M. is affected least by flucluation of sampling.

Q.16 The A.M. is not a stable average.

Q.16 The A.M. can not be located graphically.

Q.17 In case of single value missing, the A.M. can not be obtained.

Q.18 The A.M. is very much affected by extreme values.

Q.19 In case of extreme values present in data, the A.M. gives a distorted picture of the distribution.

Q.20 The A.M. may lead to a wrong inference if the details of the corresponding data is not given.

Q.21 The A.M. can not be calculated if the extreme class is open i.e. of type < 10 or > 100.

Q.22 In case of extremely asymmetrical (skewed) distribution, usually the A.M. is not a suitable measure of location.

Q.23 If wi is attached as weight to the value xi, then the weighted mean $\bar{x}_w$ is defined as

$$\bar{x}_w = \sum_i^n wi\,xi \Big/ \sum_i^n wi$$

Q.24 Median of a distribution is the value of the variable which devides it into two equal parts.

Q.25 Median of a set of values is the value which exceeds and is exceeded by the same number of values.

Q.26 The median of a set of value is the value that the number of values above it is equal to the number of values below it.

Q.27 In case of even number of values, infact any value lying between two middle values can be taken as medium.

Q.28 In case of even number of values, by convention, we take the mean of two middle values as the medium.

Q.29 Median of a set of observations is the positional average.

Q.30 Median can be calculated for a distribution with open end classes.

Q.31 In case of even number of values in the data, the median can not be determined exactly.

Q.32 Median of a set of values, is not based on all values.

Q.33 Median is not amenable to algebraic treatment.

Q.34 Median is affected much more by fluctuation of sampling as compared with mean.

Q.35 To deal with the qualitative data (honesty, beauty, intelligence, etc.) median is the only average to be used.

Q.36 Median is used to determine the typical value in the problems concerning wages, distribution of wealth etc.

Q.37 For a continuous frequency distribution, the class corresponding to the c.f. just $> \frac{N}{2}$ is known as the median class.

Q.38 In case of continuous frequency distribution, the median can be calculated using the formula

$$Md = l_1 + \frac{h}{f_{md}}\left(\frac{N}{2} - cf_p\right) = l_1 + \left(\frac{N}{2} - cf_p\right)\frac{(l_2 - l_1)}{f_{md}}$$

where l_1, l_2, cf_p & f_{md} have usual meanings.

Q.39 Most frequently occurred value in a given set of values, is known as the "Mode" of the distribution or data.

Q.40 If the maximum frequency is repeated in a given set of discreate values, the method of grouping is applied to get the value of mode.

Q.41 If the maximum frequency occurs in the very beginning or at the end of distribution the suitable method to determine the mode is the "method of grouping".

Q.42 In case if there are irregularities in the distribution then mode is determined by the method of grouping.

Q.43 In case of continuous frequency distribution, the mode is determined by the formula

$$\text{Mode} = l_1 + \frac{(f_{mo} - f_o)(l_2 - l_1)}{(2f_{mo} - f_1 - f_2)}$$

where (l_1,l_2), and (f_{mo}, f_1, f_2) are the (lower and upper limits of model class) and (frequencies of model class, class preceeding to model class and succeeding class to the model class).

Q.44 In case of irregularities in the distribution or maximum frequency being repeated or maximum frequency occuring in the very beginning or at the end of distribution, the model class is determined by the method of grouping and formula written in Q(43) is applied to get the value of mode.

Q.45 In case if the distribution is moderately asymmetrical, then for mean, median & mode, the following empirical relation reported by karl pearson among them exists. Mode = 3 Median - 2 Mean

Q.46 Mode is not affected by extreme values at all.

Q.47 Mode can be obtained easily from open and classes.

Q.48 Some distributions may come across with two modes and are said to be bi-model.

Q.49 Mode is affected to a greater extent, by fluctuation of sampling as compared with mean.

Q.50 Mode is the average to be used to find the ideal size in business forecasting, in manufacture of ready made garments, shoes, etc.

Q.51 Geometric mean of a set of n observations say, $x_1, x_2 \ldots\ldots x_n$ is defined as

$$G=\left(\prod_{i=1}^{n} xi\right)^{\frac{1}{n}}$$

Q.52 Let (n_1 & n_2) be the sizes and (G_1 & G_2), the geometric means of two series respectively, then the G.M. of the combined series is given by

$$\text{Log}\,G=\frac{\left(n_1 \log G_1 + n_2 \log G_2\right)}{\left(n_1 + n_2\right)}$$

Q.53 The G.M. is not affected much by fluctuation of sampling.

Q.54 G.M. is suitable for further mathematical treatment.

Q.55 G.M. gives more weight to small items.

Q.56 If one value of a set of values becomes equal to zero then G.M. of that data will be equal to zero.

Q.57 If one value of a set of values becomes negative then the G.M. of such data will be imaginary regardless of the magnitude of other items.

Q.58 The harmonic mean (H) of n values say $x_1, x_2 \ldots\ldots x_n$ is defined as

$$H=\frac{1}{\frac{1}{n}\sum_{i}^{n}(1/xi)}$$

Q.59 In case of frequency distribution, (fi, xi, i = 1, 2, ... n), the H.M. is defined as

$$H=\frac{1}{\frac{1}{n}\sum_{i}^{n}(fi/xi)},\quad N=\sum_{i=1}^{n} fi$$

Q.60 60 If (xi, i = 1, 2, ..., n) are values with weights (Wi, i = 1, 2, ..., n) respectively then weighted

$$H = \frac{\sum^{n} wi}{\sum_{i=1}^{n}\left(\frac{wi}{xi}\right)}$$

Q.61 H.M. is not affected much by fluctuation of sampling. If equal distances are covered per unit of time with speeds equal to V_1, V_2, ... V_n say, then average speed is given by the H.M. of V_1 V_2, ... V_n i.e.

$$\text{Average speed} = \frac{n}{\sum_{i}^{n}\left(\frac{1}{vi}\right)}$$

Q.62 Let us suppose that different distances, say S_1, S_2 ...... , S_n are travelled with different speeds, say, V_1, V_2, ... , Vn respectively then average speed is given by weighted H.M. of speed, the weights being different distances covered.

$$\text{Average speed} = \frac{\sum_{i}^{n} Si}{\sum_{i=1}^{n}\left(\frac{Si}{Vi}\right)}$$

Q.63 The H.M. is also used in averaging the ratios like price per unit, km/ hr, work done/hr, under certain conditions.

Q.64 The properties of an ideal measure of central tendency (i.e. Average) have been laid down by Prof. Yule.

Q.65 The A.M. may be regarded as best of all the averages because it satisfies all the properties of an ideal average.

Q.66 The three points which divide the series into four equal parts are called "quartiles".

Q.67 The first quartile Q1 exceeds 25% of the observations and is exceeded by 75% of the observations.

Q.68 The nine points which divide the series into ten equal parts, are called "deciles".

Q.69 The percentiles (pi, i = 1, 2, ... 99) divided the series into 100 equal parts.

Q.70 The partion values viz, (Qi, i =1, 2, 3), (Di, i = 1, 2, ..., 9) and (Pi, i =1, 2, ..., 99) can be located conveniently with the help of a curve known as ogive.

Q.71 The partition points, (Q2, Median, D5 & P50) exceed and are exceeded by 50% of the observations.

Secion-B : Fill in the Blanks

Q.1 Measures of .... give an idea about the concentration of values in the central part of the distribution.

Q.2 Measures of central tendency should be based .... values.

Q.3 Arithematic mean & geometric mean are .... for further mathematical treatment.

Q.4 Measures of central tendency should be affected .... possible by fluctuation of sampling.

Q.5 Measures of central tendency should not .... much by extreme values.

Q.6 The A.M. $\bar{x}_n$ of a set of values say x_1, x_2 ...., x_n is defined as $\bar{x}_n$ = ...

Q.7 In case of frequency distribution (fi, xi, i = 1, 2, ... k), the A.M. is defined by the expression
$\bar{x}$ = ...

Q.8 In case of grouped or continuous frequency distribution, the values of x are taken as the .... of the corresponding classes.

Q.9 If di = (xi - A) then $\bar{x}$ = ...

Q.10 In question "9", if A = $\bar{x}$ then value of $\bar{d}$ will be equal to
....

Q.11 In case of continuous frequency distribution if we take di = (xi - A)/h,
then $\bar{x}$ = ...

Q.12 The S.S. of deviations of a set of values is .... when taken about A.M.

Q.13 The A.M. of K composite series (xi, ni, i = 1, 2, ... , k), can be worked out from the expression written as
$\bar{x}$ = ...

Q.14 Out of all the measures of central tendency, the least affected measure by fluctuation of sampling is the ....

Q.15 The most stable measure of central tendency is the ....

Q.16 The A.M. can not be located by ....

Q.17 The very much affected measure of central tendency by the extreme value is the ....

Q.18 In case of extreme values present in data, the A.M. gives a .... of distribution.

Q.19 The A.M. may lead to a wrong inference if the details of the corresponding data is ....

Q.20 The A.M. of a continuous distribution (grouped data), in case of .... class can not be calculated.

Q.21 In case of extremely .... distribution, usually the A.M. is not a suitable measure of ....

Q.22 If wi is the weight attached to the value Xi, (i = 1, 2, ..., n) then weighted mean $\bar{x}_w$ is defined as
$\bar{x}_w$ = ...

Q.23 Median of a distribution is the value of the variable which divides it into .... parts.

Q.24 The median of a set of observations is the value which exceeds and is exceeded by the .... of values.

Q.25 The median of a. set of observations is the value such that the number of values above it is equal to the .... values .... it.

Q.26 In case of even number of values, infact, any value lying between two .... can be taken as median.

Q.27 In case of an even number of values, by convention, we take mean of two .... as the median.

Q.28 The median of a given set of values is the .... average.

Q.29 Median can be worked out for a distribution with an .... classes.

Q.30 Median of a set of values is .... on all values.

Q.31 Median is not amenable to .... treatment.

Q.32 Median is affected much more by .... as compared to A.M.

Q.33 To deal with the qualitative data, the only average to be used, is the ....

Q.34 To determine the typical value of measure of central tendency in the problems concerning wages, distribution of wealth etc. is the ....

Q.35 in case of continuous frequency distribution, the class corresponding to the c.f. just $>\frac{N}{1}$ is known as .... class.

Q.36 In case of continuous distribution, the median can be calculated with the following formula

$$Md = 1_1 + (\ldots\ldots - cf_p)\frac{(\ldots\ldots\ldots\ldots)}{\ldots\ldots\ldots\ldots}$$

Q.37 Most frequently occured value in a given set of values is known as the "................... " of the distribution or data.

Q.38 If the maximum frequency is repeated in a given set of discreate values, "the method of .... " is applied to get the value of " ....".

Q.39 If the most frequent value occurs either in the beginning or in the end of distribution, the suitable method to determine the mode is the " .... ".

Q.40 If there is irregularity , in the occurance of values of the given data, the suitable method to get the value of mode is the " .... "

Q.41 In case of continuous frequency distribution with (1_1, 1_2, f_o, f_1, f_2, f_{mo}) as notations, the mode can be worked out from the formula given below

$$Mo = 1_1 + \frac{(\ldots\ldots\ldots\ldots)\ (\ldots\ldots\ldots\ldots)}{(2f_{mo}\ \ldots\ldots\ldots\ldots)}$$

Q.42 For the moderately .... distribution, the following empirical relation among mean, median and mode exists Mode = (3 .... -2 ....)

Q.43 The empirical relation Mode = (3 Median - 2 Mode) has been reported by ....

Q.44 Mode is not affected by " .... " at all.

Q.45 Distributions which come across with two mode values, are known as ....

Q.46 Distributions which come across with more than two modes, are known as ....

Q.47 Mode is affected to a greater extent by .... as compared with A.M.

Q.48 To assess about the ideal size in business forecasting, in manufacture of ready made garments, the average measure used is the " ...." .

Q.49 The geometric mean of a set of n observations say $x_1, x_2, ..., x_n$ is defined as

G = ...

Q.50 Let (n_1 & n_2) and (G_1 & G_2) be the sizes and G.M. of two series respectively then expression to workout the G.M. of combined series is written as

$$\text{Log G} = \frac{(\text{..................})}{(n_1 + n_2)}$$

Q.51 The average not affected much by the fluctuation of sampling is the ....

Q.52 The average which gives more weights to small values is the ....

Q.53 The G.M. of a set of observations having one value equal to zero will always be equal to ....

Q.54 For a data having one value negative, the G.M. of that data will be of .... nature.

Q.55 The harmonic mean H.) of n values say x_1, x_2 .... x_n is defined as
H = ...

Q.56 In case of frequency distribution, (fi, xi, i =1, 2, ... n), the H.M. is defined as
H = ...

Q.57 If (xi, wi, i = 1, 2, ..., n) are the n values with corresponding weightds attached with them, H.M. is defined as
H = ...

Q.58 If equal distances are covered per unit of time with speed equal to V_1, V_2, V_3, ... , V_n (say) by a car then the average speed of the car can be calculated by the formula
Average speed = ...

Q.59 Let the distances (Si, i = 1, 2, ... , K) are travelled by a car with speeds (Vi, i = 1, 2, ... , K) respectively, the average speed of car can be calculated by
Average speed = ...

Q.60 To work out the average of the ratios, like-price/unit, km/hr, and work done/hr, under certain conditions, the suitable average used is ....

Q.61 The properties of an ideal measure of central tendency have been laid down by Prof. ....

Q.62 The three points which divide the series into four equal parts, are known as " ...." .

Q.63 The first quartile Q1 exceeds .... percent of the observatrions and is exceeded by .... percent of the observations.

Q.64 The nine points which divide the given distribution into ten equal parts, are known as " ...." .

Q.65 The 99 points (pi, i = 1, 2, ... , 99) divide the given distribution into .... equal parts.

Q.66 The partition values eg. quartiles, deciles & percentiles can be located will the help of a curve known as " ...." .

Q.67 The partition values eg. Q_2, median, D_5 & P_{50}, exceed and are exceeded by .... percent, of values of the given series .

Section-C : Multiple Choice Questions

Q.1 Measures of central tendency give an idea about the concentration of values, in the
(a) Central part of the distribution
(b) Middle part of the distribution
(c) Upper side of the distribution
(d) Both (a) & (b)

Q.2 The measures of central tendency should
(a) Be based on all the values
(b) Not be based on all the values
(c) Be based on central parts of the values
(d) Be based on some repetative values

Q.3 Measures of central tendency are
(a) Suitable for further mathematical treatment
(b) Are not suitable for further mathematical treatment
(c) Not considered for further mathematical treatment
(d) Always considered for further mathematical treatments

Q.4 Measures of central tendency should be least affected by
(a) Fluctuation of sampling (b) Extreme values of data
(c) Both (a) & (b) (d) None

Q.5 In case of frequency distribution, (fi, xi, i = 1, 2, ... , n), the A.M. is defined as

(a) $\bar{x} = \frac{1}{n}\sum_{i=1}^{n} fi\ xi$ (b) $\bar{x} = \frac{1}{N}\sum_{i}^{n} fi\ xi$

(c) $\bar{x} = \frac{1}{\sum_{i}^{n} fi}\sum_{i}^{n} fi\,xi$ (d) $\bar{x} = \frac{1}{\sum_{i}^{n} fi}\sum_{i}^{n} xi\,fi$

Q.6 In case of grouped frequency distribution, the values of variable X is taken as
(a) Mid values of the classes
(b) Lower limits of the classes
(c) Average values of lower & upper limits of the classes
(d) Both (a) & (c)

Q.7 If we define di = (xi - A) for A being some arbitrary constant then we have

(a) $\bar{x} = \bar{d} - A$ (b) $\bar{d} + A$

(c) $A + \bar{d}$ (d) Both (b) & (c)

Q.8 Let di be defined as di = $\frac{xi - A}{h}$, then we have

(a) $\bar{x} = h\bar{d} - A$ (c) $\bar{x} = A + h\bar{d}$

(b) $\bar{x} = A + ha$ (d) Both (b) & (c)

Q.9 The algebraic sum of deriations of a set of values from their A.M.

(a) Is equal to zero

(b) Is not equal to zero

(c) May or may not be equal to zero

(d) None of these

Q.10 The sum of square of deviations of a set of values taken about their A.M.

(a) Is minimum (b) Is maximum

(c) May or may not be minimum (d) None

Q.11 Out of all the averages, the least affected average by the fluctuation of sampling is the

(a) G.M. (b) H.M.

(c) Median (d) A.M.

Q.12 The measure which can not be located graphically is the

(a) Mode (b) Median

(c) Q_3 (d) A.M.

Q.13 In case of one value missing, the following measure can not be calculated

(a) A.M. (b) G.M.

(c) H.M. (d) All the three

Q.14 If the extreme class of a continous distribution is of open nature i.e < 10 or > 100 the following measure can not be calculated

(a) A.M. (b) Mode

(c) Median (d) Q2

Q.15 In case of extremely asymmetrical distribution, the following is not a suitable measure of location

(a) Mode (b) Median

(b) A.M. (d) None

Q.16 If (xi, wi, i = 1, 2, ..., n) are n values along with their weight, then weighted mean $\bar{x}_w$ can be calculated from

(a) $\sum_i^n wi\,xi \Big/ \sum_i^n wi$ (b) $\sum_i^n (xi / wi)$

(c) $\sum_i^n (wi / xi)$ (d) $\frac{1}{n}\sum_i^n wi\,xi$

Q.17 Let there be (N+1) values in data, then median is the value which exceeds and is exceeded by the

(a) Same number of values (b) $\frac{N}{2}$, number of values

(c) 50% of N values (d) All the three

Q.18 In case of even number of values in data, the median can be taken as

(a) Any value lying between two middle values

(b) A.M. of two middle values

(c) A.M. of lowest and highest value of data

(d) Both (a) & (b)

Q.19 Median of a set of observations is the

(a) Positional average of the values

(b) Half of the range of the data

(c) Value which divides the distribution into two equal parts

(d) Both. (a) & (c)

Q.20 In case of even number of values in data, the median

(a) Can not be determined exactly

(b) Can be determined exactly

(c) Can be considered any value lying in the two middle values

(d) Both (a) & (c)

Q.21 The median of a set of values is based on

(a) All values values (b) Not all values

(c) Only on few middle (d) None of the above

Q.22 Median is

(a) Not amenable to mathematical treatment
(b) Amenable to mathematical treatment
(c) Not suitable for mathematical treatment
(d) Both (a) & (c)

Q.23 By fluctuation of sampling, the median is affected

(a) Much more than A.M.
(b) Less than the A.M.
(c) Less than the G.M.
(d) None of the three

Q.24 To deal with the qualitative data, the suitable measure of central tendency is the

(a) A.M.
(b) G.M.
(c) Median
(d) Mode

Q.25 To deal with the problems related to wages, distribution of wealth etc., the suitable measure of central tendency is

(a) Median
(b) Mode
(c) A.M.
(d) G.M.

Q.26 Median class corresponds to the cumulative frequency

(a) Just $< \frac{N}{2}$
(b) Just $> \frac{N}{2}$
(c) Just $= \frac{N}{2}$
(d) None

Q.27 In case of continuous frequency distribution with usual notations, the median can be worked out from the formula

(a) $Md = l_1 + \left(\frac{N}{2} - Cf_p\right)\frac{(l_2 - l_1)}{f_{md}}$

(b) $Md = l_2 - \left(\frac{N}{2} - Cf_p\right)(l_2 - l_1) / f_{md}$

(c) $Md = l_1 + \left(\frac{2N}{4} - Cf_p\right)(l_2 - l_1) / f_{md}$

(d) Both (a) & (c)

Q.28 Mode of a distribution corresponds to the

(a) Least frequent value (b) Most frequent value
(c) Moderate frequent value (d) None

Q.29 If in the data of discrete values, the maximum frequency occurs for more than one value, then mode can be obtained using the

(a) Method 'of grouping (b) Method of inspection
(c) Method of using model class (d) None

Q.30 If the maximum frequency occurs either in, the beginning or in the end of distribution then suitable method to workout the mode is the

(a) Method of inspection
(b) Method of grouping
(c) Method of applying model class
(d) None

Q.31 If the distribution of values is of irregular nature, the suitable method to find the mode is the

(a) Method of applying model class
(b) Method of inspection
(c) Method of groUping
(d) Both (a) & (b)

Q.32 For continuous frequency distribution in usual notations, the mode can be found by using the formula

(a) $Mo = l_1 + \frac{(f_{mo} - f_o)(l_2 - l_1)}{(2f_{mo} - f_1 - f_2)}$

(b) $Mo = l_2 - \frac{(f_{mo} - f_o)(l_2 - l_1)}{(2f_{mo} - f_1 - f_2)}$

(c) $Mo = l_1 + \frac{(f_o - f_{mo})(l_2 - l_1)}{(2f_{mo} - f_1 - f_2)}$

(d) $Mo = l_1 + \frac{(f_{mo} - f_o)(l_2 - l_1)}{(2f_o - f_{mo} - f_2)}$

Q. 33 For moderately asymmetrical distribution, the following relation is true

(a) Mode = 3 Median - 2 Mean
(b) Mode = 2 Median - 3 Mean
(c) Mode = - (2 Mean - 3 Median)
(d) Both (a) & (c)

Q.34 Mode can be obtained easily

(a) From open end classes
(b) From continuous classes
(c) From discrete distributions
(d) All the three

Q.35 Any biomodel distribution has

(a) One mode
(b) Two modes
(c) Three modes
(d) None of the above

Q.36 By fluctuation of sampling, the mode affected

(a) Much more than A.M.
(b) Less than A.M.
(c) Less than G.M.
(d) None of the three

Q.37 To forecast about the ideal size of an item in business or in redimade garments, shoes, etc., the most suitable average applicable is

(a) A.M.
(b) G.M.
(c) Mode
(d) Median

Q.38 The G.M. of a set of n values, say $x_1, x_2, \ldots\ldots\ldots, x_n$ is defined as

(a) $G = \left(\prod_{i=1}^{n} xi \right)^{n}$
(b) $G = \left(\prod_{i=1}^{n} xi \right)^{\frac{1}{n}}$
(c) $G = \sqrt[n]{\prod_{i=1}^{n} xi}$
(d) Both (b) & (c)

Q.39 Let (n_1, n_2) & (G_1, G_2) be the sizes and geometric means of two set of values then G.M. of the combined set can be obtained from

(a) $\text{Log } G = \dfrac{(n_1 \log G_1 + n_2 \log G_2)}{(n_1 + n_2)}$
(b) $\text{Log } G = (n_2 \log G_2 + n_1 \log G_1) / (n_1 + n_2)$
(c) $\text{Log } G = (n_1 \log G_1 + n_2 \log G_2) / (n_1 + n_2 + 2)$
(d) Both (a) & (b)

Q.40 Geometric mean gives more weight to

(a) Large values (b) Small values

(c) Median values (d) All the three

Q.41 If one value of a set of n observations is equal to zero then the value of G.m. becomes

(a) Equal to zero (b) Negative

(c) Imaginary (d) Can not be calculated

Q.42 If one value of a set of observations becomes negative then the G.M. of that data becomes

(a) Equal to zero (b) Imaginary

(c) Very small (d) None of the above

Q.43 The harmonic mean of n values say (xi, i = 1, 2, ..., n) is defined as

(a) $H = \dfrac{1}{\frac{1}{n}\sum_{i}^{n} xi}$ (b) $H = \dfrac{1}{\left(\frac{1}{n}\sum_{i}^{n} \frac{1}{xi}\right)}$

(c) $H = \dfrac{n}{\left(\sum_{i}^{n} \frac{1}{xi}\right)}$ (d) Both (b) & (c)

Q.44 In case of freqency distribution (fi, xi, i = 1, 2, ..., K), the H.M. can be defined as

(a) $H = \dfrac{\sum^{k} fi}{\sum_{i}^{k} (fi / xi)}$ (b) $H = \dfrac{\sum_{i}^{k} fi}{\sum_{i}^{k} (xi / fi)}$

(c) $H = \dfrac{1}{\frac{1}{\sum^{n} fi} \cdot \sum_{i}^{k} \left(\frac{fi}{xi}\right)}$ (d) Both (a) & (c)

Q.45 If (xi, wi, i = 1, 2, ..., K) are k values along with their weights attached with them respectively then H.M. is defined as

(a) $H = \dfrac{\sum^{k} wi}{\sum^{k}\left(\dfrac{wi}{xi}\right)}$ (b) $H = \dfrac{1}{\dfrac{1}{\sum_{i}^{k} wi}\cdot\left(\sum_{i}^{k}\dfrac{wi}{xi}\right)}$

(c) Both (a) & (b) (d) None of the above

Q.46 If equal distances are covered by a car with speed equal to V_1, V_2, ..., V_n, then average speed V of car will be given by

(a) $V = \dfrac{n}{\sum_{i}^{k}\left(\dfrac{1}{vi}\right)}$ (b) $V = \dfrac{1}{\dfrac{1}{n}\sum^{n}(1/vi)}$

(c) $V = \dfrac{n}{\sum_{i}^{n} vi}$ (d) None of the above

Q.47 Let (Si, Vi, i = 1, 2, ... K) be the different K distances and speeds respectively then average speed V is defined as

(a) $V = \dfrac{\sum^{k} Si}{\sum^{k}\left(\dfrac{Si}{vi}\right)}$ (b) $V = \dfrac{1}{\dfrac{1}{\sum_{i}^{k} Si}\cdot\sum_{i}^{k}\left(\dfrac{Si}{vi}\right)}$

(c) $V = \sum_{i}^{k} ViSi \Big/ \sum_{i}^{k} Si$ (d) Both (a) & (b)

Q.48 To find the average of the ratios like price/unit, km/hr, work done/hr under certain conditions, the suitable measure of central tendency applicable is

(a) A.M. (b) G.M.
(c) H.M. (d) Median

Q.49 The properties of an ideal measure of central tendency have been reported by

(a) Prof. R.A. Fisher (b) Prof. C.R. Rao
(c) Prof. Karl Pearson (d) Prof. Yule

Q.50 The first quartile Q1

(a) Exceeds 25% of values
(b) Is exceeded by 75% of values
(c) Is exceeded by 50% of values
(d) Both (a) & (b)

Q.51 The partition values can be located conveniontly, with the help of a curve known as

(a) Ogive curve (b) Frequency polygon
(c) Cumulative frequency (d) Both (a) & (c)

Q.52 The partition values (Q 2, Median, D5 & P50)

(a) Exceed 50% of observations
(b) Are exceeded by 50% of observations
(c) Are exceeded by more than 50% of values
(d) Both (a) & (c)

Key Answers

Section-A : True False

1.	False	2.	True	3.	False
4.	True	5.	True	6.	True
7.	True	8.	True	9.	False
10.	False	11.	False	12.	False
13.	True	14.	True	15.	True
16.	False	17.	True	18.	True
19.	True	20.	True	21.	True
22.	True	23.	True	24.	True
25.	True	26.	True	27.	True
28.	True	29.	True	30.	True
31.	True	32.	True	33.	True
34.	True	35.	True	36.	True
37.	True	38.	True	39.	True
40.	True	41.	True	42.	True
43.	True	44.	True	45.	True
46.	True	47.	True	48.	True
49.	True	50.	True	51.	True
52.	True	53.	True	54.	True
55.	True	56.	True	57.	True
58.	True	59.	True	60.	True
61.	True	62.	True	63.	True

64. True 65. True 66. True
67. True 68. True 69. True
70. True 71. True

Section -B : Fill in the Blanks

1. Central tendency
2. On all
3. Suitable
4. As little as
5. Be affected
6. $\frac{1}{n}\sum_{i}^{n} xi$
7. $\left(\sum_{i}^{n} fixi\right) / \sum_{i}^{k} fi$
8. Middle values
9. $A + \bar{d}$
10. Zero
11. $A + h\bar{d}$
12. Minimum
13. $\sum_{i}^{k} ni\ \bar{x}i \Big/ \sum_{i}^{k} ni$
14. A.M.
15. A.M.
16. Graphically
17. A.M.
18. Distorted picture
19. Not available
20. Open
21. Asymmetrical, Location
22. $\left(\sum_{i}^{n} wixi\right) / \sum_{i}^{n} wi$
23. Two equal
24. Same number
25. Number, Below
26. Middle values
27. Middle values
28. Positional
29. Open end
30. Not based
31. Mathematical
32. Fluctuation of sampling
33. Median
34. Median
35. Median
36. $l_1 + \left(\frac{N}{2} - cf_p\right)\frac{(l_2 - l_1)}{f_{md}}$
37. Mode
38. Grouping, Mode
39. Method of grouping
40. Method of grouping
41. $\frac{(f_{mo} - f_0)(l_2 - l_1)}{(2f_{mo} - f_0 - f_1)}$
42. Asymmetrical

43. Median, Mean
44. Extreme values
45. Biomodel
46. Muttimodel distribution
47. Fluctuation of sampling
48. Mode
49. $G = \left(\prod_{i=1}^{n} xi \right)^{\frac{1}{n}}$
50. $(n_1 \log G_1 + n_2 \log G_2)$
51. G.M.
52. G.M.
53. Zero
54. Imaginary
55. $H = n \Big/ \left(\sum_{i}^{n} \frac{1}{xi} \right)$
56. $H = \sum_{i}^{n} fi \Big/ \left(\sum_{i}^{n} \frac{fi}{xi} \right)$
57. $H = \sum_{i=1}^{n} wi \Big/ \left(\sum_{i}^{n} \frac{wi}{xi} \right)$
58. $V = \frac{k}{\sum_{i}^{k} \frac{1}{Vi}}$
59. $V = \sum_{i}^{k} si \Big/ \sum_{i}^{k} \frac{si}{vi}$
60. H.M.
61. Yule
62. Quartiles
63. 50, 50
64. Desiles
65. 100
66. Ogive
67. 50

Section-C : Multiple Choice Questions

1.	(d)	2.	(a)	3.	(a)
4.	(c)	5.	(c)	6.	(d)
7.	(d)	8.	(d)	9.	(a)
10.	(a)	11.	(d)	12.	(d)
13.	(d)	14.	(a)	15.	(c)
16.	(a)	17.	(d)	18.	(d)
19.	(d)	20.	(d)	21.	(c)
22.	(d)	23.	(a)	24.	(c)
25.	(a)	26.	(b)	27.	(d)
28.	(b)	29.	(a)	30.	(b)
31.	(c)	32.	(a)	33.	(d)

34.	(d)	35.	(b)	36.	(a)
37.	(c)	38.	(d) .	39.	(d)
40.	(b)	41.	(a)	42.	(b)
43.	(d)	44.	(d)	45.	(c)
46.	(d)	47.	(d)	48.	(c)
49.	(d)	50.	(d)	51.	(d)
52.	(d)				

❑❑❑

Chapter - 4

Measures of Dispersion

Section-A : Write True / False

Q.1 Dispersion of a given set of values, gives an idea about the homogeneity or heterogeneity of the distribution of values.

Q.2 The measure of scatteredness of a set of values about an average is known as the "measure of variation or dispersion".

Q.3 The measure of dispersion must be based on all the observations.

Q.4 Measure of dispersion should be amenable to further mathematical treatment.

Q.5 Measure of dispersion should be affected as little as possible by fluctuation of sampling.

Q.6 The measures which express the spread of observations in terms of distance between the values of two selected observations, are known as "distance measures" e.g. range and interquartile range (or quartile deviation).

Q.7 The other measures of dispersion alternative to distance measures, are the average of deviations of observations from some central value, e.g. mean deviation and standard deviation.

Q.8 Range is a simplest but a crude measure of dispersion.

Q.9 Range of a set of values is based on two extreme observations which themselves are subject to chance fluctuations.

Q.10 Range based on two extreme observations of a data is not at all a measure of dispersion.

Q.11 In usual notations, the quartile deviation or semiquartile range Q is defined as

$$Q=\frac{1}{2}(Q_3 - Q_1)$$

Q.12 Quartile deviation makes use of 50% of the data and can not be regarded as reliable measure of dispersion.

Q.13 If (xi, fi, i = 1, 2, ..., n) is the frequency distribution then mean deviation from the average A is given as

Mean deviation from average $A = \frac{1}{N} = \sum_{i}^{n} fi|Xi - A|,\ \sum_{i}^{n} fi = N$

Q.14 In defining the mean deviation of a set of observation, the used average A may be mean, median or mode.

Q.15 The mean deviation being based on all observations of a set of values, is a better measure of dispersion than range or quartile range.

Q.16 In defining mean deviation, the step of ignorving the signs of deviations (xi - A) creates artificiality and renders it useless for further mathematical treatment.

Q.17 Mean deviation is least when taken about median.

Q.18 Root mean square deviation usually denoted by 's' is defined as

$s = \sqrt{\frac{1}{N}\sum_{i=1}^{k} fi(Xi - A)^2}$,for A being any arbitrary constant.

Q.19 The square of root mean square deviation (s) i.e. s^2 is known as "mean square deviation" .

Q.20 If (fi, xi, i = 1, 2, ..., K) is a frequency distribution then we have the following result

$$\text{Sum} = \sum_{i=1}^{k} fi(Xi - \bar{x}) = 0 \text{ for } \bar{x} = \text{A.M. of the distributionn}$$

Q.21 Root mean square deviation or mean square deviation is least when the deviation is taken from A.M. (i.e. $\bar{x}$).

Q.22 The minimum value of root mean square deviation (s) is equal to standard deviation (σ) for a given set of values.

Q.23 Minimum value of mean square deviation, (s^2) of a frequency distribution is equal to its variance, (σ^2).

Q.24 For a given frequency distribution (fi, xi, i = 1, 2, ..., K) the following result held good.

$$\sigma_x^2 = \frac{1}{N}\sum_{i}^{k} fi\ xi^2 - \bar{x}^2 = \sum_{i}^{k} fi\ xi^2 - \left(\frac{1}{N}\sum_{i=1}^{k} fi\ xi\right)^2$$

Q.25 For a frequency distribution (fi, xi, i = 1, 2, ..., K) and arbitrary constant A, if $di = \frac{xi - A}{h}$ then the following result holds good.

$$\sigma_x^2 = h^2 \frac{1}{N}\sum_{i}^{k} fi\left(di - \bar{d}\right)^2 = h^2\sigma_d^2$$

Q.26 For a given set of observations, the standard deviation or variance is independent of change of origin but dependent on change of scale.

Q.27 If (n_1 , n_2) are the dsizes, ($\bar{x}_1$, $\bar{x}_2$) the means, and (σ_1, σ_2) the standard deviation of the two series of values, then the variance σ^2 of the combined series of size ($n = n_1 + n_2$), is given by

$$\sigma^2 = \frac{1}{n_1 + n_2}\left[n_1\left(\sigma_1^2 + d_1^2\right) + n_2\left(\sigma_2^2 + d_2^2\right)\right]$$

$$\text{for } d_1 = \left(\bar{x}_1 - \bar{x}\right), d_2 = \left(\bar{x}_2 - \bar{x}\right) \text{and } \bar{x} = \sum_{i=1}^{2} ni\,\bar{x} \Big/ \sum_{i=1}^{2} ni$$

Q.28 The result given in Q (27) also holds good in general for more than two series of different sizes.

Q.29 For a given A.P. say, a a+d, a+2d, ... , a+2nd, we have the following result

$$\sigma^2 = \frac{n(n+1)d^2}{3}$$

Q.30 For a given set of values, the standard deviation is always greater then its mean deviation.

Q.31 To compare the variability of two series of values which differ either widely in their averages or which are measured in different units, the coefficient used is known as "coefficient of dispersion".

Q.32 For the greatest and smallest values say (A & B), of a given series, the coefficient of dispersion (C.D.) is defined as

C.D. based on range $= \frac{(A - B)}{(A + B)}$

Q.33 C.D. based on quartile deviation $= (Q_3 - Q_1) / (Q_3 + Q_1)$

Q.34 C.D. based on mean deviation

$$= \frac{\text{Mean deviation}}{\text{Average from which it is calculated}}$$

Q.35 C.D. based on standard deviation $= \frac{\sigma_x}{\bar{X}}$

Q.36 The coefficient of variation (c.v.) of a data measured on a variable x is defined as

cv. (x) $= \frac{\sigma_x}{\bar{X}} x100$

Q.37 The coefficients of dispersion or coefficient of variation are always unit less quantities.

Q.38 The C.D. as well as C.V. are useful to compare the disperion of two set of values measured in two different units of measurement.

Q.39 Out of two series say (A & B), the series having lesser C.V. than that having greater C.V., is said to be more consistent (homogeneous) than other.

Q.40 For a given frequency distribution (fi, xi, i = 1, 2, ..., K), the rth moment about mean $\bar{x}$, usually denoted by μ_r, is defined as

$$\mu_r = \frac{1}{N} \sum_{i=1}^{k} fi\ (xi - \bar{x})^2$$

Q.41 In case of defining the moments about the A.M., the following fundamental results are true.

$\mu_0 = 1, \mu_1 = 0, \mu_2 = \sigma^2$

Q.42 The rth moment of a given set of variable say x, is independent of change of origin but dependent on change of scale.

Q.43 In case of grouped frequency distribution, to calculate moment, we assume that the frequencies are concentrated at the mid points of the class intervals.

Q.44 If the distribution is symmetrical or slightly asymmetrical and the class intervals $\ngtr \frac{1}{12}$ (Range of distribution), then the assumption of concentration of frequencies in the middle of the class intervals is very nearly true.

Q.45 The assumption of concentration of frequencies in the middle of class intervals are not true in general and some error creeps into calculation of moments which is known as "Grouping error".

Q.46 If the frequency distribution is (i) continuous and (ii) the frequency tapers off to zero in both directions then the grouping error occurs.

Q.47 Sheppard suggested the .following correction in calculation of the second moment say μ_2 .

Corrected $\mu_2 = \sigma^2$ (corrected) $= \mu_2 - \frac{h^2}{12}$

where h is the width of class interval.

Q.48 Karl pearson defined the four coefficients based on first four moments

(i) $\beta_1 = \mu_3^2 / \mu_2^3,\ r_1 = \pm\sqrt{\beta_1}$

(ii) $\beta_2 = \mu_4 / \mu_2^2,\ r_2 = \beta_2 - 3$

Q.49 The four coefficients β_1, β_2, γ_1 & γ_2 defined by karl pearsion based upon first four moments, are independent of units of measurement.

Q.50 The factorial moment of order r about a points (x = b) is defined

as $\mu_r = \frac{1}{N}\sum fi(Xi-b)^{(r)}$

where $(x - b)^{(r)} = (x - b)(x - b - 1)(x - b - 2)........(x - r + 1)$

Q.51 In Q(50), if we take b = 0, then the factorial moment $\mu'_{(r)}$ is known as factorial moment about orign

Q.52 The first factorial moment about the origin is given as

$$\mu'_{(1)} = \frac{1}{N}\sum \text{fixi} = \text{Mean}\ (\bar{x})$$

Q.53 The factorial moment about origin of order 2 is given as

$$\mu'_{(2)} = \frac{1}{N}\sum_{i=1}^{k} \text{fi}\, x_1 (x_1 - 1) = \frac{1}{N}\sum \text{fi}x_i^2 - \frac{1}{N}\sum_{i}^{k} \text{fi xi}$$

$$\mu'_{(2)} = \mu'_2 - \mu'_1$$

Q.54 For the frequency distribution (xi, fi, i = 1, 2, ..., n), the rth absolute moment of the variable x, about the origin is given by

$$\frac{1}{N}\sum_{i=1}^{n} \text{fi}|\text{xi} - \bar{x}|^r$$

Q.55 The rth absolute moment of the variable x about the mean x is given by

$$\frac{1}{N}\sum_{i=1}^{n} \text{fi}|x_i^r|$$

Q.56 Skewness means "lack of symmetry".

Q.57 The skewness of a distribution gives an idea about the shape of the curve drawn for the distribution.

Q.58 If for a given distribution, we have
Mean = median = mode, then distribution is said to be skewed.

Q.59 If for a distribution, the quartiles are not at equidistant from the mediom, then distribution is said to be skewed.

Q.60 A distribution is said to be skewed if the curve , drawn with the help of data taken from the distribution is streched more in one side than the other side.

Q.61 Measure of skewness based on mean (M) and medium Md is defined as S_k = (M - Md)

Q.62 Measure of skewness based on mean (M) and mode (Mo) is defined as S_k = (M - Mo)

Q.63 Measure of skewness based on quartiles and median is defined as $Sk = (Q_3 - Md) - (Md - Q_1)$

Q.64 For comparing two set of data for skewness, the coefficient of skewness is worked out.

Q.65 The coefficients of skewness are independent of unit, of measurement.

Q.66 Karl Pearson's coefficient of skewness is defined as

$$S_k = \frac{(M - Mo)}{\sigma}$$

Q.67 If Mode is ill-defined then using the empirical relation, M0 = (3 Md - 2 M) for a moderately assymetrical distribution, we get

$$S_k = \frac{3(M - Mo)}{\sigma}$$

Q.68 The limits for karl pearson's coefficient of skewness is ± 3 , but these Its are rarely attained in practice.

Q.69 Bowely's coefficient of skewness based on quartiles, is defined as

$$S_k = \frac{(Q_3 + Q_2 - 2Md)}{(Q_3 - Q_1)}$$

and it is also known as quartile coefficient of skewness.

Q.70 Bowely's coefficient of skewness lies in the interval (-1, 1), i.e. $(-1 \leq S_k \leq 1)$

Q.71 For positively and negatively skewed distribution, we have the following figures

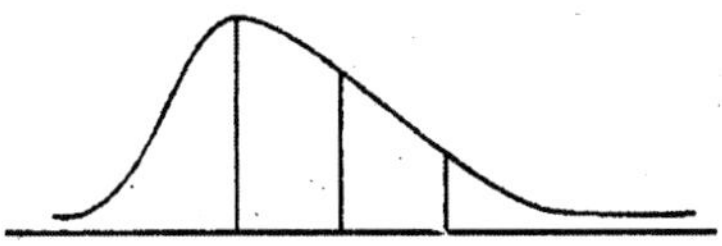

Mo Md M

+vely skewed distribution

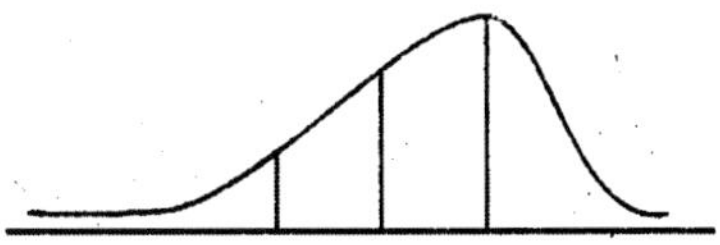

M Md Mo

-vely skewed distribution

Q.72 As reported by Prof. Karl Pearson's, the convexity of the frequency curve or kurtosis enables us to have an idea about the 'flatness or peakness' of the frequency curve.

Q.73 Kurtosis of the frequency curve is measured by the coefficient β_2 or its derivative γ_2 given as

$\beta_2 = \mu_4 / \mu^2_{2,} \gamma_2 = \beta_2 - 3$

Q.74 For normal curve (also known as mesokurtic curve), we have $\beta_2 = 3$ and $\gamma_2 = 0$.

Q.75 For platy kurtic frequency curve, we have $\beta_2 < 3$ i.e. $\gamma_2 < 0$.

Q.76 For leptokintic, we have $\beta_2 > 3$ and $\gamma_2 > 0$.

Q.77 The A.M. of 8 observations is 10 and of 2 observations is 5. The A.M. of combined set is 9.0.

Q.78 The value of 5th decide is equal to Q_2 and Median.

Q.79 The value of 50th percentile is equal to D_5 and Median.

Q.80 If the A.M. of x is 5 then A.M. of Y = (3 + 5x) is equal to 38.0.

Q.81 The difference between upper and lower limits of exclusive class intervals is known as "Midvalve".

Q.82 Median is the mathematical average.

Q.83 To measure the correct weight of an item through a faulty balance, the suitable measure of central tendency is H.M.

Section-B : Fill in the Blanks

Q.1 Dispersion of a set of values, gives an idea about the .... or .... of the values.

Q.2 The measure of scatterdness of a set of values about an average, is known as the measure of " .... or .... ".

Q.3 The measure of dispersion must be based on .... observations.

Q.4 Measure of dispersion should be applicable to further .... treatment.

Q.5 Measure of dispersion should be affected as .... as .... by fluctuation of sampling.

Q.6 Range is the .... but a .... measure of dispersion.

Q.7 Range of a set of values is based on two .... values.

Q.8 In usual notations, the quartile deviation or semi-quartile range Q is defined as

$$Q = \frac{1}{2}\,(\text{..........})$$

Q.9 Quartile deviation makes use of .... percent of data.

Q.10 Quartile deviation can not be regarded as reliable measure of ....

Q.11 Let (fi, xi, i = 1, 2, ..., K) be the frequency distribution, then mean deviation of the distribution about an arbitray average A is defined as

$$\text{M.d.} = \frac{1}{N}\sum_{i=1}^{k} fi\left|\text{..................}\right|, \text{with } N = \sum_{i=1}^{k} fi$$

Q.12 In defining mean deviation of a set of values, the arbitrary average "A" may be taken as mean, ... or mode.

Q.13 Mean deviation being based on all values, is a better measure of .... than range or .... range.

Q.14 Mean deviation is not suitable for further .... manupulation.

Q.15 Mean deviation is least when taken about ....

Q.16 Root mean square deviation (s) of a frequency distribution (fi, xi, i = 1, 2, ..., K) about a point 'A' is defined as

$$s = \sqrt{\frac{1}{N}\sum_{i=1}^{k} fi(\text{............})^2}$$

Q.17 The square of root mean square deviation is known as "..............................."

Q.18 If (fi, xi, i = 1, 2, ..., K) is a frequency distribution then

$$\sum_{i=1}^{k} fi(xi - \bar{x}) = \text{............}$$

Q.19 Root mean square deviation of a frequency distribution is minimum when deviation is taken from ....

Q.20 The minimum value of root mean square deviation is equal to .... for a given data.

Q.21 The minimum value of mean square deviation is equal to ....

Q.22 For a given frequency distribution (fi, xi, i = 1, 2, ..., K) the formula for a σ^2 x is written as

$$\sigma_x^2 = \frac{1}{N}\sum_{i=1}^{k} ....(...................)^2$$

Q.23 For a frequency distribution in term of exclusive class interval

let $di = \frac{(xi - A)}{h}$ then $\sigma_x^2 = ...\sigma_d^2$

Q.24 The standard deviation σ_x is .... of change of origin but it is .... on change of scale.

Q.25 The variance σ^2x is independent of change of .... but not of change of ....

Q.26 The variance σ^2x is .... of change of origin but it is .... on change of scale.

Q.27 Let (n_1, n_2), (X_1, X_2) and (σ_1, σ_2) be the sizes, means and standard deviations respectively of two sample data, then in usual notations, the variance σ_2 of combined sample can be calculated from the formula

$$\sigma^2 = \frac{1}{(n_1 + n_2)}\left[n_1(...........) + n_2(.........)\right]$$

Q.28 In Q(27), the A.M. ($\bar{x}$), of combined sample can be calculated from the formula written as

$$\bar{x} = \frac{(........)}{(n_1 + n_2)}$$

Q.29 For a sample values written in the arithematic progression say a, a+d, a+2d, a+3d, ... a+2nd, the variance σ^2 of the values can be expressed as $\sigma^2 = \frac{.........(.......)d^2}{3}$

Q.30 For a given data, the standard deviation is always greater than its ....

Q.31 For a given sample values, standard deviation is always than its mean deviation.

Q.32 The mean deviation of a sample values, is always .... than its standard deviation.

Q.33 The mean deviation of a population is always less than its ....

Q.34 To compare the variability (variation or dispersion) of two set of data which differ widely in their averages, the suitable coefficient used is known as "..........................".

Q.35 To compare the variability (variation, or dispersion) of two set of data measured in two different units of measurement, the siutable coefficient applicable is known as "............................".

Q.36 For a given set of data, let (A & B) be the largest and smallest values, then the coefficient of dispersion (C.D.) is defined as

C.D based on range = (A - B)/ (................)

Q.37 Coefficient of dispersion based on quartile, is defined as

C.D. = $(Q_3 - Q_1)$ / (...............)

Q.38 C.D. based on mean deviation is defined as

$$\text{C.D.} = \frac{\text{Mean deviation}}{(\ldots\ldots\ldots\ldots\ldots\ldots)}$$

Q.39 Coefficient of disperion based on standard deviation is defined as C.D. = ...

Q.40 Coefficient of variance of a random variable X is defined as C.V. (X) = (..................) x 100

Q.41 The coefficients of .... or .... are always unit less quantities.

Q.42 The C.D. as well as C.V. are useful to compare the variability of two sets of observations measured in two .... of measurement.

Q.43 In case of frequency distribution of exclusive class nature, we assume that the frequencies are concentrated at the .... the corresponding class intervals.

Q.44 Out of two set of data say A & B, the data having lesser c.v. than the other, is said to be more .... or .... than other.

Q.45 Let (fi, Xi, i = 1, 2, ..., k) be a frequency distribution then rth central moment μ_r is defined as

$$\mu_r = \frac{1}{N}\sum_{i}^{K} ....(Xi - \bar{X})^{..........} \quad for N = \sum_{i}^{K} fi$$

Q.46 Let μ be the rth central moment defined about A.M., then we have

μ_0= ...., μ_1 = ..., μ_2 = ...

Q.47 The rth (r = 2, 4, 6, ... , 2n) of a set of values, is independent of change of .... but is dependent on change of ....

Q.48 The assumption of concentration of frequencies in the middle of the class intervals is very nearly true if :

(i) The class invervals are .... $\frac{1}{12}$ (Range of distribution) and

(ii) The distribution is .... or slightly ....

Q.49 Grouping error occurs in the calculation of moment when the assumption of contraction of frequencies in the middle of .... is not true.

Q.50 In calculation of μ_2, sheppard suggested the correction as corrected $\mu_2 = \mu_2$ - (...........)

Q.51 The coefficient of skewness reported by Karl pearson based on moments are defined as

β_1 = (.......) and γ_1 = (....................)

Q.52 The coefficient of kurtosis reported by Karl Pearson are defined as β_2 = (.............), γ_2 = (...............)

Q.53 From coefficients, β_1, γ_1, β_2 & γ_2 defined by Karl Pearson are independent of "....................................".

Q.54 The factorial moment of order r about a point x = b, is defined as

$$\mu'_{(r)} = \frac{1}{N} = \sum_{i} fi(............)^{(r)}$$

Q.55 In Q (54), if we put b = 0, then factorial moment $\mu'_{(r)}$ is known as "factorial moment about (.............).

Q.56 The first factorial moment about origin is equal to the .... of the distribution..

Q.57 The rth absolute moment of the variable x about the mean $\bar{x}$ of the frequency distribution (fi, Xi, i =1, 2, ...n) is defined as

rth absolute moment = $\frac{1}{N} = \sum_{i}^{K} fi|...........|^{r}$

Q.58 The rth absolute moment of the frequency distributio (fi, Xi, i = 1, 2, 3, ..., k) about origin is defined as

$= \frac{1}{N} = \sum_{i}^{k} fi|...........|$

Q.59 Skewness means " ...."

Q.60 Skewness of a distribution gives an idea about the .... of the curve drawn from the distribution.

Q.61 For a given distribution, if Mean ≠ Median ≠ Mode, then the distribution is said to be ....

Q.62 If the quartiles of a distribution are not at equidistant from the median, then the distribution is said to be ....

Q.63 Measure of skewness based on mean (M) and median (Md) is defined as S_k = (..........)

Q.64 Measure of skewness based on mean (M) and mode (Mo) is defined as S_k = (...........)

Q.65 Measure of skewness based on quartiles and median is defined as $S_k = (Q_3$ - ....) - (Md - ....)

Q.66 Karl Pearson's coefficient of skewness is defined as S_k = (....)/σ

Q.67 For a moderately assymmetrical distribution, the coefficient of skewness S_k is defined as S_k = 3 (..........)/σ

Q.68 The limits for Karl Pearson coefficient of skewness is defined as (....... ≤ S_k ≤ ....)

Q.69 Bowley's coefficient of skewness based on quartiles is defined as

$$S_k = \frac{(Q_3 + Q_1 - \text{............})}{(\text{....................})}$$

Q.70 Bowley's coefficient of skewness lies in the interval $(\text{......} \leq S_k \leq \text{......})$

Q.71 The figure given below belongs to the .... skewed distribution.

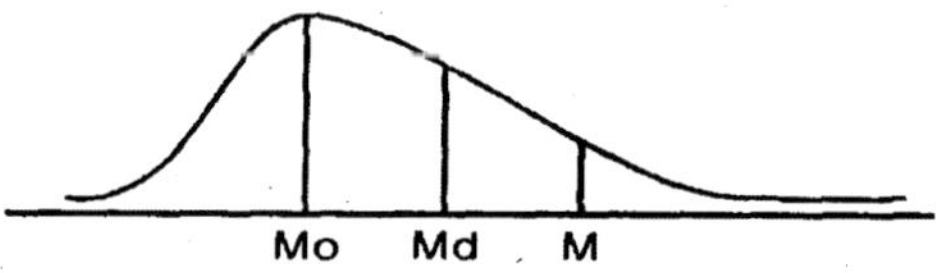

Q.72 Convexivity of the frequency curve or. kurtosis enables us to have an idea about the " .... or ...." of the frequency curve.

Q.73 Kurtosis of a frequency curve is measured by the coefficient .... or its derivative ....

Q.74 For normal curve, we have
$\beta_2 =$ ...., and $\gamma_2 =$ ....

Q.75 Normal curve is also knwon as " ...." .

Q.76 For platy Kurtile frequency curve, we have
$\beta_2 <$ .... and $\gamma_2 <$ ....

Q.77 For lapto kurtic curve, we have $\beta_2 >$ .... and $\gamma_2 >$ ....

Q.78 The A.M. of 8 observations is 10.0 and of 2 observations in 5.0. The A.M. of combined 10 observations is equal to ....

Q.79 The A.M. of a variable x is equal to 5.0. then A.M. of Y = (3 + 5x) is equal to ....

Q.80 The difference between upper and lower limits of a exclusive class interval is known as " ...."

Q.81 If w_1, w_2 are the weights of an item taken from two pans of a faulty balance, then the correct weight w of the item is given as $w = \sqrt{\text{......................}}$

Section-C : Multiple Choice Questions

Q.1 Dispersion of a data gives an idea about the
(a) Homogeneity of values in data
(b) Heterogeneity of values in data
(c) Scatterdness of values in data
(d) All the above

Q.2 Measure of scatterdness of values in the data about an average is known as
(a) Measure of variation (b) Measure of dispersion
(c) Measure of consistency (d) Both (a) & (b)

Q.3 Measure of variation must
(a) Be based on all values
(b) Not be based on all values
(c) Be amenable to further mathematical treatment
(d) Both (a) & (c)

Q.4 Measure of dispersion should
(a) Be based on all observations
(b) Not be based on all observations
(c) Be amenable to further mathematical treatment
(d) Both (a) & (c)

Q.5 Measure of disperion should be affected
(a) As little as possible (b) As much as possible
(c) Least possible (d) Both (a) & (c)

Q.6 Range of a sample values is the
(a) Simplest measure of dispersion
(b) Grude measure of dispersion
(c) Both (a) & (b)
(d) None

Q.7 Range of a sample values is based on
(a) Two extream values
(b) Lowest and highest values
(c) Both (a) & (b)
(d) Two middle values

Q.8 Two extreme values of a sample is subject to

(a) Chance fluctuations (b) Investigator

(c) Respondant (d) None

Q.9 In usual notations, the quartile deviation or semi-quartile range Q is defined

(a) $Q = -(Q_1 - Q_3)/2$ (b) $Q = (Q_3 - Q_1)/2$

(c) Both (a) & (b) (d) None

Q.10 Range based on two extreme values of a sample can

(a) Be regarded as a suitable measure of dispersion

(b) Not be regarded as suitable measure of dispersion

(c) Not be regarded as a suitable measure of variation

(d) Both (b) & (c)

Q.11 Quartile deviation Q makes use of

(a) 50 % of data (b) 25% of data

(c) Half of the data (d) Both (a) & (c)

Q.12 If (xi, fi, i = 1, 2, ..., k) is the frequency distribution then mean deviation (M.D.) is defined as

(a) $M.D. = \sum_{i}^{k} fi|xi - \bar{x}| \Big/ \sum_{i}^{k} fi$ (b) $M.D. = \sum_{i}^{k} fi|xi - A| \Big/ \sum_{i=1}^{k} fi$

(c) $M.D. = \sum_{i}^{k} fi|\bar{x} - xi| \Big/ \sum_{i=1}^{k} fi$ (d) M.D.=Both (a) & (c)

Q.13 In defining mean deviation of a sample, the average applicable may be taken as

(a) Mean (b) Median

(c) Mode (d) All the three

Q.14 Mean deviation of a sample values, is a better measure of dispersion than

(a) Range (b) Quartile range

(c) Semi quartile range (d) All the three

Q.15 Mean deviation is

(a) Useful for further mathematical treatment

(b) Not useful for mathematical treatment

(c) Better measure of dispersion than quartile range

(d) Both (b) & (c)

Q.16 Mean deviation is least when taken about

(a) Median (b) Mean

(c) Mode (d) G.M

Q.17 For a frequency distribution (fi, xi, i = 1, 2, ..., k), the mean square deviation s^2 is defined as

(a) $\sum^{k} fi(xi - A)^2 \Big/ \sum_{i}^{k} fi$ (b) $\sum_{i}^{k} fi(xi - \bar{x})^2 \Big/ \sum_{i}^{k} fi$

(c) $\sum_{i=1}^{k} fi(xi - \text{Median})^2 \Big/ \sum_{i}^{k} fi$ (d) $\sum_{i=1}^{k} fi(xi - \text{Mode})^2 \Big/ \sum_{i=1}^{k} fi$

Q.18 For any arbitrary constant A, the expression written as

$B = \sum_{i}^{k} fi(xi - A)^2 \Big/ \sum_{i}^{k} fi$ is known as

(a) Mean square deviation (b) Variance

(c) Root mean square deviation (d) None

Q.19 The function $M = \sum_{i=1}^{k} fi(xi - \bar{x}) \Big/ \sum_{i=1}^{k} fi$ defined for a frequency distribution (fi, xi, i = 1, 2, ..., k) will be equal to

(a) Variance (b) Standard deviation

(c) Zero (d) None

Q.20 Root mean square deviation or mean squared deviation is least when taken about

(a) Median (b) A.M.

(c) Mode (d) G.M.

Q.21 The minimum value of root mean quared deviation of a sample is equal to its

(a) Standard deviation (b) Coefficient of variation

(c) Coefficient of dispersion (d) Coefficient of skewness

Q.22 Mean squared deviation of a sample is equal to its

(a) Square of C.V.

(b) Variance

(c) Square of coefficient of dispersion

(d) Mean sum of square

Q.23 For a given frequency distribution (fi, xi, i = 1, 2, ... k) the following expression is true

(a) $\sigma^2 = \frac{1}{N^2}\sum_{i}^{k} fi(xi - \bar{x})^2$

(b) $\sigma^2 = \frac{1}{N}\sum_{i=1}^{k} fix_i^2 - \left(\frac{1}{N}\sum fixi\right)^2$

(c) $\sigma^2 = \frac{1}{N}\sum_{i}^{k}(xi - \bar{x})^2$

(d) $\sigma^2 = \frac{1}{k}\sum_{i}^{k} fi(xi - \bar{x})^2$

Q.24 For a given exclusive class frequency distribution, let di = (xi - A)/h then,

(a) $\sigma_x^2 = h^2\sigma_d^2$

(b) $\sigma_x = h.\sigma_d$

(c) $\sigma_x^2 = h.\sigma_d^2$

(d) Both (a) & (b)

Q.25 Let (n_1, n_2), (x_1, x_2) and (σ_1, σ_2) be the sizes, means and s.d's respectively of two samples, the variance σ^2 of combined sample in usual notations is

(a) $\sigma^2 = \frac{1}{(n_1 + n_2)}\left[n_1\sigma_1^2 + n_2\sigma_d^2\right]$

(b) $\sigma^2 = \frac{1}{(n_1 + n_2)}\left[n_1\left(\sigma_1^2 + d_1^2\right) + n_2\left(\sigma_2^2 + d_2^2\right)\right]$

(c) $\sigma^2 = \frac{1}{(n_1 + n_2)^2}\left[n_1\sigma_1^2 + n_2\sigma_2^2\right]$

(d) $\sigma^2 = \frac{1}{(n_1 + n_2)}\left[(n_1 + n_2)(\sigma_1^2 + \sigma_2^2)\right]$

Q.26 In Q (25), the mean x of the combined sample will be written as

(a) $\bar{x} = \sum_{i}^{2} ni\ \bar{xi} \Big/ \sum_{i=1}^{n} ni$

(b) $\bar{x} = \frac{(n_1 - 1)\bar{x}_1 + (n_2 - 1)\bar{x}_2}{n_1 + n_2}$

(c) $\bar{x} = \frac{(n_1 - 1)\bar{x}_1 + (n_2 - 1)\bar{x}_2}{n_1 + n_2 - 2}$

(d) $\bar{x} = \frac{n_1\bar{x}_1 + n_2\bar{x}_2}{n_1 + n_2 - 1}$

Q.27 For a given airthmatic progression say, a, a+d, a+2d, ..., a+2nd, we have

(a) $\sigma^2 = \frac{n(n+1)d^2}{3}$ (b) $\sigma^2 = \frac{n(n+1)d^2}{2}$

(c) $\sigma^2 = \frac{n(n+1)d^2}{4}$ (d) $\sigma^2 = \frac{n(n-1)d^2}{2}$

Q.28 For a given sample, its standard deviation is always
(a) Greater than its mean deviation
(b) Less tha its mean deviation
(c) Equal to its mean deviation
(d) Equal to its C.V.

Q.29 To compare the variability of two sample values measured in two different units of the maesurement, the suitable meaure is the
(a) Coefficient of variation (b) Coefficien of dispersion
(c) Standard deviation (d) Relative variance

Q.30 For the greatest and smallest values say (A & B) of a sample, the coefficient of dispersion (C.D). based on range is defined as

(a) $C.D. = \frac{(A-B)}{(A+B)}$ (b) $C.D. = \frac{(A+B)}{(A-B)}$

(c) $C.D. = \frac{(B-A)}{(A+B)}$ (d) $C.D. = \frac{(A-B)}{(A+B)^2}$

Q.31 C.D. based on quartile deviation is defined as

(a) $\frac{Q_3 - Q_1}{Q_3 + Q_1}$ (b) $\frac{Q_1 - Q_3}{Q_1 + Q_3}$

(c) $\frac{Q_2 - Q_1}{Q_1 + Q_2}$ (d) $\frac{Q_1 - Q_2}{Q_2 + Q_1}$

Q.32 C.D. based on mean deviation is defined as

(a) $\frac{\text{Mean deviation}}{\text{Average from which it is calculated}}$

(b) $\frac{\text{Mean deviation}}{\overline{x}}$

(c) $\frac{\text{Mean deviation}}{\text{Median}}$ (d) $\frac{\text{Mean deviation}}{\text{Mode}}$

Q.33 C.D. based on standard deviation is defined as

(a) $\left(\frac{\sigma_x}{\overline{x}}\right)^2$ (b) $\frac{\sigma_x}{\overline{x}}$

(c) $\frac{\sigma_x}{\overline{x}^2}$ (d) $\frac{\overline{x}}{\sigma_x}$

Q.34 Coefficient of variation of a variable x is defined as

(a) $\frac{\sigma_x}{\overline{x}}\times 100$ (b) $\left(\frac{\sigma_x}{\overline{x}}\right)^2\times 100$

(c) $\frac{\sigma_x}{\overline{x}^2}\times 100$ (d) $\frac{\overline{x}}{\sigma_x}\times 100$

Q.35 Coefficient of dispersion (C.D.) is always quantity
(a) Without unit of measurement
(b) With a unit measurement
(c) Presented in percentage
(d) To mesure the consistency of data

Q.36 Coefficient of varation (C.V.) is always a quantity to compare the variable of two set of data measured in
(a) Two different units of measurement
(b) The same unit of measurement
(c) Percentage
(d) Both (a) & (b)

Q.37 Out of two set of data say A & B, the data having lesser C.V. than that of other is said to be more
(a) Consistant (b) Heterogeneous
(c) Homogenous (d) Both (a) & (c)

Q.38 For a given frequency distribution, the rth moment μ_r about mean $\overline{x}$ is defined as

(a) $\frac{1}{N}\sum_{i}^{k} fi(xi-\overline{x})^2$ (b) $\frac{1}{N}\sum_{i}^{k} fi(xi-\overline{x})^r$

(c) $\frac{1}{N}\sum_{i}^{k} (xi-\overline{x})^r$ (d) $\frac{1}{N}\sum_{i}^{k} fi(xi-\overline{x})^{2r}$

Q.39 In case of defining moment about A.M. the fundamental results are true.

(a) $\mu_1 = 0, \mu_2 = \sigma^2$
(b) $\mu_0 = 1, \mu_1 = 0$
(c) $\mu_1 = 1, \mu_0 = 0$
(d) Both (a) & (b)

Q.40 The rth moment of random variable x is

(a) Independent of change of origin
(b) Independent of change of scale
(c) Dependent of change of scale
(d) Both (a) & (c)

Q.41 For h being the class interval of exclusive class, shepard suggested the following corrected value of μ_2

(a) $\mu_2 - \frac{h^2}{12}$
(b) $\mu_2 - \frac{h}{12}$
(c) $\mu_2 + \frac{h^2}{12}$
(d) $\mu_2 - \frac{2}{3}h^2$

Q.42 The coefficient of skewness f3 defined by Karl Pearson is written as

(a) $\beta_1 = \mu_3^2 / \mu_2^3$
(b) $\beta_1 = \mu_3 / \mu_2^2$
(c) $\beta_1 = \mu_2^2 / \mu_3^2$
(d) $\beta_1 = \mu_3 / \mu_2^2$

Q.43 The coefficient of Kurtosis 132 defined by Karl Pearson can be calculated from the function

(a) μ_4 / μ_2
(b) μ_4 / μ_2^2
(c) μ_2^2 / μ_4
(d) μ_4^2 / μ_2^2

Q.44 The four coefficients $\beta_1, \beta_2, \gamma_1$ & γ_2 defined by Karl Pearson are

(a) Independent of unit of measurement
(b) Dependent of unit of measurement
(c) Free from unit of measurement
(d) Both (a) & (c)

Q.45 The factorial moment $\mu^{i}_{(r)}$ of order r about a point x = b, can be calculated from the expression for a frequency distribution

(a) $\frac{1}{N}\sum_{i=1}^{k} fi\,(xi - b)^{(r)}$
(b) $\frac{1}{N}\sum_{i}^{k} fi\,(xi - b)^{(2r)}$
(c) $\frac{1}{N}\sum_{i}^{k} (xi - b)^{(r)}$
(d) None

Q.46 For a given frequency distribution, the first factorial moment about the origin is equal to

(a) Mean (b) Median

(c) Mode (d) G.M.

Q.47 For a frequency distribution (fi, xi, i = 1, 2, ... , k), the rth absolute moment about the mean x can be calculated from the formula

(a) $\frac{1}{N}\sum_{i}^{k} fi(xi-\bar{x})^r$ (b) $\frac{1}{N}\sum_{i}^{k} fi|xi-\bar{x}|^r$

(c) $\frac{1}{N}\sum_{i}^{k} fi(xi-\bar{x})^{(2r)}$ (d) $\frac{1}{N}\sum_{i}^{k} fi|xi-\bar{x}|^{(2r)}$

Q.48 The skewness of a distribution gives us an idea about the shape of

(a) Curve (b) Histogram

(c) Frequency polygon (d) Bar diagram

Q.49 A frequency distribution is said to be skewed if

(a) Mean ≠ Median (b) Median ≠ Mode

(c) Mean ≠ Mode (d) Mean ≠ Median Mode

Q.50 A distribution is said to be skewed if the quartiles are not at equidistant from

(a) Median (b) Mode

(c) Mean (d) G.M.

Q.51 Measure of skewness based on quartiles and median is defined as

(a) $(Q_3 - Md) + (Md - Q_1)$ (b) $(Q_3 - Md) - (Md - Q_1)$

(c) $(Md - Q_1) - (Q_3 - Md)$ (d) $(Md - Q_1) + (Q_3 - Md)$

Q.52 Karl Pearson's coefficient of skewness Sk is defined as

(a) $\frac{M-Mo}{\sigma}$ (b) $\frac{M-Md}{\sigma}$

(c) $\frac{Md-Mo}{\sigma}$ (d) $\frac{Md-M}{\sigma}$

Q.53 For a moderately asymmetrical distribution having ill-defined mode, coefficient of skewness is defined by the function

(a) 2 (M - Md)/σ (b) 3 (M - Md)/σ

(c) 3 (Md - Mo)/σ (d) 3 (Mo - M)/σ

Q.54 Limits for Karl Pearson's coefficient of skewness Sk is

(a) (-3 ≤ Sk ≤ 3) (b) (-2 ≤ Sk ≤ 2)

(c) (-1 ≤ Sk ≤ 1) (d) None

Q.55 Bowley's coefficient of skewness Sk based on quartiles is calculated from the expression

(a) $\frac{(Q_3 + Q_1 - 2Md)}{(Q_3 - Q_1)}$ (b) $\frac{(Q_3 + Q_1 - 2M)}{(Q_3 - Q_1)}$

(c) $\frac{(Q_3 + Q_1 - 2Mo)}{(Q_3 + Q_1)}$ (d) $\frac{(Q_3 + Q_1 - 2Md)}{(Q_3 + Q_1)}$

Q.56 Bowley's coefficient of skewness liesin an interval given as

(a) (-2 ≤ Sk ≤ 2) (b) (-1 ≤ Sk ≤ 1)

(c) (-1.5 ≤ SK ≤ 1.5) (d) None

Q.57 The curve indicates for the distribution to be

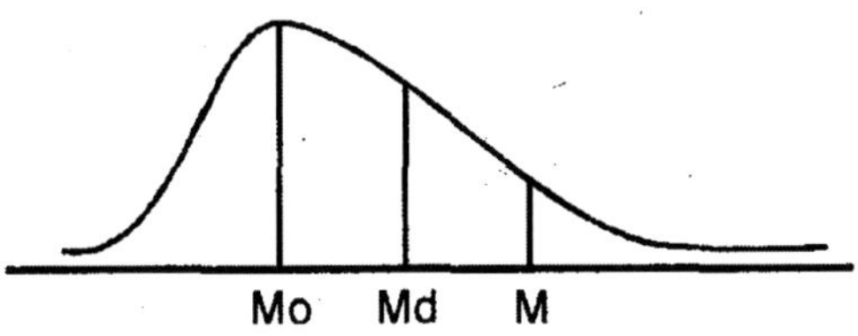

(a) Negatively skewed (b) Positively skewd

(c) Lepto - kurtic (d) Meso - kurtic

Q.58 As reported by Karl Pearson, the convexivity or Kurtosis of the frequency curve gives an idea about the curve of its

(a) Flatness (b) Skewness

(c) Symmetry (d) None

Q.59 Kurtosis of the frequency curve is measured by the coefficient

(a) β_2 (b) β_1

(c) γ_1 (d) $\beta_1 + 1$

Q.60 For normal curve, we have the result

(a) $\beta_2 = 3$ (b) $\beta_2 = 2$

(c) $\gamma_2 = \beta_2 + 2$ (d) $\gamma_2 = \beta_2 + 1$

Q.61 For platy kurtic frequency curve, we have the result

(a) $\beta_2 > 3, \gamma_2 > 0$ (b) $\beta_2 < 3, \gamma_2 < 0$

(c) $\beta_2 < 3, \gamma_2 > 0$ (d) $\beta_2 > 3, \gamma_2 < 0$

Q.62 The value of 5th Decile is equal to

(a) Q_2 (b) Median

(c) Mode (d) Both (a) & (b)

Q.63 The value of 50th percentile is equal to

(a) Median (b) Mode

(c) D_5 (d) Both (a) & (c)

Q.64 The A.M. of x is equal to 5 then A.M. of Y = (3 + 5x) is equal to

(a) 3.8 (b) 28

(c) 35.0 (d) 48.0

Q.65 The difference between upper and lower limits of an exclusive class is known as

(a) Mid value (b) Median

(c) Length of class (d) Class density

Q.66 Let (W1, W2) be the weight of an item taken from two pans of a faulty balance, then the corrected weight of the item will be equal to

(a) $(w_1 w_2)^{\frac{1}{2}}$ (b) $(w_1 + w_2)^{\frac{1}{2}}$

(c) $(W_1 + W_2) / 2$ (d) None of these

Key Answers

Section-A : True False

1.	True	2.	True	3.	True
4.	True	5.	True	6.	True
7.	True	8.	True	9.	True
10.	True	11.	True	12.	True
13.	True	14.	True	15.	True
16.	True	17.	True	18.	True
19.	True	20.	True	21.	True
22.	True	23.	True	24.	True
25.	True	26.	True	27.	True
28.	True	29.	True	30.	True
31.	True	32.	True	33.	True
34.	True	35.	True	36.	True
37.	True	38.	True	39.	True
40.	False	41.	True	42.	True
43.	True	44.	True	45.	True
46.	True	47.	True	48.	True
49.	True	50.	True	51.	True
52.	True	53.	True	54.	True
55.	True	56.	True	57.	True
58.	True	59.	True	60.	True
61.	True	62.	True	63.	True
64.	True	65.	True	66.	True
67.	True	68.	True	69.	True
70.	True	71.	True	72.	True
73.	True	74.	True	75.	True
76.	True	77.	True	78.	True
79.	True	80.	False	81.	False
82.	False	83.	False		

Section -B : Fill in the Blanks

1.	Homogenity, Heterogenity	2	Dispersion, Variation
3.	All	4.	Mathematical
5.	Little, Possible	6.	Simplest, Crude
7.	Extreme	8.	$(Q_3 - Q_1)$
9.	50	10.	Dispersion

11. $|Xi - \bar{x}|$
12. Median
13. Dispersion, Quartile
14. Mathematical
15. Median
16. (Xi - A)
17. Mean square deviation
18. Zero
19. Airthmatic Mean
21. Standard deviatior
21. Variance
22. fi $(Xi - \bar{x})$
23. h^2
24. Independent, Dependent
25. Origin, Scale
26. Independent, Dependent
27. $(\sigma_1^2 + d_1^2), (\sigma_2^2 + d_2^2)$
28. $(n_1\bar{x}_1 + n_2\bar{x}_2)$
29. n (n + 1)
30. Mean deviation
31. Greater
32. Less
33. Standard deviation
34. Coefficient of dispersion
35. Coefficient of variation
36. (A + B)
37. $(Q_3 + Q_1)$
38. The average from which it is calculated
39. $\sigma_x / \bar{x}$
40. $\left(\frac{\sigma_x}{\bar{x}}\right)$
41. Variation or Dispersion
42. Different units
43. Mid point
44. Consistent, Homogeneous
45. fi $(\ )^r$
46. 1, 0, σ^2
47. Origin, Scale
48. Not greater than, Symmetrical, Asymmetrical
49. Class interval
50. $(h^2/L2)$
51. $(\mu_3^2 / \mu_2^3), \sqrt{\beta_1}$
52. $(\mu_4 / \mu_2^2), (\beta_2 - 3)$
53. Unit of measurement
54. (xi - b)
55. (Origin)
56. Arithmatic mean
57. $|Xi - \bar{x}|$
58. $|x_i^r|$
59. "Lack of symmetry"
60. Shape
61. Skewed
62. Skewed
63. (Mean - Median)
64. (Mean - Mode)

65.	Median, Q_1	66. (Mean - Mode)
67.	(Mean - Median)	68. (-3, 3)
69.	(- 2 Md)/ $(Q_3 - Q_1)$	70. $(-1 \leq Sk \leq 1)$
71.	Positively	72. Flatness, Peakness
73.	β_2 or γ_2	74. 3, zero
75.	Miso Kurtile	76. -<3,<0
77.	->3,> 0	78. 9.0
79.	28.0	80. Class interval
81.	$\sqrt{W_1W_2}$	

Section-C : Multiple Choice Questions

1.	(d)	2.	(d)	3.	(d)
4.	(d)	5.	(d)	6.	(c)
7.	(c)	8.	(a)	9.	(c)
10.	(d)	11.	(d)	12.	(d)
13.	(d)	14.	(d)	15.	(d)
16.	(a)	17.	(a)	18.	(a)
19.	(c)	20.	(b)	21.	(a)
22.	(b)	23.	(b)	24.	(d)
25.	(b)	26.	(a)	27.	(a)
28.	(a)	29.	(a)	30.	(a)
31.	(a)	32.	(b)	33.	(a)
34.	(a)	35.	(d)	36.	(d)
37.	(b)	38.	(d)	39.	(d)
40.	(a)	41.	(a)	42.	(b)
43.	(d)	44.	(a)	45.	(a)
46.	(a)	47.	(b)	48.	(a)
49.	(d)	50.	(a)	51.	(b)
52.	(a)	53.	(b)	54.	(a)
55.	(a)	56.	(b)	57.	(b)
58.	(a)	59.	(a)	60.	(a)
61.	(a)	62.	(d)	63.	(d)
64.	(b)	65.	(c)	66.	(a)

❑❑❑

Chapter - 5

Theory of Probability

Section-A : Write True / False

Q.1 If an experiment is repeated under an identical conditions, we generally come to a result (i.e. outcome) which is unique or certain.

Q.2 In an experiment, the result (outcome) may not be unique but it is one of the several possible outcomes.

Q.3 The phenomena described in Ql, are known as "deterministic or predictable".

Q.4 In deterministic phenomena, the results can be predicted with certainity.

Q.5 A deterministic model stipulates that the conditions under which an experiment is performed, determine the ouctome of the experiment

Q.6 The phenomena which do not lead themselves to deterministic approach, are known as "Unpredictable or probabilistic" phenomena.

Q.7 The phenomena frequently occured in business, economic and social science or even in our day to day life, are known as "deterministic phenomena".

Q.8 A manufacturer can ascertain the future demand of his product with certainity.

Q.9 A sales manager can not predict with certainity about the sales target next year.

Q.10 If an electric bulb has lasted for one year, nothing can be predicted about its future life.

Q.11 Galileo (1564-1642), an Italian mathematician, was the first man to make an attempt to measure probability in a quantitative way.

Q.12 Galileo, first time dealt with some problems related to the theory of dice in gambling.

Q.13 A numerical mesure of uncertainity is provided by a very important branch of mathematics called the "The theory of probability".

Q.14 The probability theory describes certainity by the quantity 1.0.

Q.15 The theory of probability describes "Impossibility" by the quantity 0.0.

Q.16 The probability theory describes the various grades of uncertainities by a coefficient denoted by p and ranging as $(0 < p < 1.0)$

Q.17 B. Pascal (1623-62) and P. Fermat (1601-65) laid the first foundation of the "Science of probability".

Q.18 The next stalwart in the theory of probability was "Bernouillie" (1654-1705) whose "Treatise on Probability" was published post humously by his nephew, N Bernouille in 1713.

Q.19 Chebychew (1821-94) founded the "Russian School of Statisticians".

Q.20 An experiment conducted under identical conditions in which if an outcome is not unique, but may be any one of the several possible outcomes, in known as "Random experiment".

Q.21 Tossing a coin, throwing a die, selecting a card from a pack of playing cards etc are not the examples of random experiments.

Q.22 The result of a random experiment, is called an "Outcome".

Q.23 Any perticular performance of a random experiment is called a trial.

Q.24 The ouctcome or the combinations of outcomes are khown as "Event".

Q.25 An event is called simple if it corresponds to a single possible outcome of the experiment otherwise it is known as a "Compound or composite event".

Q.26 The total number of all possible outcomes of a random experiment, is known as the "exhaustive events or cases".

Q.27 In drawing two cards from a pack of cards, the exhaustive number of event is equal to 52_{C_2}.

Q.28 In throwing n dice, the exhaustive number of cases, is 6^n.

Q.29 In drawing a card from a pack of cards, the number of cases favourable to draw an ace is equal to 4.

Q.30 In throwing of two dice, the number of cases favourable to getting the sum 7 is equal to 6.

Q.31 It two or more than two events can not happen simultaneously in the same trial, then such events are known as "Mutually exclusive or incompatible" events.

Q.32 In throwing a die, all the 6 faces numbered 1 to 6 are mutually exclusive events.

Q.33 If taking into consideration all the relevant evidenes, that there is no reason to expect one outcone in preference to other outcomes, then such events are known as "Equally Likely Events".

Q.34 In throwing an unbiased die, the event of numbers from 1 to 6, to come on upper most face, are equally likely.

Q.35 If the happening or non-happening of an event is no affected by the happening or non happening of the other events then such events are known as "Independent Events".

Q.36 When a die is thrwon twice, the result of rist throw does not affect the results of second throw.

Q.37 If a random experiment or trial results in "n" exhaustive, mutually exclusive and equally likely outcomes, out of which m are favourable to the occurrence of an event "E" then the probability of happening of event "E" is given as $p = \left(\frac{m}{n}\right)$.

Q.38 The definintion of probability of happening of event E given as $P = \left(\frac{m}{n}\right)$ in situation described in Q37 was given by "James Bernouille"

Q.39 The limits of probability of happening an event E lies as $(0 \leq P(E) \leq 1)$.

Q.40 Sometimes, we say that "the odds in favour of E are m : (n - m)" or the "the odds against E are (n - m) : n".

Q.41 The non happening of an event E is called the "Complementary event of E".

Q.42 The complementary event of E is denoted by $\overline{E}$ or by E.

Q.43 The relation $P(E) + P(\overline{E}) = 1$, holds good.

Q.44 P (Happening of atleast one of the events $E_1, E_2, \ldots\ldots, E_n$) = 1 - P (None of the events $E_1, E_2, \ldots\ldots., E_n$ happens).

Q.45 Probability "p" of the happening of an event E is also known as the probability of success of even E.

Q.46 The probability of failure of an event E is denoted by $q = (1 - p)$.

Q.47 If P (E) = 1, then the event E is called sure event.

Q.48 If P (E) = 0, then the event E is called "An Impossible" event.

Q.49 The definition of probability $P(E) = \frac{m}{n}$ given in Q37, is known as (classical or "a priori" or mathematical) defnition of probability.

Q.50 If the various outcomes of the random experiment are not equally likely and exhaustive number of outcomes is infinite then the mathematical definition of probability fails.

Q.51 If in N trials an event E happens M times, then the probability of the happening of E, denoted by P (E), is given by

$$P(E) = \lim_{N \to \infty} \frac{M}{N}$$

Q.52 The probability defined in Q51, is known as "Statistical or Empirical Probability".

Q.53 If an experiment is repeated a large number of times, the experiment outcomes may not remain identical and homogenous.

Q.54 The limit in Q51 may not attain a unique value, however large N may be.

Q.55 The probability that a leap year selected at random will contain 53 sunday is equal to $\frac{2}{7}$.

Q.56 If two unbiased dice are thrown then probability that both the dice show the same number is equal to $p = 6 / 36 = \frac{1}{6}$

Q.57 If two dice are thrown then probability that the total of numbers on two dice is greater than 8 is equal to $\frac{10}{36} = \frac{5}{18}$

Q.58 The n persons are seated in n chairs at a round table. The probability that two specified persons are sitting next to each other is equal to

$$\frac{(n-2)!2!}{(n-1)!} = \frac{2}{(n-1)}$$

Q.59 A car is parked among N cars in a row not at either end. One his returned, the owner find that eactly r of the N places are still occupied. The probability that both neighbouring places are empty will be equal to $=({}^{n-3}C_{r-1})/({}^{n-1}C_{r-1})$

Q.60 Set theory was developed by the German mathematician, G. Cantor (1845-1918).

Q.61 A set is well defined as collection or aggregate of all possible objects having properties and specified according to a well defined rule.

Q.62 The objects consitituting a set 'A' are called elements, members or points of set A.

Q.63 If A is a subset of B then we write symbolically as $A \subseteq B$ or $B \geq A$

Q.64 Two sets A and B are said to be equal or identical if $A \subseteq B$ and $B \subseteq A$ and we write A = B or B = A.

Q.65 A null set or an empty set is one which does not contain any element at all and is denoted by ϕ).

Q.66 An empty set ϕ is the subset of every set.

Q.67 The union of two sets A and B denoted by $A \cup B$ and is defined as a set consisting all those elements which belongs to either A or B.

Q.68 Symbolically, $A \cap B = \{x : x \in A \text{ and } x \in B\}$ and

$\bigcap_{i=1}^{n} Ai = \{x : x \in Ai \text{ for atleast one } i = 1, 2, \ldots.., n\}$

Q.69 The intersection of two sets A and B denoted by A n B is defined as a set consisting of those elements which belongs to both A and B.

Q.70 $A \cap B = \{x : x \in A \text{ and } x \in B\}$

Q.71 $\bigcap_{i=1}^{n} Ai = \{x : x \in Ai, \text{ for all } i = 1, 2, \ldots\ldots, n\}$

Q.72 If A and B have no 'common points i.e. $A \cap B = \phi$, then sets A and B are said to be disjoint, or mutually exclusive or non overlaping.

Q.73 The difference of two sets A and B, denoted by A - B is defined as a set consisting of those elements of A which do not belong to B i.e.

$A - B = \{x : x \in A \text{ and } x \notin B\}$

Q.74 The complement or negative of any set A denoted by A is a set containing all elements of the universal set S that are not eleinent of A i.e. $\overline{A} = S - A$.

Q.75 Let A, B and C be the subsets of an universal set S, then

(a) Coinmulative laws : $A \cup B = B \cup A$
$A \cap B = B \cap A$

(b) Associative laws : $(A \cup B) \cup C = A \cup (B \cup C)$
$(A \cap B) \cap C = A \cap (B \cap C)$

(c) Distributive law : $A \cap (B \cup C) = (A \cap C) \cup (A \cap C)$
$A \cup (B \cap C) = (A \cup C) \cap (A \cup C)$

(d) Complementary laws: $A \cup \bar{A} = S,\ A \cap \bar{A} = \phi$
$A \cup S = S,\ A \cap S = A$
$A \cup \phi = A,\ A \cap \phi = \phi$

(e) Difference law : $A - B = A \cap \bar{B}$
$A - B = A - (A \cap B) = (A \cup B) - B$
$A - (B - C) = (A - B) \cup (A - C)$
$(A \cup B) - C = (A - C) \cup (B - C)$
$A - (B \cup C) = (A - B) \cap (A - C)$
$(A \cap B) \cup (A - B) = A,\ (A \cap B) \cap (A - B) = \phi$

Q.76 De-Margan's Law :

$$\left(\overline{\bigcup_{i}^{n} Ai}\right) = \bigcap_{i}^{n} Ai \quad \text{and} \quad \left(\overline{\bigcap^{n} Ai}\right) = \bigcup_{i}^{n} Ai$$

Q.77 Involution Law : $(\bar{\bar{A}}) = A$

Q.78 Indempotency Law : $A \cup A = A,\ A \cap A = A$

Q.79 Let $\{A_n\}$ be a sequence of sets in S. The limit supermum or limit superior of the sequence usually written as lim sup A_n is the set of all those elements which belong to A_n for infinitely many n.

Thus $\lim\limits_{N \to \infty} \sup A_n = \{ x : x \in A_n \text{ for infinitely many } n\}$

Q.80 The set of all those elements which belong to A_n for all but a finite number of n called limit infinimum or limit inferior of the sequence and is denoted by lim. inf. A_n.

Thus $\lim\limits_{N \to \infty} \text{inf. } A_n = \{ x : x \in A_n \text{ for all but a finite } n\}$

Q.81 The sequence $\{A_n\}$ is said to have a limit if lira. sup. A_n = lim inf. A_n and the common value gives the limit of the sequence.

Q.82 lim sup $A_n = \bigcap_{m=1}^{\infty}\left(\bigcup_{n=m}^{\infty} A_n\right)$ and lim inf $A_n = \bigcup_{m \to 1}^{\infty}\left(\bigcap_{n=m}^{\infty} A_n\right)$

Q.83 $\{A_n\}$ is a monotone (infinite) sequence of sets if either (1) $A_n C A_{n+1} \forall n$ or $A_{n+1} \subset A_n \forall n$.

Q.84 If $A_n \subset A_{n+1}$ then the sequence $\{A_n\}$ is said to non decreasing (or increasing) sequence and is usually expressed as $A_n \uparrow$.

Q.85 If $A_n \supset A_{n+1}$ then the sequence $\{A_n\}$ is said to be non-increasing (or decrasing) sequence and is denoted by $A_n \downarrow$.

Q.86 For a monotone sequence (non-increasing or non decreasing) the limit always exists and we write

$$\lim_{n \to \infty} A_n = \begin{cases} \bigcup_{n \to 1}^{\infty} A_n \text{ in case } A_n \uparrow \\ \bigcap_{n=1}^{\infty} A_n \text{ in case } A_n \downarrow \end{cases}$$

Q.87 A ring R is a non-empty class of sets which is closed under the formation of finite 'union or 'difference' i.e.

if $A \in R$, $B \in R$,then $A \cup B \in R$, and $A - B \in R$

Q.88 A field F (or an algebra) is a non empty class of sets which is closed under formation of finite union and complementation. Thus

(i) $A \in F, B \in F \Rightarrow A \cup B \in F$ and (ii) $A \in F \Rightarrow \bar{A} \in F$

Q.89 A a ring is a non empty class of sets which is closed under the formation of countable union and difference. Thus

(i) $Ai \in C, i = 1, 2, \ldots.. \Rightarrow \bigcup_{i=1}^{\infty} Ai \in C$

(ii) $A \in C, B \in C \Rightarrow A - B \in C$

Q.90 The axiomatic definition of probability includes both the classical and statistical definitions as particular cases and overcomes the deficiency of each of them.

Q.91 The axiomatic approach of mathematical theory of probability relies entirely upon the "logic of deduction".

Q.92 The results of a trial in a random experiment is called an outcome or an elementary event or a sample point.

Q.93 The totality all possible outcomes of a random experiment constitues the "sample space".

Q.94 Tha sample space of a random experiment is usually denoted by S or Ω and the outcomes of the experiment are denoted by e_1, e_2, e_3, ... or by w_1, w_2, w_3, ...

Q.95 A sample space is called "discreted" if it contains only finitely or infinitely many points which can be arranged into a simple sequence w_1, w_2, ... while a sample space containing nondenumerable number of points is called a "continous sample space".

Q.96 Of all possibloe outcomes in the sample space of an experiment, some outcomes statisfy a specified discription and we call it as an event.

Q.97 An empty set ϕ is a sub set of S and ϕ is also an event, known as impossible event.

Q.98 P (A) is the probability function defined on a σ field B of evets if the following properties (axioms) hold.
1. For each A ∈ B, P (A) is defined, is real and P (A) ≥ 0
2. P (S) = 1
3. If $[A_n]$ is any definite or infinite sequence of disjoint events in B then $P\left(\bigcup_{i=1}^{n} Ai\right)=\sum_{i}^{n} P(Ai)$

Q.99 The set function P defined on σ field B, taking its value in real line and satisfying the above three axioms, is called the "Probability measure".

Q.100 The triplet {S, B, P} is often called probability space.

Q.101 The σ algebraB is taken to be the collection of all subsets of S.

Q.102 The classical probability may be regarded as a special case of axiomatic probability.

Q.103 The statistical or frequency approach of probability may be regarded as a special case of axiomatic probability.

Algebra of Events

Q.104 For the events A, B, C we have
$A \cup B = \{w \in S; w \in A \text{ or } w \in B\}$

Q.105 $A \cap B = \{w \in \Omega; w \in A \text{ and } w \in B\}$

Q.106 $\overline{A}$ (A complement) = $\{w \in S : w \notin A\}$

Q.107 $A - B = \{w \in \Omega; w \in A \text{ but } w \notin B\}$

Q.108 $A \subset B \Rightarrow$ forevery $w \in A$, $w \in B$

Q.109 A = B if and only if A and B have same domen i.e. if A B and $B \subset C$

Q.110 A and B are disjoint (mutually exclusive) if $A \cap B = \phi$ (null set).

Q.111 We have $A \cap B = A + B$ if A and B are disjoint.

Q.112 Since all the events are subsets of S, all the laws of set theory hold for algebra of events.

Q.113 Atleast one of the events A or B occurs $\rightarrow w \in A \cup B$

Q.114 Neither A nor B occurs $\Rightarrow w \in \overline{A} \cup \overline{B}$

Q.115 Event A occurs and B does not occur $\Rightarrow w \in A \cup \overline{B}$

Q.116 Probability of complementary event $\overline{A}$ of A is gfiven by $P(\overline{A}) = 1 - P(A)$

Q.117 $P(\overline{A} \cap B) = P(B) - P(A \cap B)$

Q.118 If $B \subset A$ then $P(A \cap \overline{B}) = P(A) - P(B)$

Q.119 If A and B are two events and are not disjoints then
$P(A \cup B) = P(A) + P(B) - P(A \cap B)$

Q.120 For three non mutually exclusive events A, B & C
$P(A \cup B \cup C) = P(A) + P(B) + P(C) - P(A \cap B) - P(B \cap C) - P(C \cap A) + P(A \cap B \cap C)$

Q.121 The probabilities $P(A_1), P(A_2), \ldots\ldots, P(A_n)$ of mutually exclusive forms of A, are known as "partial probabilities"

Q.122 According to Boole's inequality for n events A1, A2, ... , A_n, we have

$$P\left(\bigcup_{i=1}^{n} Ai\right) \geq \sum_{i=1}^{n} P(Ai) - (n-1)$$

Q.123 $P\left(\bigcup_{i=1}^{n} Ai\right) \leq \sum_{i=1}^{n} P(Ai)$

Q.124 For n events A1, A2, ..., A_n

$$P\left(\bigcup_{i=1}^{n} Ai\right) \geq \sum_{i=1}^{n} P(Ai) - \sum_{1 \leq i \leq j \leq n} P(A_i \cap A_j)$$

Q.125 For two events A and B, we have

$$P(A \cap B) = P(A).P(B / A), P(A) > 0$$
$$\text{or} = P(B).P(A / B), P(B) > 0$$

where P (B/ A) represents conditional probability of occurence of B when the event A has already happend.

Q.126 An event A is said to be independent of another event B if P (A/B) = P (A) if P (B) $\neq$ 0.

Q.127 If the events A and B are such that P (A) $\neq$ 0, P (B) $\neq$ 0 and A is independent of B then B is independent of A if $P(A \cap B) = P(A)p(B)$

Q.128 For a fixed event B, with P (A) $\neq$ 0 we have

$$P(S/B) = \frac{P(S \cap B)}{P(B)} = \frac{P(B)}{P(B)} = 1$$

Q.129 $P(A \cup B / C) = P(A / C) + P(B / C) - P(A \cap B / C)$

Q.130 If A and B are independent events then A and $\overline{B}$, $\overline{A}$ and B, and $\overline{A}$ & $\overline{B}$ are also independent.

Q.131 **Pair wise Independent Events :**

The events A1, A2, ..., A_n are said to be pair wise independent if

$$P(A_i \cap A_j) = P(Ai)P(Aj), i \neq j = 1,2,......., n$$

Q.132 **Mutually Independent Events :**

The events A1, A2, ..., A_n are said to be mutually independent if

$$P(A_{i1} \cap A_{j2} \cap A_{i3} \cap ....A_{ik}) = P(A_{i1})P(A_{j2})......P(A_{ik}),$$

k = 2, 3, ..., n

Q.133 Total number of conditions for mutual independent of A_1, A_2, ..., A_n is equal to $(2^n - n - 1)$

Q.134 If A, B, C are mutually indepedent events then $A \cup B$ and C are also independent.

Q.135 If A, B and C are random events in a sample space and if A, B, C are pairwise independent and A is independent of $B \cup C$ then A, B and C are mutually independent.

Q.136 P (Happening of at least one of the events A_1, A_2, ..., A_n) = 1 - P (None of the events A_1, A_2, ..., A_n, happens)

Q.137 For any two events A and B, we have
$P(A \cap B) \leq P(A) \leq P(A \cup B) \leq P(A) + P(B)$

Q.138 **Bayes's Theorem :** Thomas Bayes was a British Mathematician who gave his well known "Bayes Theorem) in 1763.

Q.139 Bayes theorem states, that if E_1, E_2, ..., E_n are mutually disjoint events with (PEi) ≠ 0, (i = 1, 2, ..., n) then for any arbitrary event A which is subset of $\bigcup_{i=1}^{n} Ei$ such that P (A) > 0, we have

$$P(Ei / A) = \frac{P(Ei)P(A / Ei)}{\sum_{i=1}^{n} P(Ei)P(A / Ei)} = \frac{P(Ei)P(A / Ei)}{P(A)} \quad i = 1, 2, ..., n$$

Q.140 To deduce the probability from mathematical concepts, some statements are made in terms of properties of some concepts. These propertires are often known as "Axioms" or "Rules" of Probability theory.

Section-B : Fill in the Blanks

Q.1 An outcome of an experiment is one of the several ....

Q.2 A deterministic model states that the condition under which an experiment is performed, determines the outcome of an ....

Q.3 A phenomena which do not lead to a deterministic approach, are known as "Unpredictable or phenomena.

Q.4 The phenomena frequently occured in business, economic and social science or even in our day to day life are know as " .... " phenomena.

Q.5 If an electric bulb has lasted for one year, " .... " can be predicted about its future life.

Q.6 Galileo (1564-1642) an Italian mathematician, was the first man to make an attempt to measure " .... " in a quantitative way.

Q.7 Galileo, first time dealt with some problems related to the theory of dice in " ...." .

Q.8 A numerical measure of uncertainity is provided by the theory of " ...." .

Q.9 The probability theory describes the certainity of an event by a quanity equal to ....

Q.10 The theory of probability describes the impossibility of happening of an event, equal to ....

Q.11 The probability theory describes the various grades of uncertainities by a coefficient denoted by p which lies in limits (.....................)

Q.12 B pascel (1623-62) and P Fermat (1601-65) laid the first foundation of the "Science of ...." .

Q.13 The next stal wart in the theory of probability was "J. ...." .

Q.14 Chebechev founded the "Russian school of ...." .

Q.15 An experiment conducted under identical conditions in which if an outcome is not unique but may be anyone of the several possible outcomes, is known as " ...." .

Q.16 Tossing of a coin or throwing of a dice is an example of a "................ experiment".

Q.17 The result (observation) occured in a random experiment, is called an " ...." .

Q.18 Any particular performance of a random experiment is called a " ...."

Q.19 The outcomes or the combination of outcomes of a random experiment, are known as " ...." .

Q.20 An event of a random experiment which is the combination of two or more outcomes, is known as "Composite or ...." event.

Q.21 Total number of all possible outcomes of a random experiment is known as " .... " number of events or cases.

Q.22 In drawing two cards from a pack of playing cards, the exhaustive number of events is equal to ....

Q.23 In throwing n dice, the exhauctive number of events is equal to ....

Q.24 In throwing two dice, the number of cases favourable to get the sum 7 is equal to ....

Q.25 If two or more events can not happen simultaneously in the same trial, then such events are known as " .... or incomptible" events.

Q.26 In throwing a die, all the 6 faces numbered 1 to 6 are "....................... exclusive" events.

Q.27 If all the outcomes of a random experiment, have some probability to happen then such events are known as "................ likely events".

Q.28 In throwing an unbiased die, the six faces numbered from 1 to 6 to come on upper most face, are " .... " likely.

Q.29 If the happening of an event is not affected by the happening of other events then such events are known as " ...." .

Q.30 When a die is thrown twice, the results of first throwing is "........................" of the results of second throw.

Q.31 The classical definition of probability of happening of an event "E" was given by " .... "

Q.32 Some times we say that "odds in favour of E are m : (n - m)" or odds against E are .... " .

Q.33 The non happening of an event E is called the " .... event of E".

Q.34 The complementary event of E is denoted by ....

Q.35 The relation P (E) + P $(\overline{E})$ = ...

Q.36 P (Happening of atleast one of the events E_1, E_2, ..., E_n) = 1 - P (............. of the events E_1, E_2, ..., E_n) happens.

Q.37 The probability "p" of the happening of an event E is also known as the " .... of success" of event E.

Q.38 If "p" is the probability of success of an event E then the probability of failure of the event is denoted by q and q = ...

Q.39 If P (E) = 1.0 then the event E is known as ....

Q.40 If P (A) = 0, then the event A is known as an (...............) event.

Q.41 Definition of probability P (E) = $\frac{m}{n}$, (where m is the number of favourable events to E and n is the exhaustive number of event) is known as classical or "prior or .... " definition of probability.

Q.42 If the various outcomes of a random experiment are not equally likely and exhaustive number of events are not finite then the " ...." definition of probability fails.

Q.43 If in N trials, an event E happens M times then the probability of happening of E, denoted by P (E) is defined as P (E) = ...

Q.44 The definition of "probability" given in Q (43) is known as " .... or .... probability".

Q.45 If an experiment is repeated a large number of times, then experimental outcomes may not remain " .... or homogeneous".

Q.46 The probability that a leap year selected at random will contain 53 sunday is equal to ....

Q.47 If two unbiased dice are thrown simulteneously then the probability that both dice will show the same number on its upper most face is equal to ....

Q.48 If two dice are thrown then the probability that the total of numbers on upper most faces is greater than 8 is equal ....

Q.49 The n persons are seated in n chairs at a round table. The probability that two specified persons are sitting next to each other is equal to ....

Q.50 The probability that a ceiling fan in the class will fall is equal to ....

Q.51 Set theory was developed by "German-mathematician" named " .... " during (1845-1918).

Q.52 The objects consistituting a set A, are called " .... or member or points" of the set A.

Q.53 If A is a subset of another set B then we write it symbolically as A .... B or B .... A.

Q.54 Two sets A and B are said to equal or identical if A .... B and B .... A and we write A = B or B = A.

Q.55 A null set or "an .... set" is one which does not contain any " .... " at all and is denoted by ....

Q.56 An empty set ϕ is the subset of " .... " set.

Q.57 The union of two sets A and B is denoted by .... and is defined as the set containing all those elements which belong to " .... ".

Q.58 Symbolically, $A \cup B$ = { x:.....................}

Q.59 Symbolically $\bigcup_{i=1}^{n} A_i$ = {x:x $\in$ Ai for .... one i=1,2,..........,n}

Q.60 The intersection of two sets A and B is denoted as ....

Q.61 An interaction $A \cap B$ of two sets A and B is defined a a set consisting of those elements which belongs to ....

Q.62 $A \cap B = \{w : w \in$ ....$\}$

Q.63 $\bigcup_{i=1}^{n} Ai = \{w : w \in Ai$ for ....$\}$

Q.64 If A and B have no common elements then $A \cap B =$ ....

Q.65 If $A \cap B = \phi$, then the two sets A and B are said to be " ...." or non overlaping or " .... " sets.

Q.66 The difference of two sets A and B are denoted by ....

Q.67 The difference of two sets A and B, is defined as a set A - B contaning of those elements of " .... " which do not belong to ....

Q.68 Symbolically $A - B = \{w : w \in$ .... and ω........$B\}$

Q.69 The complement of any set A is denoted by .... or ....

Q.70 The $\overline{A}$ is a set containing all elements of the universal set S that are .... of A.

Q.71 Symbolically, we have $\overline{A} = S$ ....

Q.72 Let A, B be the subsets of an universal set S, then according to commutative law, and

$A \cup B =$ .... and $A \cap B =$ ....

Q.73 For three sets A, B, C,the associative law says that $(A \cup B) \cap C =$ ....

Q.74 For three set A, B, C, the associative law states that $(A \cap B) \cap C =$ ....

Q.75 For three sets A, B, C, the distributive law state that $A \cup (B \cap C) = (A \cup B)$ ....

Q.76 For three sets A, B, C, the distributive law states that $A \cap (B \cup C) =$ .... $\cap (A \cap C)$

Q.77 For a set A and universal set S, the complementary law states that $A \cup \overline{A} =$ ....,and $A \cap \overline{A} =$ ....

Q.78 For a set A and universal set S, the complementary law says that $A \cup S =$ ...., and $A \cap S =$ ....

Q.79 For a null set ϕ and other set A, the complementary law says that $A \cup \phi =$ ...., and $A \cap \phi =$ ....

Q.80 For two sets A and B, the difference law says that $A - B = A \cap$............

Q.81 For two sets A and B, the difference law says that $A - B = A - ($.........$) = (A \cup B)$ ....

Q.82 For three sets A, B, C, the "Difference law" states that $A - (B - C) = (A - B)$ .... $(A - C)$

Q.83 For three sets A, B, C the "Difference law" states that $(A \cup B) - C = (A - C)$.....$(B - C)$

Q.84 For three sets A, B, C, the "Difference law" states that $A - (B \cup C) = (A - B)$....$(A - C)$

Q.85 For three sets A, B, C, the "Difference law" states that $(A \cap B) \cup (A - B) =$ ....and $(A \cap B) \cap (A - B) =$ ....

Q.86 De-Margon's Law says that

$$\left(\overline{\bigcup_{i=1}^{n} Ai}\right) = \bigcap_{i=1}^{n} \bar{A}i \text{ and } \left(\overline{\bigcap_{i=1}^{n} Ai}\right) = \ldots\ldots\ldots\ldots$$

Q.87 Involuation law states that $(\bar{\bar{A}}) =$ ....

Q.88 Idempotency law says that

$A \cup A =$ ...., and $A \cap A =$ ....

Q.89 Let $\{A_n\}$ be a sequence of sets in S. The limit supremum or limit superior of the sequence is written symbolically as

$\lim_{N \to \infty} \sup A_n = \{ w : w \in An$.......$\}$

Q.90 The set of all those elements which belongs to A_n for all but a finite number of n is called : lim .... or .... inferior, of the sequence.

Q.91 Limit infinimum .... of a sequence of set $\{A_n\}$ is symbolically written as

$\lim_{n \to \infty} \inf A_n = \{ w : w \in An \text{ for } ....\}$

Q.92 The sequence $\{A_n\}$ is said to have a limit iff :
$\lim_{n \to \infty} \sup A_n =$ .... and the common value gives the
" ...."

Q.93 Lim sup An and lim inf An can be written symbolically as

lim sup An = $\bigcap_{m=1}^{\infty}$ (..........) and lim inf An = $\bigcup_{m=1}^{\infty}$ (..........)

Q.94 The sequence {An} of sets is a monotone (infinite) sequence of sets if
$A_n \subset A_{n+1}$ .... or $A_n \supset$ .... $\forall$ n

Q.95 If $A_n \subset$..........., then sequence {An} is said to non decreasing (..............)sequence and is usually expressed as ....

Q.96 If $A_n \supset$ ...., then the sequence {An} is said to be non increasing (or ....) sequence and is denoted by An $\downarrow$

Q.97 For a monoton sequence, the limit always exists and we write

$$\lim_{n\to\infty} A_n = \begin{cases} \text{......................in case } A_n \uparrow \\ \text{......................in case } A_n \downarrow \end{cases}$$

Q.98 A ring R is a non empty class of sets which is closed under formation of finite " .... " or " ...."

Q.99 A ring R is a " .... " set of class say A and B if A $\in$ R,B $\in$ R then .... $\in$ R. and A - B ....

Q.100 A field F (or an algebra) is a " .... " class of sets which is closed under the formation of finite " .... and ...."

Q.101 If A, B are two sets and F is a field then we have
A $\in$ F, B $\in$ F $\Rightarrow$ A $\cup$ B $\in$ ....

Q.102 For a set A and field F, when A $\in$ F $\Rightarrow$ $\overline{A}$................

Q.103 A σ ring is a " .... " class of sets which is closed under the formation of countable " .... and .... " .

Q.104 Let Ai, i = 1, 2, ... be a sequence of countalbe sets, and C is a σ ring then we have

Ai ϵ C,i =1,2,....... ⇒ .... ϵ C

Q.105 Let A and B be two sets and C is a σ ring then we have

A ϵ C, B ϵ C ⇒ A - Bϵ ....

Q.106 The axiomatic definition of probability includes both " .... and .... " definitions of probability as particular cases and overcomes the defficiency of " .... " .

Q.107 The axiomatic approach of mathematical theory of probability relies entirely upon the " .... "

Q.108 The result of a trial in a random experiment is called an " .... " or an " .... " event or a sample point.

Q.109 The totality of all possible outcomes of a random experiment constitutes the " .... " .

Q.110 The sample space of a random experiment is usually denoted by " .... or .... "

Q.111 A smaple space contaning only finitely or infinitely many points which can be arranged into a simple sequence w_1, w_2, w_3, ..., is called " .... " sample space.

Q.112 A sample space containing non-denomuerable number of points, is called " .... " sample space.

Q.113 Out of all possible outcomes of a sample space, some outcomes staisfy a specified discriptions which are known " .... " .

Q.114 An empty set ϕ i.e. null set is also known as " .... " event.

Q.115 Let P (A) be a probability function defined on a σ field B of events then

A ϵ B, P (A) is real and P(A) ≥ ....

Q.116 Let S be a sample space and P (A) is the probability function then P (S) = ...

Q.117 If {An} is an infinite or finite sequence of disjoint events in s field B, then we have $P\left(\bigcup_{i=1}^{n} Ai\right) =$ ....

Q.118 The set function P defined on σ field B, taking its value in real line and satisfying some (three) axiom, is known as :........... "

Q.119 The triplet {S, B, P} is often known as " .... "

Q.120 The σ algebra B is taken to be the collection of all "subsets of .... "

Q.121 The statistical or frequency approach of probability may be regarded as a special case of " .... "

Q.122 The properties of some statements made about the mathematical concept of probability are often known as " .... or .... "of probability theory.

Q.123 For the events A, B, C and sample space S, we can write symbolically

$A \cup B = \phi$ {w ϵ S : w ϵ A or ....}

Q.124 $A \cap B$ = {x ϵ Ω : x ϵ A ....}

Q.125 $\overline{A}$(A complement) = { w ϵ Ω : ....}

Q.126 A – B = {w ϵ Ω : w ϵ A but.........}

Q.127 $A \subset B \Rightarrow$ for every w ϵ A, ...

Q.128 A = B if $A \Rightarrow B$ and ....

Q.129 A and B are mutually exclusive if $A \cap B =$ ....

Q.130 We have $A \cup B = A+B$ if A and B are ....

Q.131 Atleast one of the events A or B occurs can be symbolically written as : w ϵ ....

Q.132 Neither A nor B occurs can be symbolically written as w ϵ ....

Q.133 Event A occurs and B does not occur can be symbolically written as w ϵ ....

Q.134 Probability of complementary event $\overline{A}$ of A is given by P $(\overline{A})$ = 1 - ....

Q.135 $(\overline{A} \cap B)$ = P(B) — P(.........)

Q.136 if B⊂ A then P(A ∩ $\overline{B}$) = P(A) — P(........)

Q.137 If A and B are not disjoint events then
P(A ∪ B) = P(A) + P(B) — P(.......)

Q.138 For not mutually exclusive events A, B and C, we have
P(A∪B∪C)=P(A)+P(B)+P(C)—P(A∩B)—P(B∩C)—P(.......)+P(......)

Q.139 The probabilities P (A_1), P (A_2), ..., P (A_n) of mutually exclusie forms of event A, are known as (.............).

Q.140 For n events A_1, A_2, ..., A_n Boole's inequality states that

$$P\left(\bigcap_{i=1}^{n} Ai\right) \geq \ldots\ldots\ldots\ldots$$

Q.141 $P\left(\bigcup_{i=1}^{n} Ai\right) \ldots\ldots\ldots\ldots \sum_{i=1}^{n} P(Ai)$

Q.142 For n events A_1, A_2, ..., A_n

$$P\left(\bigcup_{i=1}^{n} Ai\right) \geq \ldots\ldots\ldots\ldots - \sum_{1 \leq i \leq j \leq n} P(Ai \cap Aj)$$

Q.143 For two events A and B, we have
P(A ∩ B) =P (A)..P (.........) forP(A) ≠ 0
=P (B)..P(..........) forP(B) ≠ 0

Q.144 An event A is said to independent of another event B if
P (A/B) = P (........) for P (B) ≠ 0

Q.145 For P (A) ≠ 0 and P (B) ≠ 0, the two events A and B are said to be independent if
P(A ∩ B) = ...

Q.146 For a fixed event B with P (A) ≠ 0, we have

$$P(S/B) = \frac{P(S \cap B)}{P(B)} = \frac{P(\ldots)}{P(B)} = \ldots\ldots\ldots$$

Q.147 $P(A \cup B / C) = P(A/C) + P(B/C) - P(\text{...........})$

Q.148 If A and B are independent events then A & $\overline{B}$, $\overline{A}$ & B and $\overline{A}$ & $\overline{B}$ are also ....

Q.149 The sequence of events $A_1, A_2, \ldots\ldots, A_n$ are said to be pairwise independent iff

$P(Ai \cap Aj) =$ .... for $i \neq j = 1, 2, \ldots, n$

Q.150 The events $A_1, A_2, \ldots\ldots, A_n$ are said to be mutually independent if

$P(Ai_1 \cap Aj_2 \cap \ldots\ldots Ai_k) =$ .... for $k = 2, 3, \ldots\ldots, n$

Q.151 The total number of conditions for mutual independence of $A_1, A_2, \ldots\ldots, A_n$ is equal to = ...

Q.152 If A, B & C are mutually independent event then $A \cup B$ & C are ....

Q.153 If A, B & C are pairwise independent events in a sample space S and A is independent of $B \cup C$ then A, B& C are ....

Q.154 P (At least one of the events $A_1, A_2, \ldots, A_n$ happens) = 1 - P (.........................)

Q.155 For any two events A and B we have

$P(A \cap B) \leq P(A) \leq \ldots\ldots \leq P(A) + P(B)$

Q.156 Thomas Bayes was a British Mathematician who gave his well known theorem " .... " in 1763.

Q.157 Bayes theorem states that if $E_1, E_2, \ldots\ldots, E_n$ are mutually disjoint events with $P(Ei) \neq 0$, $(i = 1, 2, \ldots, n)$ then for any arbitrary event A which is subset of $\bigcup_{i=1}^{n} Ei$ such that $P(A) \neq 0$, we have

$$P(Ei/A) = \frac{P(Ei).P(\ldots\ldots\ldots)}{P(A)}$$

Section-C : Multiple Choice Questions

Q.1 In an experiment repeated under an identical conditions, the result obtained is generally
(a) Unique (b) Certain
(c) May be one of the several possible outcomes
(d) Both (a) & (b)

Q.2 The phenomena described in Ql, are known as
(a) Deterministic (b) Predictable
(c) Nonpredicatable (d) Both (a) & (b)

Q.3 In deterministic phenomena, the results can be predicted with
(a) Certainity (b) Non Certainity
(c) With probability one (d) Both (a) & (c)

Q.4 A deterministic model states that the conditions under which is an experiment performed, the outcome of the experiment is
(a) Determined (b) Not determined
(c) Probabilistic (d) None

Q.5 The phenomena not leading themselves to deterministic approach, are known as
(a) Probabilisitic (b) Non-predictable
(c) Uncertain Phenomena (d) All the three

Q.6 The phenomena frequently occured in business, economic & social science or even in our day to day life, are known as
(a) Deterministic (b) Undeterministic
(c) Probabilistic (d) Bot (b) & (c)

Q.7 A manufacturer can ascertain the future demand of his product
(a) With Certainity (b) With non certainity
(c) In probabilistic way (d) Both (b) & (c)

Q.8 Sale target next year by a sale manager
(a) Can be predicted with certainity
(b) Can not be predicted with surity
(c) Can be predicted in a probabilistic way
(d) Both (a) & (c)

Q.9 If an electric bulb has lasted for one year, then its future life can
(a) Be predicted with certainity
(b) Not be predicted with surity
(c) Be predicted in probabilistic
(d) Both (b) & (c)

Q.10 Who did first time deal with some problems related to the theory of dice in gambling
(a) Galieo (b) Bernouille
(c) Prof. R.A. Fisher (d) Kempthorn

Q.11 A numerical measure of uncertainity is provided by a very important branch of mathematics known as
(a) The theory of large number
(b) The theory of small number
(c) The theory of probability
(d) The theory of point estimation

Q.12 The probability theory describes the certainity with the value
(a) 0.5 (b) 0.0
(c) 1.00 (d) 100.00

Q.13 The theory of probability describes the "Impossibility" with the value equal to
(a) 0.0 (b) 0.1
(c) 1.00 (d) 0.5

Q.14 The probability theory describes the various grades of uncertainity by a coefficient denoted by p and it ranges as
(a) $-1.0 < p < 0.0$ (b) $-1.0 < p < 1.0$
(c) $0.0 < p < 1.0$ (d) $-0.5 < p < 0.5$

Q.15 The next stal wart in the theory of probability is
(a) J. Bernouille (b) N. Bernouille
(c) Karl Pearson (d) D. Moiver

Q.16 Who founded the "Russion School of Statisticians" ?
(a) Chebychov b) Karl Pearson
(c) Karl Mark (d) Brejnev

Q.17 Tossing a coin, throwing of a dice & selecting a card from a pack of playing cards, are
(a) The examples of random experiments
(b) Not the examples of random experiments
(c) The examples of complicated experiments
(d) Nothing can be said with surity

Q.18 The result of a random experiment is known as
(a) An outcome (b) An event
(c) Trial (d) Both (a) & (b)

Q.19 The outcome or combination of outcomes of a random experiment, is known as
(a) Event (b) Trial
(c) Result (d) None

Q.20 An event which corresponds to a single outcome of an experiment, is known as
(a) Simple event (b) Compound event
(c) Composite event (d) Both (b) & (c)

Q.21 Total number of all possible outcomes of a random experiment is knonw as
(a) Exhaustive events (b) Exhaustive cases
(c) Conditional events (d) Both (a) & (b)

Q.22 In throwing n dice, the exhaustive number of events is equal to
(a) n (b) 6n
(c) 6^n (d) None

Q.23 The number of cases favourable to draw an ace from a pack of playing card is equal to
(a) 4 (b) ${}^{52}C_4$
(c) ${}^{13}C_4$ (d) 4x ${}^{52}C_4$

Q.24 The number of case favourable to get a sum 7 in throwing of two dice is equal to
(a) 5 (b) 6
(c) 8 (d) None

Q.25 If two or more events can not occure simultaneously in the same experiment, then such events are known as

(a) Mutually exclusive (b) Favourable events

(c) Incompatible events (d) Both (a) & (c)

Q.26 In throwing a die, all the 6 faces numbered from 1 to 6, are

(a) Mutually exclusive (b) Incompatible events

(c) Independent events (d) All the three

Q.27 If there is no reasons to expect one event in preference to other events in a trial, then such events are known as

(a) Equally likely events (b) Equally probable events

(c) Dependent events (d) Both (a) & (b)

Q.28 In throwing a balanced die, the numbers from 1 to 6 to come on upper most face, are

(a) Equally probable (b) Equally likely

(c) Both (a) & (b) (d) None

Q.29 If the happening/non - happening of an event is not affected by the happening / non-happening of other event then such events are knonw as

(a) Dependent events (b) Independent events

(c) Mutually exclusive events (d) Equally likely events

Q.30 When a die is thrown twice, then the results of both throw are

(a) Independent (b) Mutually exclusive

(c) dependent (d) Both (a) & (b)

Q.31 The definition of probability of happening of an event E writen as P (E) = (m/n) was given by

(a) James Bernouille (b) N. Bernouille

(c) D. Moiver (d) Galileo

Q.32 Sometimes, we can state that

(a) Odds in favour of E are m : (n - m)

(b) Odds against E are (n - m) : m

(c) Odds in favour of $\overline{E}$ = (n — m): m

(d) All the three are correct

Q.33 Non happening of an event E is knonw as
(a) Complementary of event E
(b) Mutually exclusive of event E
(c) Independent of event E
(d) Dependence on event E

Q.34 Complementary of an event E is denoted by
(a) $\overline{E}$ (b) E^c
(c) E' (d) Both (a) & (b)

Q.35 The following relation is true
(a) $P(E) + P(\overline{E}) = 1$ (b) $P(E) + P(E^c) = 1$
(c) $P(E) = 1 - P(\overline{E})$ (d) All the three

Q.36 P (Atleast one of the vents, Ei, i = 1, 2, ..., n happens) is equal to
(a) 1 - P (None of the events, Ei, i = 1, 2, ..., n, happens)
(b) $P\left(\bigcup_{i=1}^{n} Ei\right)$
(c) $P\left(\bigcap_{i=1}^{n} Ei\right)$
(d) None

Q.37 Probability "p" of happening of an event E, is also known as
(a) Probability of success of E
(b) P (Failure of E)
(c) P (Complementary of E)
(d) P (Independent of E)

Q.38 The probability of failure of an event E is denoted by
(a) $q = (1 - p)$ (b) $P(\overline{E})$
(c) $P(E^c)$ (d) $P(E \cap \overline{E})$

Q.39 If P (E) = 1, then the event 'E' isknown as
(a) Sure event (b) Certain event
(c) Rare event (d) Both (a) & (b)

Q.40 If P (A) = 0, then the event 'A' is known as

(a) Rare event (b) Sure event

(c) Impossible event (d) Both (a) & (c)

Q.41 For (m, n) as the favourable number of event and exhaustive number of event, the definition of probability of an event E given as P (E) = (m/n) is known as

(a) Classical probability (b) Apriori probability

(c) Mathematical probability (d) All the three

Q.42 If (N, M) denote the number of (trials, favourable event E) respectively then $P(E) = \lim_{N \to \infty} \left(\frac{M}{N}\right)$ is known as

(a) Statistical probability (b) A priori probability

(c) Empirical probability (d) Both (a) & (c)

Q.43 Probability that a leap year selected at random will contain 53 sunday is equal to

(a) $\frac{3}{7}$ (b) $\frac{2}{7}$

(c) $\frac{1}{7}$ (d) None

Q.44 In case of throwing two dice, the probability to get same number on upper most face, is equal to

(a) $\frac{1}{3}$ (b) $\frac{1}{6}$

(c) $\frac{1}{2}$ (d) None

Q.45 In case of throwing two dice, the probability of getting total of two numbers occured on upper most faces is greater than 8 is equal to

(a) $\frac{6}{18}$ (b) $\frac{7}{18}$

(c) $\frac{5}{18}$ (d) $\frac{11}{18}$

Q.46 Set theory was developed by the German mathematician

(a) G. Cantor (b) D. Moiver

(c) Draper (d) Pascal

Q.47 The objects constituting a set are called

(a) Elements (b) Members

(c) Points (d) All the three

Q.48 Let A be the subset of B then we write symbolically as

(a) $A \subseteq B$ (b) $B \supseteq A$

(c) $A \in B$ (d) Both (a) & (b)

Q.49 Two sets A & B are said to be equal or identical if

(a) $A \subseteq B$ and $B \subseteq A$ (b) $A \in B$ and $B \in A$

(c) A = B or B = A (d) Both (a) & (c)

Q.50 A null set is

(a) An empty set (b) ϕ

(c) That which does not contain any element at all

(d) All the three

Q.51 An empty set ϕ

(a) Is the subset of every set (b) Has no element at all

(c) Is a null set (d) All the three

Q.52 $A \cup B$ is defined as a set containing all those elements which

(a) Belong to either A or B (b) Belong to A and B

(c) Belong to A and $\overline{B}$ (d) Belong to B and $\overline{A}$

Q.53 Symbolically $A \cup B$ is written as

(a) $\{x : x \in A \text{ or } x \in \overline{B}\}$ (b) $\{x : x \in A \text{ and } x \in B\}$

(c) $\{x : x \in A \text{ or } x \in \overline{B}\}$ (d) $\{x \in B \text{ or } x \in \overline{A}\}$

Q.54 Symbolically, $A \cap B$ can be written as

(a) $\{x : x \in A \text{ or } x \in B\}$ (b) $\{x : x \in A \text{ and } x \in B\}$

(c) $\{x : x \in A \text{ or } x \in \overline{B}\}$ (d) $\{x : x \in \overline{A} \text{ and } x \in B\}$

Q.55 Symbolically, $\bigcap_{i=1}^{n} Ai$ can be Written as

(a) $\bigcap_{i=1}^{n} Ai = \{x : x \in Ai \ \forall\, i = 1, 2, \ldots\ldots\ldots, n\}$

(b) $\bigcap_{i=1}^{n} Ai = \{x : x \in Ai \text{ for somi } Ai, i = 1, 2, \ldots\ldots, n\}$

(c) $\bigcap_{i=1}^{n}$ Ai = {x : x ϵ Ai fornotall i =1,2,....., n}

(d) $\bigcap_{i=1}^{n}$ Ai = {x : x ϵ Ai or Aj, i ≠ j = 1,2,....., n}

Q.56 For two sets A & B if $A \cap B = \phi$, then A & B are said to be

(a) Mutually exclusive sets (b) Disjoint sets

(c) Non overlaping sets (d) All the three

Q.57 The difference of two sets A & B can be defined as

(a) A - B = {x : x ϵ A and x ∉ B}

(b) A - B = {x : x ∉ B and x ϵ B}

(c) A - B = {x : x ϵ A and x ϵ $\overline{B}$}

(d) A - B = {x : x ϵ $\overline{A}$ and x ϵ $\overline{B}$}

Q.58 Let S be a universal set then complement or negative of any set A denoted by $\overline{A}$ is defined as

(a) $\overline{A}$ =S - A (b) $\overline{A} = S \cap A$

(c) $\overline{A} = S \cup A$ (d) $\overline{A} = S \cup A^{-1}$

Q.59 Let A & B be two subsets of an universal set S then commutative law states that

(a) $A \cup B = B \cup A$ (b) $A \cap B = B \cap A$

(c) $A \cup B = B \cap \overline{A}$ (b) Both (a) & (b)

Q.60 Let A, B & C be the subsets of an universal set S, then associative law states that

(a) $(A \cup B) \cup C = A \cup (B \cup C)$

(b) $(A \cup B) \cap C = A \cap (B \cup C)$

(c) $(A \cap B) \cap C = A \cap (B \cap C)$ (d) Both (a) & (c)

Q.61 For three sets A, B & C being the subsets of an universal set S, the distributive law states that

(a) $A \cup (B \cap C) = (A \cup B) \cap (B \cup C)$

(b) $A \cap (B \cup C) = (A \cap B) \cup (B \cap C)$

(c) $A \cap (B \cup C) = (A \cap B) \cup (B \cap C)$ (d) Both (a) & (b)

Q.62 For a set A and an universal set S, the complementary laws states that

(a) $A \cup \overline{A} = S$ (b) $A \cap \overline{A} = \phi$

(c) $A \cup S = S$ (d) All the three

Q.64 For any set A, involution law says that

(a) $(\overline{\overline{A}}) = A^{-1}$ (b) $(\overline{\overline{A}}) = A$

(c) $(\overline{\overline{A}}) = \overline{A}$ (d) $(\overline{\overline{A}}) = A^{-1}$

Q.65 For any set A, the cidempontency law states that

(a) $A \cup A = A$ (b) $A \cap A = \phi$

(c) $A \cap A = A$ (d) Both (a) & (c)

Q.66 Let $\{A_n\}$ be a sequence of sets in S. Then we can write

(a) $\lim_{n \to \infty} \sup A_n = \{x; x \in A_n \text{ for inf initely many } n\}$

(b) $\lim_{n \to \infty} \sup A_n = \{x; x \in \overline{A}_n \text{ for inf initely many } n\}$

(c) $\lim_{n \to \infty} \sup A_n = \{x; x \in A_n \text{ for all but finite } n\}$

(d) $\lim_{n \to \infty} \sup A_n = \{x; x \in \overline{A}_n \text{ for all but finite } n\}$

Q.67 For $\{A_n\}$ being a sequence of sets in S, limit infinimum of the sequence can be written as

(a) $\lim_{n \to \infty} \inf An = \{x : x \in A_n \text{ for all but finite } n\}$

(b) $\lim_{n \to \infty} \inf An = \{x : x \in \overline{A}_n \text{ for all but finite } n\}$

(c) $\lim_{n \to \infty} \inf An = \{x : x \in A_n \text{ for inf initely many } n\}$

(d) $\lim_{n \to \infty} \inf An = \{x : x \in \overline{A}_n \text{ for inf initely many } n\}$

Q.68. The sequence $\{A_n\}$ is said to have a limit iff

(a) $\lim \sup A_n = \lim \inf A_n$ (b) $\lim \sup A_n = \lim \inf \overline{A}_n$

(c) $\lim \sup \overline{A}_n = \lim \inf \overline{A}_n$ (d) None of the above

Q.69 For any sequence of set {An}, we can write

(a) lim sup An = $\bigcap_{m=1}^{\infty}\left(\bigcup_{n=m}^{\infty} An\right)$

(b) lim sup An = $\bigcap_{m=1}^{\infty}\left(\bigcup_{n=m}^{\infty} \bar{A}n\right)$

(c) lim sup An = $\bigcup_{m=1}^{\infty}\left(\bigcup_{n=m}^{\infty} An\right)$

(d) lim sup An = $\bigcup_{m=1}^{\infty}\left(\bigcup_{n=m}^{\infty} \bar{A}n\right)$

Q.70 For any sequence $\bigcap_{m=1}^{\infty}\left(\bigcup_{n=m}^{\infty} An\right)$ n), we can write

(a) lim inf An = $\bigcap_{m=1}^{\infty}\left(\bigcup_{n=m}^{\infty} \bar{A}n\right)$

(b) lim inf An = $\bigcup_{m=1}^{\infty}\left(\bigcap_{n=m}^{\infty} \bar{A}n\right)$

(c) lim inf An = $\bigcap_{n=1}^{\infty}\left(\bigcap_{n=m}^{\infty} \bar{A}n\right)$

(d) lim inf An =

Q.71 {An} is a monotone sequence of set if

(a) Either $A_n \subset A_{n+1} \forall n$
(b) or $A_n \supset A_{n+1} \forall n$
(c) Either $A_n \supset \overline{A}_{n+1} \forall n$
(d) Both (a) & (b)

Q.72 If $A_n \subset A_{n+1}$, then the sequence {An} is said to be

(a) Non decreasing
(b) Non increasing
(c) Increasing
(d) Both (a) & (c)

Q.73 If $A_n \supset D\ A_{n+1}$, then the sequence {An} is said to be

(a) Non increasing
(b) Non decreasing
(c) Decreasing
(d) Both (a) & (c)

Q.74 For a monotone sequence {An}, the limt always exists and we write it as

(a) $\lim An = \bigcup_{n=1}^{\infty} An$ if An↑ (b) $\lim An = \bigcap_{n=1}^{\infty} An$ if An ↓

(b) $\lim An = \bigcap_{n=1}^{\infty} An$ if An ↑ (d) Both (a) & (b)

Q.75 Let R be a ring of non-empty class of sets A & B

(a) If $A \in R, B \in R$ then $A \cup B \in R$

(b) If $A \in R, B \in R$ then $A - B \in R$

(c) If $A \in R, B \in R$ then $A \cup \overline{B} \in R$

(d) Both (a) & (b)

Q.76 Let F be a "field" of non-empty class of sets A and B if

(a) $A \in F, B \in F \Rightarrow A \cup B \in F$ (b) $A \in F \Rightarrow A^{-1} \in F$

(c) $A \in F \Rightarrow \overline{A} \in F$ (d) Both (a) & (c)

Q.77 Let C be a a ring and is non-empty class of sets then

(a) $Ai \in C$ fori=1,2,..... $\Rightarrow \bigcup_{i=1}^{n} A_i \in C$

(b) $A \in C, B \in C \quad A \cap B \in C$

(c) $A \in C, B \in C \quad A - B \in C$ (d) Both (a) & (c)

Q.78 Axiomatic definition of probability

(a) Includes classical definition of probability

(b) Includes statistical definition of probability

(c) Overcomes the deficiency of both definitions

(d) All the three are true

Q.79 The totality of all possible outcomes of a random experiment constitute the

(a) Sample space (b) Field

(c) Real line (d) Universal set

Q.80 The sample space of a random experiment is usually denoted by

(a) S (b) Ω

(c) $\overline{S}$ (d) Both (a) & (b)

Q.81 The outcomes of a random experiment are denoted by

(a) $e_1, e_2, e_3, ...,$ (b) $w_1, w_2, w_3, ...$

(c) $\overline{e}_1, \overline{e}_2, \overline{e}_3, ...$ (d) Both (a) & (b)

Q.82 A sample space S is called "Discrete" if it contains

(a) Finitely many points (b) Infinitely many points

(c) Infinitely any number of points (d) Both (a) & (b)

Q.83 An empty set ϕ is

(a) Subset of S (b) An event

(c) An impossible event (d) All the three

Q.84 Let P (A) be a probability function defined on a σ field B of events if

(a) For A ϵ B, P(A) is defined and P (A) ≥ 0

(b) P (S) = 1

(c) For {An} being a finite or infinite sequence of disjoint set in B,

$$P\left(\bigcup_{i=1}^{n} Ai\right) = \sum_{i=1}^{n} P(Ai)$$

(d) All the three

Q.85 The probability function P (A) defined on a a field B of event in Q 84 is called

(a) Probability measure (b) Probability space

(c) Sample space (d) None

Q.86 The triplet (S, B, P} is called the

(a) Probability space (b) Probability measure

(c) Sample space (d) Universal set

Q.87 The σ field B is the collection of all

(a) Subsets of S (b) Elements of S

(c) Events of S (d) None

Algebra of Events

Q.88 For the events A, B, C we have

(a) $A \cup B = \{w \in S, w \in A \text{ or } w \in B\}$

(b) $A \cup B = \{w \in S, w \in A \text{ and } w \in B\}$

(c) $A \cup B = \{w \in S, w \in A \text{ or } w \in \overline{B}\}$

(d) $A \cup B = \{w \in S, w \in A \text{ and } w \in \overline{B}\}$

Q.89 For the events A, B, C we have

(a) $A \cap B = \{w : w \in A \text{ and } w \notin B\}$

(b) $A \cap B = \{w : w \in A \text{ and } w \in B\}$

(c) $A \cap B = \{w : w \in A \text{ and } w \in \overline{B}\}$

(d) $A \cap B = \{w : w \in A \text{ or } w \in B\}$

Q.90 For the set A and sample space S, we have

(a) $A\text{-}B = \{w : w \in S, w \in A \text{ and } w \in B\}$

(b) $A\text{-}B = \{w : w \in S, w \in A \text{ but } w \notin B\}$

(c) $A\text{-}B = \{w : w \in S, w \in A \text{ or } w \in B\}$

(d) $A\text{-}B = \{w : w \in S, w \in A \cap B\}$

Q.91 For a set A and sample S, we can write

(a) $\overline{A}$(Complement)= $\{w \in S, w \in A\}$

(b) $\overline{A} = \{w \in S, w \notin A\}$

(c) $\overline{A} = \{w \in S, w \notin A^{-1}\}$

(d) $\overline{A} = \{w \in A, w \notin S\}$

Q.92

(a) $A \subset B \Rightarrow \{w \in A, w \in B\}$ (b) $A \subset B \Rightarrow \{w \in A, w \notin B\}$

(c) $A \subset B \Rightarrow \{w \in A \text{ and } w \in B\}$ (d) Both (a) & (c)

Q.93 A and B are said to be mutually exclusive events if

(a) $A \cap B = \{w \in A \text{ and } w \notin B\}$

(b) $A \cap B = \{w \in A \text{ and } w \in B\}$

(c) $A \cap B = \phi$ (d) Both (a) & (c)

Q.94 We have

(a) $A \cup B = A + B$, if A and B are disjoint

(b) $A \cup B = A + B - A \cap B$ if A and B are not disjoint

(c) $A \cup B = A + \overline{B}$ (d) Both (a) & (b)

Q.95 At least one of the events A or B occurs if

(a) $w \in A \cup B$ (b) $w \in A \cap B$

(c) $w \in A \cup B$ (d) $w \in A \cap \overline{B}$

Q.96 Neither A or B occurs

(a) $w \in A \cup B$ (b) $w \in \overline{A} \cap \overline{B}$

(c) $w \in A \cap B$ (d) $w \in \overline{A} \cup \overline{B}$

Q.97 Event A occurs and B does not occur

(a) $w \in A \cap B$ (b) $w \in A \cap \overline{B}$

(c) $w \in A \cup B$ (d) $w \in \overline{A} \cup \overline{B}$

Q.98 $P(\overline{A})$ is given by

(a) $P(\overline{A}) = 1 - P(A)$ (b) $P(\overline{A}) = 1 - P(A^{-1})$

(c) $P(\overline{A}) = 1 + P(A)$ (d) None

Q.99 For two events A and B we have

(a) $P(\overline{A} \cap B) = P(B) - P(A \cap B)$ (b) $P(\overline{A} \cap B) = P(A \cup B)$

(c) $P(\overline{A} \cap B) = P(B) - P(A \cup B)$ (d) None

Q.100 If $B \subset A$ then

(a) $P(A \cap \overline{B}) = P(A) - P(B)$ (c) $P(A \cap \overline{B}) = P(A) - P(\overline{B})$

(b) $P(A \cap \overline{B}) = P(A) - P(\overline{\overline{B}})$ (d) Both (a) & (c)

Q.101 For three non mutually exclusive events A, B, C we have

(a) $P(A \cup B \cup C) = P(A)+P(B)+P(C)- P(A \cap B)-P(B \cap C)-P(C \cap A)+P(A \cap B \cap C)$

(b) $P(A \cup B \cup C) = P(A)+P(B)+P(C)+ P(A \cap B)+P(B \cap C)+P(C \cap A)-P(A \cap B \cap C)$

(c) $P(A \cup B \cup C)=P(A)+P(B)+P(C)- P(A \cap B)-P(B \cap C)-P(C \cap A)+P(\overline{A} \cap \overline{B} \cap \overline{C})$

(d) $P(A \cup B \cup C)=P(A)+P(B)+P(C)- P(A \cup B)-P(B \cup C)-P(C \cup A)+P(\overline{A} \cap \overline{B} \cap \overline{C})$

Q.102 For Pi (i = 1, 2, ..., n) be the probabilities of mutually exclusive and Ai (i = 1, 2, ..., n) are known as

(a) Partial probabilities (b) Conditional probabilites
(c) Statistical probabilities (d) Priori probabilities

Q.103 For n events say Ai (i = 1, 2, ..., n), Boole's inequally states that

(a) $P\left(\bigcap_{i=1}^{n} Ai\right) \geq \sum_{i=1}^{n} P(Ai) - n$

(b) $P\left(\bigcap_{i=1}^{n} Ai\right) \geq \sum_{i=1}^{n} P(Ai) - (n-1)$

(c) $P\left(\bigcap_{i=1}^{n} Ai\right) \geq \sum_{i=1}^{n} P(Ai) - (n^2-1)$

(d) None

Q.104 For n events say A A A A we have

(a) $P\left(\bigcup_{i=1}^{n} Ai\right) \geq \sum_{i=1}^{n} P(Ai) - \sum_{1 \leq i \leq j \leq n}^{n} P(Ai \cap Aj)$

(b) $P\left(\bigcup_{i=1}^{n} Ai\right) \leq \sum_{i=1}^{n} P(Ai) - \sum_{1 \leq i \leq j \leq n}^{n} P(Ai \cap Aj)$

(c) $P\left(\bigcup_{i=1}^{n} Ai\right) \geq \sum_{i=1}^{n} P(Ai) + \sum_{1 \leq i \leq j \leq n}^{n} P(Ai \cap Aj)$

(d) None

Q.105 For two events A and B we have

(a) $P(A \cap B) = P(A).P(B / A)$, for $P(A) > 0$

(b) $P(A \cap B) = P(B).P(A / B)$, for $P(A) > 0$

(c) $P(A \cap B) = P(B).P(A / B)$, for $P(B) > 0$

(d) Both (a) & (c)

Q.106 For a fixed event B, with $P(A) \neq 0$, we have

(a) $P(S / B) = \frac{P(S \cap B)}{P(B)}$ (b) $P(S / B) = 1.0$

(c) $P(S / B) = \frac{P(S \cup B)}{P(B)}$ (d) Both (a) & (b)

Q.107 For three events A, B, C we have

(a) $P(A \cup B / C) = P(A / C) + P(B / C) + P(A \cap B / C)$

(b) $P(A \cup B / C) = P(A / C) + P(B / C) - P(A \cap B / C)$

(c) $P(A \cup B / C) = P(A / C) + P(B / C) + P(A \cup B / C)$

(d) None

Q.108 If A and B are independent events then

(a) A & $\overline{B}$ are independent (b) $\overline{A}$ &B are independent

(c) $\overline{A}$ & $\overline{B}$ are independent (d) All the three

Q.109 Total number of conditions for mutual independence of Ai (i = 1, 2, ..., n) is equal to

(a) $(n^2 + n - 1)$ (b) $(n^2 - n + 1)$

(c) $(n^2 - n - 1)$ (d) $(n^2 - 2n - 1)$

Q.110 If A, B, C are mutually independent events then

(a) $A \cup B$ and C are independent

(b) $A \cup B$ and C are dependent

(c) $A \cup C$ and B are independent

(d) Both (a) & (c)

Q.111 For two events A and B, we have

(a) $P(A \cap B) \leq P(A)$ (b) $P(A \cap B) \leq P(A \cup B)$

(c) $P(A \cap B) \leq P(A) + P(B)$ (d) All the three

Q.112 Thomas Bayes who gave his well konw "Bayes theorem in 1763 was a

(a) British mathematician (b) Italian mathematician

(c) Chiness mathematician (d) U.S. mathematician

Q.113 If Ei (i = 1, 2, ..., n) are mutually disjoint events with P (Ei) ≠ 0, then for any arbitrary event A $\subset \bigcup_{i=1}^{n}$ Ei and P (A) > 0, Bayes Theorem states that :

(a) $P(Ei/A) = \frac{P(Ei)P(A/Ei)}{\sum_{i=1}^{n} P(Ei)P(A/Ei)}$

(b) $P(Ei/A) = \frac{P(Ei)P(A/Ei)}{P(A)}$

(c) $P(Ei/A) = \frac{P(Ei)P(A/Ei)}{\sum_{i=1}^{n} P(Ei)}$

(d) Both (a) & (b)

Key Answers

Section-A : True False

1.	False	2.	True	3.	True
4.	True	5.	True	6.	True
7.	False	8.	False	9.	True
10.	True	11.	True	12.	True
13.	True	14.	True	15.	True
16.	True	17.	True	18.	True
19.	True	20.	True	21.	False
22.	True	23.	True	24.	True
25.	True	26.	True	27.	True
28.	True	29.	True	30.	True
31.	True	32.	True	33.	True
34.	True	35.	True	36.	True
37.	True	38.	True	39.	True
40.	True	41.	True	42.	True
43.	True	44.	True	45.	True
46.	True	47.	True	48.	True

49.	True	50.	True	51.	True
52.	True	53.	True	54.	True
55.	True	56.	True	57.	True
58.	True	59.	True	60.	True
61.	True	62.	True	63.	True
64.	True	65.	True	66.	True
67.	True	68.	True	69.	True
70.	True	71.	True	72.	True
73.	True	74.	True	75.	True
76.	True	77.	True	78.	True
79.	True	80.	True	81.	True
82.	True	83.	True	84.	True
85.	True	86.	True	87.	True
88.	True	89.	True	90.	True
91.	True	92.	True	93.	True
94.	True	95.	True	96.	True
97.	True	98.	True	99.	True
100.	True	101.	True	102.	True
103.	True	104.	True	105.	True
106.	True	107.	True	108.	True
109.	True	110.	True	111.	True
112.	True	113.	True	114.	True
115.	True	116.	True	117.	True
118.	True	119.	True	120.	True
121.	True	122.	True	123.	True
124.	True	125.	True	126.	True
127.	True	128.	True	129.	True
130.	True	131.	True	132.	True
133.	True	134.	True	135.	True
136.	True	137.	True	138.	True
139.	True	140.	True		

Section-B : Fill in the Blanks

1. Possible outcomes
2. Experiment
3. Probabilitistic
4. Probabilistic
5. Nothing
6. Probability
7. Gambling
8. Probability

9.	1.0	10. Zero
11.	$(0 \leq P \leq 1)$	12. Probability
13.	Bernouille	14. Statistician
15.	Random Experiment	16. Random
17.	Event	18. Trial
19.	Event	20. Compound
21.	Exhaustive	22. $^{52}C_2$
23.	6^n	24. 6
25.	Mutually Exclusive	26. Mutually
27.	Equally	28. Equally
29.	Independent events	30. Does not affect
31.	J. Bernoulle	32. (n - m) : m
33.	Complemenatry	34. $\overline{E}$
35.	1.0	36. None
37.	Probability	38. (1 - p)
39.	Certain or sure	40. Impossible
41.	Mathematical	42 Classical
43.	M/N	44. Classical, Ariori
45.	Identical	46. 2/7
47.	$\frac{1}{6}$	48. 5/18
49.	2/(n-1)	50. Zero
51.	G. Contor	52. Element
53.	$\subset$ or $\supset$	54. $\subset$, $\subset$
55.	Empty, Element 4	56. Every
57.	$A \cup B$, Either A or B	58. {x ϵ A or x E ϵ B}
59.	Atleast	60. $A \cap B$
61.	A and B	62. { w ϵ A and w ϵ B}
63.	{All i = 1, 2, ..., n)	64. ϕ
65.	Disjoint, Mutually Exclusive	66. A - B
67.	A, B	68. {w ϵ A and ∉ B}

69. $\overline{A}$ or A^c

70. Not the members

71. -A

72. $B \cup A$, $B \cap A$

73. $A \cup (B \cup C)$

74. $A \cap (B \cap C)$

75. $\bigcap (A \cup C)$

76. $A \cap B$

77. S, ϕ

78. S, A

79. A, ϕ

80. $\overline{B}$

81. $A \cap B, - B$

82. U

83. U

84. U

85. A, ϕ

86. $\bigcup_{i=1}^{n} \overline{A}i$

87. A

88. A, A

89. {for infinitely many n}

90. Infinimum, lira

91. {All but a finite n}

92. $\lim_{n \to \infty} \inf An$, the limit of the sequence

93. $\left(\bigcup_{n=m}^{\infty} An\right), \left(\bigcap_{n=m}^{\infty} An\right)$

94. $\forall$ n, A_{n+1}

95. A_{n+1}, Increasing

96. A_{n+1}, Decreasing

97. $\left(\bigcup_{n=1}^{\infty} An\right)$ in case $A\uparrow$

$\left(\bigcup_{n=1}^{\infty} An\right)$ in case $A\downarrow$

98. Union or Difference

99. Non-empty $A \cup B \in R$

$A - B \in R$

100. Non-empty, Union

complementation

101. F

102. $\in$ F

103. Union, Difference

104. $\bigcup_{i=1}^{\infty} Ai$

105. C

106. Classical, Statistical, Each of them

107. Logic of deduction	108. Outcome, Elementary
109. Sample space	110. S or Ω
111. Discrete	112. Continous
113. Event	114. Impossible
115. ≥ 0.0	116. $= 1.0$
117. $\sum_{i=1}^{n} P(Ai)$	118. Probability Measure
119. Probability space	120. S
121. Axiomatic Probability	122. Axioms Rules
123. $w \in B$	124. and $X \in B$
125. $\{w \notin A\}$	126. $\{w \notin B\}$
127. $\{w \in B\}$	128. $B \subset C$
129. $= \phi$	130. Independent
131. $w \in A$ or $w \in B$	132. $w \in \overline{A}$ and $w \in \overline{B}$
133. $\cap \overline{B}$	134. P (A)
135. $(A \cap B)$	136. P (B)
137. $(A \cap B)$	138. $C \cap A$, $A \cap B \cap C$
139. Partial Probabilities	140. $\sum_{i=1}^{n} P(Ai) - (n - 1)$
141. $\leq$	142. $\sum_{i=1}^{n} P(Ai)$
143. (B/A), (A/B)	144. P (A)
145. P (A). P (B)	146. P (B), 1.0
147. $P(A \cap B/C)$	148. Independent
149. P (Ai) P (Aj)	150. $\prod_{K=2}^{n} P(Aik)$
151. $(2^n - n - 1)$	152. Independent
153. Independent	154. P (None of events happens)
155. $\leq p(A \cup B)$	156. Bayes Theorem
157. (A/Ei)	

Section-C : Multiple Choice Questions

1. (c) 2. (d) 3. (d)
4. (a) 5. (d) 6. (d)
7. (d) 8. (d) 9. (d)
10. (a) 11. (c) 12. (c)
13. (a) 14. (c) 15. (a)
16. (a) 17. (a) 18. (d)
19. (a) 20. (a) 21. (d)
22. (c) 23. (a) 24. (b)
25. (d) 26. (d) 27. (d)
28. (c) 29. (b) 30. (d)
31. (a) 32. (d) 33. (a)
34. (d) 35. (d) 36. (a)
37. (a) 38. (a) 39. (d)
40. (d) 41. (d) 42. (d)
43. (b) 44. (b) 45. (c)
46. (a) 47. (d) 48. (d)
49. (d) 50. (d) 51. (d)
52. (a) 53. (a) 54. (a)
55. (a) 55. (d) 56. (d)
57. (a) 58. (a) 59. (d)
60. (d) 61. (d) 62. (d)
63. (d) 64. (b) 65. (d)
66. (a) 67. (a) 68. (a)
69. (a) 70. (b) 71. (d)
72. (d) 73. (d) 74. (d)
75. (d) 76. (d) 77. (d)
78. (d) 79. (a) 80. (d)
81. (d) 82. (d) 83. (d)
84. (d) 85. (a) 86. (a)
87. (a) 88. (a) 89. (b)
90. (b) 91. (b) 92. (d)
93. (d) 94. (d) 95. (a)
96. (b) 97. (b) 98. (a)
99. (a) 100. (d) 101. (a)
102. (a) 103. (b) 104. (a)
105. (d) 106. (d) 107. (b)
108. (d) 109. (c) 110. (d)
111. (d) 112. (a) 113. (d)

❑❑❑

Chapter - 6

Random Variables and Distribution Functions

Section-A : Write True / False

Q.1 A random variable (r.v) is a real number x connected with the outcomes of a random experiment.

Q.2 In tossing of a balanced die at one time, the number coming on the upper most face of the die is a random variable x which takes any one out of 6 values taken as 1, 2, 3, 4, 5, 6.

Q.3 Let s be the sample space associated with a given random experiment. A real valued function defined on s and taking values in R $(-\infty, \infty)$ is called one dimensional random variable.

Q.4 In general a n dimensional random variable is simply a function whose domain is S and whose range is a collection of n-tuples of real members (vectors in n-space).

Q.5 A random variable is a function x(w) with domain S and range R,$(-\infty < R < \infty)$ such that for every real number a, the event $\{w: x(w) \leq a\} = B$.

Q.6 We say that the function x(w) is the measurable real function on (S, B) where $B \subset R$.

Q.7 One dimensional random variables are donated by capital letters, X, Y, Z, ... etc.

Q.8 A typical outcomes of the experiment is denoted by w or e.

Q.9 X (w) represents the real number which the r.v. X associates with the outcome w.

Q.10 The values which X, Y, Z, ... etc. can assume, are denated by lower case letters viz. x, y, z, ... etc.

Q.11 We can write $P(x \leq a) = P [w : x (w) \in (-\infty , a)]$

Q.12 We can write that, $P (a < x \leq b) = P [w: x (w) \in (a, b)]$.

Q.13 The analogus meaning to $P(x = a \text{ or } x = b) = P\{(x = a) \cup (x = b)\}$

Q.14 The analogus meaning to $P(x = a \text{ and } x = b) = P\{(x = a) \cap (x = b)\}$

Q.15 The triplet (S, B, P) is known as probability space.

Q.16. A function x(w) from S to $R (-\infty,\infty)$ is a r.v. iff for real a, we have $\{w: x (w) \leq a)\} \in B$.

Q.17 If (x_1 & x_2) are r.v's and c is a constant then C x_1, $x_1 + x_2$, and x_1, x_2, x_n are also r.v's.

Q.18 If (x_1, x_2) are two random variables then, max (x_1, x_2) and Min. (x_1, x_2) are also random variables.

Q.19 If x is a r. v. and f(x) is a continuous function, then f(x) is a r.v.

Q.20 If x is a r.v. and f(x) is an increasing or decreasing function of x then f(x) is also a r.v.

Q.21 Let x be a r.v. The function F defined for all real x by $F(x) = P(x \leq x) = P\{w: x (w) \leq x\}, -\infty < x < \infty$ is called distribution function of r.v. x.

Q.22 A distribution function is also known as the "cumulative distribution function".

Q.23 The domain of the distribution function is $(-\infty, \infty)$ and its range is [0, 1].

Q.24 If F is the d.f of the r.v. x and if a < b then we have $P\{a < x \leq b\} = F(b) - F(a)$

Q.25 If $P(x = a) = 0$ and $P(x = b) = 0$ then the four events. $a \le x \le b$, $a \le x < b$, $a < x \le b$, and $a < x < b$ have the same probability and is equal to $F(b) - F(a)$.

Q.26 All the distribution functions are monatonically non-decreasing (ie increasing) functions and lie in the limt (0, 1).

Q.27 Let F be a d.f. of a r.v. x then

$$F(-\infty) = \lim_{x \to -\infty} f(x) = 0 \text{ and } \lim_{x \to \infty} f(x) = 1$$

Q.28 A variable which can take only a countable number of real values and these values depend on chance, is called a "Discreate random variable or Discreate stochaustic variable or Discrete chance variable".

Q.29 A real values function defined on a discrete sample space, is called a "Discrete random variable".

Q.30 The number of telephonic calls/unit of time, the number of axidents taken place/week, the number of heads taken place in n bernoullian trials, etc. are the example of discrete random variable.

Q.31 If x is a discrete random variable with distince values (xi, i = 1, 2, ………, xn,) then the function P (x) defined as

$$P_x(x) = \begin{cases} P(X = xi) = pi \text{ for } x = xi \\ 0 \qquad , \text{for } x \ne xi, i = 1,2,\ldots.. \end{cases}$$

is called probability mass function (p.m.f.) of r.v. X.

Q.32 The set of the values (xi, pi, i = 1, 2, ……, n) taken by r.v. x, is known as the spectrum of the random variable x.

Q.33 The set of the values (xi, pi = 1, 2, …., n) taken by r.v. x, is known as the spectrum of the random variable x.

Q.34 For discrete r.v., a knowledge of (p.m. f) makes it possible for us to work out the probabilities of arbitrary events. i.e. if E is a set of real numbers then we have

$$P(X \in E) = \sum_{x \in E \cap S} P(x), \text{ where S is the sample space.}$$

Q.35 We can write $p(xi) = P(x = xi) = F(xi) - F(x_{i-1})$ where F is the d.f. of X.

Q.36 A random variable X is said to be continuous, if it can take all possible values between certain limits say $(a \le x \le b)$.

Q.37 A random variable X is said to be continous when its different values can not be put in 1 to 1 correspondence with a set of positive integers.

Q.38 The height and weight of students of a class can be taken as the continous random variables.

Q.39 Let f (x) be any continous function of X defined in a small interval (x, x + dx) of length dx so that f (x) dx represents the probability that X falls in the infinitesimal (x, x + dx).

Q.40 For the probability density function (p.d.f) of a random continuous variable X, symbolically, we can write
$P(x \le X \le x+dx) = f_x(x)dx$

Q.41 The f (x) dx represents the area bounded by the curve y = f (x), x axis and ordinate at the points (x & x + dx). Then function f (x) is known as probability density function (p.d.f) or simply the density function of r.v. X.

Q.42 The expression f (x) dx usually denoted as d F (x), is known as the probability differential.

Q.43 The curve y = f (x) defined as in Q (41), is known as the "probability density curve" or simply "probability curve".

Q.44 The p.d.f., f (x) of the r.v. X is defined as

$$f(x) = \lim_{\delta x \to 0} \frac{P(x \le X \le x + dx)}{\delta x}$$

Q.45 The probability for a variate value to lie in the interval dx is f (x) dx.

Q.46 The probability for a variate value to fall in the finite interval [α, β] is denoted by
$P(\alpha \le x \le \beta = \int_{\alpha}^{\beta} f(x)dx$

Q.47 The integral $\int_{\alpha}^{\beta} f(x)dx = P(\alpha \le x \le \beta)$ represents the area between the curve y = f (x), x axis and the ordinate at (x = α & x + dx = β).

Q.48 If [a, b] is the range of random variable X then $\int_{a}^{\beta} f(x)dx = 1.0$

Q.49 The range of a continuous random variable x may be finite or infinite.

Q.50 In case of a continuous r.v., X, the probability at apoint x = c is always equal to zero for all possible values of c.

Q.51 For a continuous random variable X, it does not matter whether we include the end points of the interval (α, β) i.e.
$P(\alpha \le x \le \beta)=P(\alpha < x \le \beta) = P(\alpha \le x < \beta) =P(\alpha< x <\beta)$

Q.52 Let X be a random variable defined as (a ≤ x ≤ b) and f (x) be the p.d.f of a r.v. X, then arithmatic mean $\bar{x}$ is defined as

$$\bar{X} = \int_a^b x\, f(x)\, dX$$

Q.53 The harmonic mean H is given by

$$\frac{1}{H} = \int_a^b \frac{1}{x} f(x)dx$$

Q.54 The geometric mean G is given by

$$\log G = \int_a^b \log x\, f(x)dx$$

Q.55 The rth raw moment of r.v. X about origin is defined as

$$\mu_r = \int_a^b x^r f(x)dx$$

Q.56 The variance σ^2 of a r.v. X is defined as

$$\sigma^2 = \mu_2 = \int_a^b x^2 f(x)dx - \left(\int_a^b xf(x)dx\right)^2$$

Q.57 Let Md denotes the median point. Then Md can be obtained by solving

$$\int_a^{Md} f(x)dx = \int_{Md}^b f(x)dx = \frac{1}{2}$$

Q.58 Mean deviation can be defined as

$$M.D. = \int_a^b |x - \bar{x}|\, f(x)dx$$

Q.59 The quartile Q_3 is defined as

$$Q_3 = \int_a^{Q_3} f(x)dx = \frac{3}{4}$$

Q.60 The ith decile Di can be defined as

$$Di = \int_a^{Di} f(x)dx = \frac{i}{10}, i = 1, 2, ..., 9$$

Q.61 The mode Mo of a r.v. X can be obtained from the solution of f' (x) = 0 and f" (x) < 0 provided (a ≤ Mo ≤ b)

Q.62 Let X be a continous r.v. with p.d.f f (x), then the function

$$F_x(x) = P(X \le x) = \int_{-\infty}^{x} f(t)dt, -\infty < t < \infty$$

is called the distribution function (d.f.) or sometimes cumulative distribution function (c.d.f.) of the r.v. X.

Q.63 We have $F'(x) = \frac{d}{dx} F(x) = f(x) \ge 0$ and we states that F (x) is non decreasing function of X.

Q.64 We have the following result for c.d.f $F_F(x)$.
$0 \le F_x(x) \le 1.0$

Q.65 $F(-\infty) = \lim_{X \to -\infty} F(x) = \lim_{X \to -\infty} \int_{-\infty}^{x} f(x)dx = 0$

Q.66 $F_x(+\infty) = \lim_{X \to \infty} F(x) = \lim_{X \to \infty} \int_{-\infty}^{x} f(x)dx = \int_{-\infty}^{\infty} f(x)dx = 1.0$

Q.67 The function $F_x(x)$ is a continous function of X on the right.

Q.68 The discontinuties of F (x) are at the most countable.

Q.69 Let (X & Y) be two random variables defined on the same sample space S, then the function (x, y) that assigns a point in R^2 (R x R), is called two dimensional random variable.

Q.70 Let w є S then the value of (x, y) at w is given by the pair of real numbres {x (w), y (w)}.

Q.71 The notation {x ≤ a, y ≤ b} denotes the events of all elements w є S such that x (w) ≤ a & y (w) ≤ b.

Q.72 The probability of the event $\{x \le a, y \le b\}$ is denoted by $P\{x \le a, y \le b\}$

Q.73 Let $A = \{a < x \le b\}$, $B = \{c < y \le d\}$ be two events. Then $P(a < x \le b, c < y \le d) = P(A \cap B)$

Q.74 Two dimenstional r.v. is said to be discrete if it takes atmost a countable number of points in R^2.

Q.75 Two random variable X and Y are said to be jointly distributed if they are defined on the same probability space.

Q.76 If the joint probability function is denoted by $P_{xy}(x, y)$, then the probability of certain event E is given by
$P_{xy}(x, y) = P[(x, y) \in E]$

Q.77 Let (X, Y) be two random variables on a sample space S with respective image sets X (S) = {xi, i = 1, 2, ..., n} and y (S) = {yj, j = 1, 2, ..., m}. Then, we make the product set as

$$S(x) \times S(y) = \{x_1, x_2, \ldots X_n\}\{y_1, y_2, \ldots, y_m\}$$

Q.78 We deonte the probability of the ordered pair (xi, yj) as $P(X = x_i, Y = y_i)$.

Q.79 The joint probability function P on X (S) Y (S) is defined as $p_{ij} = P = (X = x_i \cap Y = y_i) = p(x_i, y_i)$

Q.80 Let (X. Y) be two dimensional r.v. then their joint probability mass function denoted by $P_{xy}(x, y)$'is defined as

$$P_{X,Y}(xi, yj) = \begin{cases} P(X = x_i, Y = y_j \text{ for a value}(x_i, y_j) \text{ of } (X, Y)) \\ 0, \quad \text{otherwise} \end{cases}$$

Q.81 For the joint probability mass function $P_{xy}(xi, yj)$ we can write that

$$\sum_{i=1}^{n}\sum_{j=1}^{m} P(xi, yj) = 1.0$$

Q.82 Let P (xi, yj) be the joint p.m.f of two r.v.'s (X, Y). Then p.d.f. of X alone can be determined as

$P_x(xi) = \sum_{j=1}^{m} p(x_i, y_j) = P_i$ and is known as marginal p.m.f or discrete p.m.f. of X.

Q.83 From two dimensional p.m.f say P_{xy} (xi, yj), the conditional discrete p.m.f. of X, given Y = y, is dienfed as

$$P_{x/y}(X/Y)=\frac{P(X=x, Y=y)}{p(Y=y)}, \text{ provided } P(Y=y)\neq 0$$

Q.84 From the conditional p.m.f P_{xy} (xi, /y) we can write that

$$\sum_{i=1}^{n}\frac{P(xi/y)}{p(Y=y)}=\frac{P(Y=y)}{p(Y=y)}=1.0$$

Q.85 Λ ncccssary and sufficicnt condition for the two discrete r.v. (X, Y) to be independent is $P_{x,y}(X = xi.\ Y = yj) = P_x(X = xi).\ P_y(Y = yj)$ for all values of (X, Y) i.e. for (xi, yj).

Q.86 The distribution function of the two dimensional random variables (X, Y) is a real valued function F defined for all real (x, y) by the relation $F_{xy}(x, y) = P(X = x, Y = y)$

Q.87 For real numbers a_1, b_1, a_2 and b_2

$P(a_1 < x \leq b_1, a_2 < y \leq b_2) =$

$F_{xy}(b_1, b_2) + F_{xy}(a_1, a_2) - F_{xy}(a_1, b_2) - F_{xy}(b_1, a_2)$

Q.88 Let $a_1<a_2$, $b_1<b_2$ then we have $F(b_1, a_2)\geq F(a_1, a_2)$ and $F(a_1, b_2) \geq F(a_1, a_2)$ and it shows that F (x, y) is monotonic non decreasing function.

Q.89 $F(-\infty, y) = F(x, -\infty) = 0$ and $F(-\infty, \infty) = 1$

Q.90 If density function f (x, y) is continuous at (x, y) then

$$\frac{\partial^2 F(x, y)}{\partial x \partial y} = f(x, y)$$

Q.91 In case of jointly contionuous random variables, the marginal distribution functions of (X & Y) are $F_x(x) = \sum_y p(X \leq x, Y = y)$ and $F_y(y) = \sum_X P(X = x, Y \leq y)$

Q.92 In case of joint continous r.v. (X, Y) the marginal distribution fucntions of (X & Y) are

$$F_X(x) = \int_{-\infty}^{x} \left\{ \int_{-\infty}^{\infty} f(x, y)\, dy \right\} dx \text{ and } \int_{-\infty}^{y} \left\{ \int_{-\infty}^{\infty} f(x, y)\, dx \right\} dy$$

Q.93 The joint density function $f_{xy}(x, y)$ is defined as

$$f_{xy}(x,y) = \frac{\partial^2 F(x, y)}{\partial x \partial y} = \lim_{\partial x \to 0, \partial y \to 0} \frac{P(x \le x \le x+dx, \le y \le y+dy)}{\partial x \partial y}$$

Q.94 The probability that the,point (x, y) will lie in the infinitesimal rectangular region dx dy is given by

$$p\left(x - \frac{dx}{2} \le X \le x + \frac{dx}{2}, y - \frac{dy}{2} \le Y \le y + \frac{dy}{2}\right) = f(x,y)dxdy$$

Q.95 The marginal p.d.f of X for discrete variable is given as

$$f_x(x) = \sum_y P(x, y)$$

Q.96 The marginal p.d.f of X for continous r.v. is given by

$$f_x(x) = \int_{-\infty}^{\infty} f_{xy}(x,y)dy$$

Q.97 From the marginal distributions of two jointaly distributed r.v's we can not determine the joint distribution of these two r.v.'s.

Q.98 The conditional distribution function $F_{Y/X}(y/x)$ denotes the distribution function of Y when X has already assumed the particular value x.

Q.99 Mathematically, we can write $F_{Y/X} < (y/x) = P(Y \le y / X = x)$

Q.100 We can write $f_{Y/X}(y/x) = \dfrac{f(x, y)}{f(x)}$ for $f(x) > 0$

Q.101 The two r.v.'s X and Y with joint p.d.f., f (x, y) and marginal p.d.f's f (x) and g (y) respectively are said to be stochastically (ramdomly) independent iff : $f(x, y) = f(x).\, g(y)$

Q.102 If the two random variables X and Y are stochastically independent then for possible paris of real numbers (a_1, b_1) & (a_2, b_2) where ai < bi, we can write
$P(a_1 \le x \le b_1) \cap P(a_2 \le y \le b_2) = P(a_1 < x \le b_1)P(a_2 < y \le b_2)$

Q.103 Let y = g (x) be any continous function and is also a r.v. Then the distirbution of Y = g(x) is uniquely determined.

Q.104 Let X be a continous r.v. with p.d.f. f (x). Let y = g(x) be strictly a monotonic function of x. Let g (x) be differentiable ∀ x. Then p.d.f of r.v., y is given by

$$h(y) = f(x). \left|\frac{dy}{dx}\right|$$

Q.105 Let two r.v.'s (X & Y) be transformed to new variables as u = u(x, y) and v = v (x, y), where (u, v) are continuously differentiable function. Then jacobiam of transformation :

$$j = \frac{\partial(x,y)}{\partial(u,v)} = \begin{vmatrix} \frac{\partial x}{\partial u} & \frac{\partial y}{\partial u} \\ \frac{\partial x}{\partial v} & \frac{\partial y}{\partial v} \end{vmatrix}$$

Q.106 The joint p.d.f g (u, v) of the transformed variates u and v is given as g (u, v) = f (x, y) | j | where | j | is the Jacobian of transformation and f (x, y) is expressable in terms of u and v.

Q.107 Let f (x, y) be joint p.d.f of independent continuous r.v. of (X, Y) and let make the transformation u = X + Y. Then p.d.f. of u say h (u) is given by

$$h(u) = \int_{-\infty}^{\infty} f(v)\, f_Y(u - v)\, dv$$

Q.108 In Q (107), the function h (u) is said to be the convalution of $f_x(x)$ and $f_Y(y)$ and we write it as h (•) = f $(x)^x$ f (y)

Q.109 Let (X, Y) be two independent continuous r.v.'s then p.d.f. of U = (X - Y) is given by

$$h(u) = \int_{-\infty}^{\infty} f(x,x-u)dx = \int_{-\infty}^{\infty} f(x)f(x-u)dx$$

Q.110 Let (X, Y) be two independent continuous r.v. Then the p.d.f. of their product u = xy is given by

$$h(u) = \int_{-\infty}^{\infty} f_{XY}\left(x, \frac{u}{x}\right)\frac{dx}{|x|}$$

Q.111 The p.d.f. of the quotient z = x/y, is given by

$$g_1(z) = \int_{-\infty}^{\infty} g\,(vz,v)|\,v\,|dv = \int_{-\infty}^{\infty} f_x(vz) f_Y\,(v)|\,v\,|dv$$

where v = y ⇒ x = vz and new range of the variable z is determined from the range of the variables (X & Y) using transformation z = (x/y).

Section- B : Fill in the Blanks

Q.1 A random variable is a .... number x connected with the outcomes of a .... experiment.

Q.2 In tossing of a balanced die at one time, the number of coming up on the upper most face of the die is a .... variable x.

Q.3 A real valued function defined on a sample space, S and taking values in R (......), is known as one .... random variable.

Q.4 In general a one dimensional random variable is simply a function whose .... is S, and whose range is a collection of n triples of .... numbers.

Q.5 A random variable is a .... x (w) with domain S and range R(.....) such that for every real number 'a', the event {w : x (w) ...... a} = B.

Q.6 We can say that the function x (w) is the .... real function on (...............).

Q.7 A typical outcome of an experiment is generally denoted by .... Or ....

Q.8 X (w) represents, the .... which the random variable x associates with the outcome w.

Q.9 The probability theory describes the certainity of an event by a quanity equal to ....

Q.10 The analogus meaning of P (x = a or x = b) can be written in set theory as P {(x = a) .... (x = b)}

Q.11 The analogus meaning of P (x = a and x = b) can be written in set theory as P {(x = a) .... (x = b)}

Q.12 The triplet (S, B, P) is known as ....

Q.13 A functin x (w) from S to R ($-\infty < R < \infty$) is a random variable iff for real 'a' we have $\{w : x(w) < a\} \in B$

Q.14 If (X_1, X_2) are random variables and c is a constant then CX_1, X_1X_2 and ($X_1 + X_2$), are also ....

Q.15 If (X_1, X_2) are two random variables then Max (X_1, X_2) and Min (X_1, X_2) are also

Q.16 If X is a r.v. and f (x) is a continous function of x then f (x) is also a ....

Q.17 If X is a r.v. and f (x) is an increasing or decreasing function of X then f (x) is also a ....

Q.18 A function F defined for all real x by

$F(x) = P(X \leq x) = P\{w : x(w) \leq x\}, -\infty < x < \infty$ is known as .... random variable x.

Q.19 A distribution function is also known as .... distribution function.

Q.20 The domain of a distribution function F (x) is (.......................) and its range is [.......................].

Q.21 If F (x) is a distribution function of x and if a < b, then we can write

$p(a < x \leq b) =$ ....

Q.22 If P (X = a) = 0 and P (X = b) = 0 then four events $a \leq x \leq b$, $a < x \leq b$, $a \leq x < b$ and $a < x < b$ have the same probability and is equal to ....

Q.23 All the distribution functions are monotnically non .... functions and lie in the limits (.............).

Q.24 If F (x) is a distribution function of x, then F ($-\infty$) .... and F (∞) = ...

Q.25 A variable x which can take only a countable number of real values and these values depend on chances, is called a (Discrete ....) or (Discrete ....) or (discrete chance variable).

Q.26 A real valued function defined on a discrete sample space S, is called a " .... random variable".

Q.27 The function P (x) defined for a discrete random variable x as

$P_x(x) = P(x = xi) = pi$ for $x = xi$

$= 0$ for $x \neq xi$, i = 1, 2, ..., n is called the "............................" function.

Q.28 The set of ordered pairs {xi, p (xi), i = 1, 2, ..., n} is known as the "Probability .... " of the r.v. x.

Q.29 The set of values (xi, i =1, 2, ..., n) taken by the random variable X, is known as the " .... " of x.

Q.30 A knowledge of p.m.f of X, for an arbitrary event E of set of real numbers and a sample space S, faselitates us to work out

$$P(x \in E) = \sum_{x \in \ldots\ldots} p(x)$$

Q.31 The relation between p (X = x) and distribution function F (X = x) = F (x) - F (.........).

Q.32 A random variable X is said to be continuous, if it can take all possible values between certain ...., say $(a \leq x \leq b)$.

Q.33 A random variable X is said to be continuous when its different values can not be put into 1 to 1 correspondence with a set of ....

Q.34 The f (x) being a continuous function of x, the f (x) dx represents the probability that x falls in the infinitesimal interval $(\ldots\ldots \leq x \leq \ldots\ldots)$.

Q.35 The f (x) dx represents the area bounded by the curve y = f (x), x axis and the ordinate at the points (.......................).

Q.36 The expression f (x) dx usually denoted by d F (x), is known as the " .... ".

Q.37 The curve y = f (x) defined in Q (35) is known as the "probability density curve" or simply " .... ".

Q.38 The p.d.f, $f_x(x)$ of random variable X is defined as

$$f(x) = \lim_{\partial x \to 0} \frac{P(\ldots\ldots \leq x \leq \ldots\ldots)}{\partial x}$$

Q.39 The probability that a random variable x will lie in the interval "dx" is equal to " .... " .

Q.40 The probability that a variate X will lie in an interval [α, β] can be writeen as $P(\alpha \le x \le \beta) = \int_{\alpha}^{\beta}$ ....

Q.41 If [a, b] is the range of random variable X then $\int_a^b f(x)dx =$ ....

Q.42 The range of a continous random variable X may be finite or ....

Q.43 In case of a continuous random variable X, the probability at a point c is always equal to .... for all possible values of X.

Q.44 Let f (x) be a p.d.f. of a r.v., X defined in (........), then arithmetic mean $\bar{x}$ is defined as

$$\bar{x} = \int_a^b \text{.......} dx$$

Q.45 The harmonic mean H is defined as $\frac{1}{H} = \int_a^b \text{........} dx$

Q.46 The geometric mean G of a random variable X is defined as

$$\log G = \int_a^b \log x \text{ ....}$$

Q.47 The rth raw moment μ_r' of a r.v. X is defined as

$$\mu_r' = \int_a^b \text{.......} f(x)dx$$

Q.48 The variance σ_x of a r.v. X is defined as

$$\sigma_x^2 = \int_a^b \text{........} dx - \left(\int_a^b x f(x)dx\right)^2$$

Q.49 Let Md denotes the median point of a r.v. X then Md can be determined by solving

$$\int_{a}^{Md} f(X)dx = \int_{Md}^{b} \text{.............} = \frac{1}{2}$$

Q.50 Mean deviation M.D. of a r.v., X can be defined as

$$\text{M.D.}= \int_{a}^{b} |\text{...........}| f(x)dx$$

Q.51 The 3rd quartile Q_3 can be obtained from the equation

$$Q_3 = \int_{a}^{b} \text{.........}dx = \frac{3}{4}$$

Q.52 The ith decile Di (i = 1, 2, ..., 9) can be obtained by solving

$$Di = \int_{a}^{Di} \text{.......} = \frac{i}{10}, \ i = 1, 2, \text{....}, 9$$

Q.53 Let X be a r.v. defined as $(a \le x \le b)$ the moe Mo can be obtained from the solution of f'(x) = 0 and f"(x) .... provided $(a \le Mo \le b)$

Q.54 Let X bea continuous r.v. with p.d.f. f(x) then the function Fx (x) defined as $Fx(x) = P(X \le x) = \int_{-\infty}^{x} f(t)\,dt, \ (-\infty < t < \infty)$ is known as " .... " or comulative distribution function of X.

Q.55 The distribution function Fx (x) is a "non .... " function of X.

Q.56 The distribution function Fx (x) has the limits defined as (...... ≤ Fx (x) ≤.......)

Q.57 For the distribution function Fx (x), we have Fx (—∞) =

$$\lim_{x \to -\infty} \int_{-\infty}^{x} f(x)dx = \text{...................}$$

Q.58 $F_x(+\infty) = \lim_{x \to \infty} \int_{-\infty}^{x} f(x)dx = \text{..................}$

Q.59 The function Fx (x) is a .... function of X on the right.

Q.60 The discontinuities of F(x) are at the most ....

Q.61 Let (X, Y) be two r.v.'s defined on same sample space S, then the function (x, y) which assigns a point in R^2 (R x R), is called .... dimensional r.v.

Q.62 Let w ϵ S, then the value of (X, Y) at w is given by the pair of real numbers (.........., ...) .

Q.63 The notation {x ≤ a, y ≤ b} denotes the events that all elements w ϵ S, such that x(w) ≤ a, ... ≤ b.

Q.64 Let A = {a < x ≤ b}, B = {c < y ≤ d} be two events then P (a < x ≤ b, c < y ≤ d) = ...

Q.65 Two dimensional r.v. (X, Y) is said to be discrete if it takes " .... " of points in R^2.

Q.66 Two r.v's (X, Y) are said to be jointly distributed if they are defined on the same probability ....

Q.67 Pxy (x, y) being joint probability function of (X, Y), then the probability of certain event E is given by P(x, y) = P [(.....) ϵ E]

Q.68 Let X(S) = {xi, i = 1, 2, ..., n} and Y(S) = {yj, j = 1, 2, ..., m} be two image sets on a sample space S. Then joint probability function P on X(S) x Y(S) is defined as pij = p(xi, yj) = P (........)

Q.69 Let (X, Y) be two dimensional discrete r.v, then their joint p.m.f. P(x,y) is defined as

$$P\,(xi,\, yi) = \begin{cases} P\,(\ldots\ldots) & \text{for a value } (xi,\, yi) \\ 0 & \text{, otherwise} \end{cases}$$

Q.70 For joint p.m.f. P(xi, yj), we have the result that

$$\sum_{i=1}^{n}\sum_{j=1}^{m} P\,(xi,\, yi) = \ldots\ldots$$

Q.71 From Joint p.m.f. (P(xi, yj), we can determine the p.m.f. of r.v. X alone as

$$P(xi) = \sum_{\ldots\ldots}^{\ldots\ldots} P\,(xi\,,\, Yi) = pi\,,\, i = 1, 2, \ldots, n, j = 1, 2, \ldots, m$$

Q.72 The p.m.f. of r.v. X alone obtained from joint p.m.f. P(x, y) is called .... p.m.f. or discrete .... of r.v. X.

Q.73 From joint p.m.f. say P(x, y), the conditional p.m.f. of X for given Y = y, is defined as

$$P\,(X/Y = y) = \frac{P\,(..........)}{P\,(..........)}\text{ , (provided } P(y) \neq 0$$

Q.74 From the conditional p.m.f. say P(xi / Y = y), we can write that

$$\sum_{i=1}^{n} \frac{P\,(xi\,/\,y)}{P\,(Y = y)} = \,..........$$

Q.75 For two discrete r.v. (X, Y) to be independent, the necessary and sufficient condition is that
P(X, Y) = P (....) P (....) values of (xi, yj)

Q.76 The d.f. of two r.v. (X, Y) is a real valued function F defined for all (....................) by the relation _
F (x, y) = P(X = ...,Y = ...)

Q.77 For real numbers (a_1, b_1) and (a_2, b_2) for two r.v. (X, Y), we can write the relation -

$$P\,(a_1 < x \leq b_1, a_2 < y \leq b_2) = F_{xy}\,(a_1, a_2) + F_{xy}\,(b_1, b_2) - F_{xy}\,(..................) - F_{xy}\,(..................)$$

Q.78 The joint distribution function F (x, y) is monotonically (...........) function of (X, Y).

Q.79 For joint distribution function F_{xy} (x, y), we have the result
F (—∞ , y) = F(x, —∞) = ...

Q.80 For joint distribution function F(x, y), we have F(—∞ , ∞) = ...

Q.81 If density function f(x,y) is continuous at (x,y) then we have

$$\frac{.........}{\partial x\, \partial y} = f(x,y)$$

Q.82 In case of jointly continuous r.v. (X, Y), the marginal distribution function of (X) is given as $F_x\,(x) = \sum_{..........} P(X \leq x, ...)$

Q.83 In case of joint continuous r.v. (X, Y), the marginal distribution function of X is given as $F_x(x) = \int_{-\infty}^{x}\left\{\int_{-\infty}^{x} f(x, y)dy\right\}$

Q.84 In case of joint p.d.f. f (x, y), the marginal distribution function of Y is given as

$$F_Y(y) = \int_{-\infty}^{y}\left\{\int_{-\infty}^{\infty} f(\ldots\ldots\ldots)\ldots\right\}\ldots.$$

Q.85 The joint density function f(x,y) is defined as

$$f(x\ y) = \lim_{\partial x \to 0, \partial y \to 0} \frac{p\ (x \le X \le \ldots\ldots\ldots, y \le Y \le \ldots\ldots)}{\partial x\ \partial y}$$

Q.86 The probability that the point (x, y) will lie in the infinitesimal rectangular region dx dy is given by. f(x, y) dx dy =

$$p\left(\ldots. \le X \le x + \frac{dy}{2}, y - \frac{dy}{2} \le Y \le \ldots..\right)$$

Q.87 The marginal p.d.f. of x for discrete r.v. is given as f (x) =

$$\sum_{y} P\ (\ldots\ldots)$$

Q.88 The marginal p.d.f. of Y in case of continuous joint p.d.f. f(x,y), is given by

$$f(y) = \int_{-\infty}^{y} \ldots\ldots..dx$$

Q.89 From two marginal distributions of two r.v. (X, Y) we can not determine the " .... " of these two r.v. (X, Y).

Q.90 From joint density function, f(x, y) we can determine the two " .... " f(x) and f(y).

Q.91 The conditional density function f(y/x) denotes the density function of y when .... has already assumed the particular value ....

Q.92 Mathematically, we can write
F (y/x) = P(Y ≤ .... / X = ...)

Q.93 We can write mathematically that

$$f(y/x) = \frac{\text{........}}{f(x)}, \text{ for } f(x) > 0$$

Q.94 Two r.v. (X, Y) with joint p.d.f. f(x, y) and marginal pdf as f(x) & f(y), are said to stochastically (randomly) independent iff. f(x, y) = ...

Q.95 Let (X, Y) be the stochastically independent two r.v's then for possible pairs of real numbers (a_1, b_1) & (a_2, b_2) where ai < bi, we can write

$P(a_1 < x \leq b_1) \cap P(a_2 < y \leq b_2) = P(\text{...........}) P(\text{............})$

Q.96 If the distribution of Y = g(x) is uniqualy determined then the function. Y = g(x) is a .... function of x.

Q.97 Let X be continuous r.v. with p.d.f. f(x) and Y = g(x) be a strictly monotonic function of x. Let g(x) be differentiable V x. Then p.d.f. of Y say h(y) is given by

$$h(y) = f(x) \left| \frac{\text{........}}{\text{........}} \right|$$

Q.98 Two r.v's (X, Y) are transformed to new variables as U = U (x, y) and V = v (x, y) where (U, V) are continuously differentiable function.

Then Jacobian of transformation J is given by

$$j = \frac{\partial(\text{.........})}{\partial(\text{.........})} = \begin{vmatrix} \frac{\partial x}{\partial u} & \frac{\partial y}{\partial u} \\ \frac{\partial x}{\partial v} & \frac{\partial y}{\partial v} \end{vmatrix}$$

Q.99 The joint p.d.f. say g(u, v) of the transformed variates U and V is given by

g(u, v) = f(........) where | j | is the Jacobian of transformation.

Q.100 If (X, Y) are independent continuous r.v's, then p.d.f. of U = X + Y is given by

$h(u) - \int f_x(v) f_y(\text{...............}) dv$ where v = x, and y = (u-v)

Q.101 In Q(100), the function h(u) is said to be the convaluation of $f_x(x)$ and $f_Y(y)$ and we write it as h(.) = f(x)* f(....)

Q.102 Let (X, Y) be two independent continuous r.v's. Then the p.d.f. of their product u = XY is given by $h(u) = \int_{-\infty}^{\infty} f_{xy}(x, ...) \frac{dx}{|x|}$

Section -C : Multiple Choice Questions

Q.1 A r.v. is a .... x connected with the outcomes of a random experiment.
(a) Real number (b) Integer number
(c) An event (d) Counting number

Q.2 In tossing of a balanced die at one time, the number coming on the upper most face is a -
(a) A random variable (b) An event
(c) An outcome of the experiment (d) All the three

Q.3 Let S be a sample and X be a r.v. defined on S, then X will takes the values in -
(a) $(-\infty < R < b)$ (b) R $(a < R < \infty)$
(c) Integers (d) Both (a) & (b)

Q.4 In general the n dimensional r.v. Xis defined on a sample space S and whose range is n tuples of ....
(a) Real numbers (b) Counting numbers
(c) Integers (d) Both (a) & (b)

Q.5 L etX (w) be a function with domain S and range R, $(-\infty < R < \infty)$ such that the event. $\{w : x(w) \leq a\} = B$ then 'a' is
(a) An integer (b) Real number
(c) Positive number (d) None of the three

Q.6 The function X (w) is the measurable function on
(a) (S, B) (b) (B, S)
(c) S, (d) Ω

Q.7 X (w) is associated with the outcome 'w' of a random experiment and it represents the -
(a) Integer number (b) Real number
(c) Positive integer number (d) None

Q.8 We can write $P(X \leq a)$ as

(a) $P\{w : x(w) \in (-\infty, a]\}$
(b) $P\{w: x(w) \in (-\infty , \infty)\}$
(c) $P\{w : x(w) \in (\infty, a)\}$
(d) $P\{w :x(w) \in (0,a)\}$

Q.9 We can write that $P(a < x < b)$ as

(a) $P\{w : x (w) \in (a, b)\}$
(b) $P\{w : x (w) \in (a, b]\}$
(c) $P\{w : x (w) \in [a, b]\}$
(d) $P\{w : x (w) \notin (a,b)\}$

Q.10 The analogus meaning to P (x = a, or x = b) is

(a) $P\{w : (x = a) \cap (x= b)\}$
(b) $P\{w : (x = a) \cup (x = b)\}$
(c) $P\{w : (x \neq a) \cup (x = b)\}$
(d) $P\{w : (x = a) \cup (x \neq b)\}$

Q.11 The analogus meaning to P (x = a and x = b) is equal to

(a) $P\{w : (x = a) \cap (x = b)\}$
(b) $P\{w : (x = a) \cup (x = b)\}$
(c) $P\{w : (x \neq a) \cup (x = b)\}$
(d) $P\{w : (x = a) \cup (x \neq b)\}$

Q.12 The triplet (S, B, P) is known as -

(a) Probability space
(b) Sample space
(c) σ field
(d) Borel field

Q.13 A function X (w) from S to R (...) is a random variable iff for a real 'a', we have

(a) $\{w : x(w)< a\} \in B$
(b) $\{w : x(w)< a\} \notin B$
(c) $\{w : x(w)< a\} \subset B$
(d) $\{w : x(w)< a\} \not\subset B$

Q.14 If (X_1 & X_2) are r.v's then $X_1 + X_2$ and $X_1 X_2$ are

(a) Random variables
(b) Constants
(c) Not random variables
(d) Both (b) & (c)

Q.15 If (X_1, X_2) are r.v's them Max (X_1, X_2) and Min (X_1, X_2) are -

(a) Constants
(b) Random variables
(c) Stochastic variables
(d) Both (b) & (c)

Q.16 If X is a r.v. and f(x) is a continuous function, then f(x) is a

(a) Constant
(b) r.v.
(c) Stoohastic variable
(d) Both (b) & (c)

Q.17 Let X be a r.v. and f(x) is monotonic function of X then f(x) is a
(a) Stochastic variable (b) Constant
(c) Random variable (d) Both (a) & (c)

Q.18 Let X be a r.v. and F(x) is defined for all real x as $F(x) = P(X \leq x) = P\{w : x(w) \leq x\}, -\infty < x < \infty$. Then F(x) is known as
(a) Distribution function of r.v. X (b) Density function of r.v. X
(c) Non decreasing function of X (d) Both (a) & (c)

Q.19 A distribution function F(x) is also known as
(a) Cumulative distribution function
(b) Probability distribution function
(c) Probability density function
(d) Both (a) & (b)

Q.20 The domain of the distribution function is
(a) $[-\infty, -\infty]$ (b) $[\infty,\infty)$
(c) $(-\infty, \infty)$ (d) None

Q.21 The range of a distribution function is
(a) (O, R) (b) [0, 1]
(c) [0, 1) (d) (0,1]

Q.22 If F(x) is the distribution function of r.v. X and if a < b then we can write that F(b) - F(a) is equal to
(a) $P\{a < x < b\}$ (b) $P\{a < x \leq b\}$
(c) $P\{a \leq x \leq b\}$ (d) $P\{a \leq x < b\}$

Q.23 If $P(x = a) = 0$ and $P(x = b) = 0$, then F(b) - F(a) is equal to
(a) $P\{a \leq x \leq b\}$ (b) $P\{a \leq x \leq b\}$
(c) $P\{a \leq x < b\}$ (d) All the three

Q.24 All the distribution functions are monotonically -
(a) Non-increasing (b) Non decreasing
(c) Increasing (d) Both (b) & (c)

Q.25 Let F(x) be a distribution function of X then $F(-\infty) = \lim_{x \to -\infty} F(x)$ is equal to
(a) Zero (b) ≥ 0
(c) 1.0 (d) None

Q.26 Let F(x) is a distribution function then $\lim_{x \to \infty} F(x)$ is equal to -

(a) 0.0 (b) 1.0

(c) ≥ 1.0 (d) < 1.0

Q.27 A variable which can take only a countable number of real values and these values depend on chance, is called a

(a) Discrete random variable

(b) Discrete stochaustic variable

(c) Discrete chance variable

(d) All the three

Q.28 A real valued function defined on a discrete sample space S, is called -

(a) Discrete chance variable (b) Discret random variable

(c) Discrete stochaustic variable (d) All the three

Q.29 The number of telephonic call/unit of time, the number of accidents taken place/minute, the number of heads taken place in n bernouille's trials is an example of -

(a) Discrete random variable

(b) Discrete stochastic variable

(c) Continuous random variable

(d) Both (a) & (b)

Q.30 Let X be a discrete r.v. with distinct values (xi, i = 1, 2, n) then the function P(x) defined as

$$p(x) = \left\{ \begin{array}{l} P\ (X = xi) = pi,\ \ \text{for } x = xi \\ \qquad\qquad\ \ = 0 \quad \text{for } x \neq xi, i = 1,2,.....n \end{array} \right\} \text{ is called}$$

(a) Probability mass function

(b) Probability density function

(c) Probability distribution function

(d) Indicator function

Q.31 The set of ordered pairs (xi, pi, i = 1, 2, 3, ..., n ...) is known as-

(a) Probability density function of X

(b) Probability distribution function of X

(c) Probability mass function of X

(d) Probability generating function

Q.32 The set of all possible values (xi, i = 1, 2, ... n, ...) taken by a randondom variable x, is known as

(a) Sample space (b) Spectrum of r.v. x

(c) Probability of r.v. (d) Both (a) & (b)

Q.33 Let F(x) denotes the distribution function of r.v. X, then F(x) - F(xi-1) is equal to -

(a) P (x = xi) (b) p(xi)

(c) P(x $\leq$ xi) (d) Both (a) & (b)

Q.34 A r.v. X is said to be continuous when its different values can not be put in to 1 to 1 correspondence with a set of ....

(a) Integers (b) Positive integers

(c) Real numbers (d) None

Q.35 Let f(x) be any continuous function of X defined in an interval (x, x + dx). Then f(x) dx gives the probability for x to lie in the interval -

(a) (x — dx, x + dx) (b) $\left(x - \frac{dx}{2}, x + \frac{dx}{2}\right)$

(c) (x, x + dx) (d) Both (a) & (b)

Q.36 For a continuous p.d.f. f(x) of a r.v. X, we can state that f(x) dx is equal to

(a) P(x $\leq$ x $\leq$ x+ dx) (b) $P\left(x - \frac{dx}{2} \leq x \leq x + \frac{d_2}{2}\right)$

(c) p(x—dx $\leq$ x $\leq$ x) (d) None

Q.37 If f(x)dx represents the area bounded by the curve y = f(x), x axis and the ordinate at points (x, x + dx) then f(x) is known as-

(a) Probability density function of X

(b) Distribution function of x

(c) Density function of x (d) Both (a) & (c)

Q.38 The expression $f(x)dx = \frac{dF(x)}{dx}$ is known as

(a) Probability differential of x

(b) Probability density function of x

(c) Probability distribution function of x

(d) Probability measure of x

Q.39 The curve y = f (x) defined in Q (37) is known as

(a) Probability curve (b) Probability differential

(c) Probability density curve (d) Both (a) & (c)

Q.40 The p.d.f. f (x) of r.v. X is defined as

(a) $\lim_{\delta x \to 0} \frac{P(x \le X \le x + dx)}{\partial x}$

(b) $\lim_{\delta x \to 0} \frac{P\left(x - \frac{\partial x}{2} \le X \le x + \frac{\partial x}{2}\right)}{\partial x}$

(c) $\lim_{\delta x \to 0} \frac{P(x - \partial x \le X \le x)}{\partial x}$ (d) P (X = x)

Q.41 The probability for a variate value to lie in the internal dx is equal to -

(a) f(x) (b) f(x)dx

(c) F(x) (d) $\frac{2}{dx} F(x)$

Q.42 The probability for a variate value to fall in the finite interval [α,β] is given by

(a) P (α ≤ x ≤ β) (b) P(α ≤ x ≤ β)

(c) $\int_{\alpha}^{\beta} f(x)dx$ (d) Both (a) & (c)

Q.43 The variance σ_x^2 of r.v. X is equal to -

(a) $\int_{-\infty}^{\infty} xf(x)dx - \left(\int_{-\infty}^{\infty} x^2 f(x)dx\right)^2$

(b) $\int_{-\infty}^{\infty} x^2 f(x)dx - \left(\int_{-\infty}^{\infty} xf(x)dx\right)^2$

(c) $\int_{-\infty}^{\infty} x^2 F(x)dx - \left(\int_{-\infty}^{\infty} xF(x)dx\right)^2$ (d) Both (b) & (c)

Q.44 Let Md denotes the median point of the r.v. X. Then Md can be obtained from the equation

(a) $\int_{a}^{Md} f(x)dx = \int_{Md}^{b} f(x)dx = \frac{1}{2}$

(b) $\int_a^{Md} x.f(x)dx = \int_{Md}^{b} x f(x)dx = \frac{1}{2}$

(c) $\int_a^{Md} f^2(x)dx = \left(\int_a^b f(x)dx\right)^2 = \frac{1}{2}$

(d) $\left(\int_{Md}^{b} f(x)dx\right)^2 = \left(\int_a^{Md} f(x)dx\right)^2 = \frac{1}{2}$

Q.45 Mean deviation of a r.v. X can be defined as

(a) $\int_a^b |x-\bar{x}| f(x)dx$

(b) $\int_a^b |x-\bar{x}|^2 f(x)dx$

(c) $\left[\int_a^b |x-\bar{x}|^2 f(x)dx\right]^{\frac{1}{2}}$

(d) None of these

Q.46 The quartile Q3 can be obtained from -

(a) $\int_a^{Q_3} f(x)dx = \frac{3}{4}$

(b) $\int_a^{Q_3} f(x)dx = \frac{1}{3}$

(c) $\int_a^{Q_3} xf(x)dx = \frac{3}{4}$

(d) $\int_a^{Q_3} x.f(x)dx = \frac{1}{3}$

Q.47 The ith decile Di (i = 1, 2, ... 9), can be obtained from -

(a) $\int_a^{Di} f(x)dx = \frac{i}{10}$

(b) $\int_a^{Di} x f(x)dx = \frac{i}{10}$

(c) $\int_a^b x^i f(x)dx = \frac{i}{10}$

(d) $\int_a^{Di} f(x)dx = \frac{10}{i}$

Q.48 The mode Mo of a r.v. X for (a < Mo < b) can be obtained from the solution of -

(a) f' (x) = 0 and f" (x), > 0

(b) f' (x) = 0 and f" (x) < 0

(c) f" (x) = 0 and f' (x) < 0

(d) f" (x) = 0 and f' (x) > 0

Q.49 Let X be the continuous r.v. with p.d.f. f(x) then the distribution function of X is given by -

(a) $F_X(x)$

(b) P(x = x)

(c) $\int_{-\infty}^{x} f(t)\, dt$

(d) All the three

Q.50 If F(x) is a non decreasing function of r.v. X then

(a) F' (x) ≥ 0 (b) $\frac{d}{dx}F(x) \geq 0$

(c) f (x) ≥ 0 (d) All the three

Q.51 The distribution function F_x (x) possess the following property

(a) $0 \leq F_x(x) \leq 1$ (b) $-1 \leq F_x(x) \leq 0$

(c) $-1 \leq F_x(x) \leq 1$ (d) $-\frac{1}{2} \leq F_x(x) \leq \frac{1}{2}$

Q.52 Let F(x) be the distribution function of r.v. X then we have the following result

(a) $F(-\infty) = 0$ (b) $\lim_{x \to -\infty} F(x) = 0$

(c) $\lim_{x \to -\infty} \int_{-\infty}^{x} f(x)dx = 0$ (d) Al the thre

Q.53 Let F(x) be the distribution function of r.v. X then the following is true

(a) $\lim_{x \to -\infty} F(x) = 1.0$ (b) $\lim_{x \to \infty} \int_{-\infty}^{x} f(x)dx = 1.0$

(c) $\int_{-\infty}^{\infty} f(x)dx = 1.0$ (d) All the three

Q.54 The function F(x) is continuous function of X on the ....

(a) Right side of x (b) Left side of x

(c) At point x (d) None

Q.55 A function f(x, y) defined on a sample space S, assigns a point in R^2 (R x R) then (X, Y) are random variables of

(a) One dimension (b) Two dimension

(c) Three dimension (d) None

Q.56 Let w ϵ S then the value of (X, Y) at the point w is given by the pair of -

(a) Real numbers {x(w), y (w))

(b) Real numbers {y(w), x (w)}

(c) Integer numbers {x(w), y (w))

(d) Integer numbers {(y(w), x(w)}

Q.57 The notation $\{x < a, y < b\}$ denotes the events of all elements $w \in S$, such that -

(a) $x(w) \leq a, y(w) \leq b$ (b) $x(w) = a, y(w) = b$

(c) $y(w) \leq b, x(w) \leq a$ (d) $y(w) = b, x(w) = a$

Q.58 Let $A = \{a < x < b\}$, $B = \{c < y < d\}$ be two events then $P\{a < x < b, c < y < d\}$ is equal to the

(a) $P(A \cup B)$ (b) $P(A \cap B)$

(c) $P(\overline{A} \cap \overline{B})$ (d) None

Q.59) Two r.v's (X, Y) are said to be discrete, if it takes atmost the -

(a) The uncountable number of points in R^2

(b) The countable number of points in R^2

(c) The countable number of points in R

(d) The uncountable number of points in R

Q.60 Two r.v's (X, Y) are said to be jointly distributed if they are defined on the

(a) Same probability space (b) Different samle space

(c) Same sample space (d) Both (a) & (c)

Q.61 If joint probability function is denated by P (x, y) then the probability of certain event E is given by

(a) $P\{(x, y) \in E\}$ (b) $P\{(x,y) \in \overline{E}\}$

(c) $P\{(y, x) \in E\}$ (d) $P\{(y,x) \in \overline{E}\}$

Q.62 The joint probablity function P on X(S) is defined as -

(a) $p_{ij} = P(X = xi \cap Y = yi)$ (b) $p_{ij} = P(X = xi \cup Y = yi)$

(c) $p_{ij} = P(xi, yi)$ (d) $p_{ij} = 1)(X = xi, \text{ or } Y = yi)$

Q.63 For joint p.m.f. P(xi, yj), we can write

(a) $\sum_{i=1}^{n} \sum_{j=1}^{m} p(xi, yi) = 1.0$ (b) $\sum_{i=1}^{n} \sum_{j=1}^{m} p(xi, yi) > 1.0$

(c) $\sum_{i=1}^{n} \sum_{j=1}^{m} p(xi, yi) \neq 1.0$ (d) $\sum_{i=1}^{n} \sum_{j=1}^{m} P(xi, yi) = 0.5$

Q.64 Let P (xi, yi) be the joint p.m.f. of two r.v.'s (X, Y). Then the p.m.f. of X alone can be determined as

(a) $\sum_{j=1}^{m} P(x_i, y_i) = P_{i\cdot}$ (b) $\sum_{j=1}^{m} P(x_i, y_i) = P_{\cdot j}$

(c) $\sum_{j=1}^{m} P(x_i, y_i) = P_{i\cdot}$ (d) $\sum_{j=1}^{m} P(x_i, y_i) = P_{\cdot j}$

Q.65 For joint p.m.f. P(x, y), the conditional p.m.f. of X given Y = y, is defined as

(a) $P(x/y) = \frac{P(X=x, Y=y)}{P(Y=y)}$ for $P(y) \neq 0$

(b) $P(x/y) = \frac{P(X=x, Y=y)}{P(X=x)}$ for $P(x) \neq 0$

(c) $P(x/y) = \frac{P(X=x, Y=y)}{P(X \geq xi)}$ for $P(x) \neq 0$

(d) $P(x/y) = \frac{P(X=x, Y=y)}{P(Y \geq y)}$ for $P(y) \neq 0$

Q.66 A necessary and sufficient condition for two discrete r.v's (X, Y) to be independent is that

(a) $P(X = xi, Y = yj) = P(X = xi).P(Y = yj)$
(b) $P(X = xi, Y = yj) \neq P(X = xi).P(Y = yj)$
(c) $P(X = xi, Y = yj) = P(X = xi) + P(Y = yj)$
(d) $P(X = xi, Y = yj) = P(X = xi) \cap P(Y = yj)$

Q.67 For real numbers a_1, b_1, a_2, b_2, we have

(a) $P(a_1 < X \leq b_1, a_2 < Y \leq b_2) =$
$F(a_1, a_2) + F(b_1, b_2) - F(a_1, b_2) - F(b_1, a_2)$
(b) $P(a_1 < X \leq b_1, a_2 < Y \leq b_2) =$
$F(a_1, a_2) + F(b_1, b_2) + F(a_1, b_2) + F(b_1, a_2)$
(c) $P(a_1 < X \leq b_1, a_2 < Y \leq b_2) =$
$F(a_1, b_2) + F(b_1, a_2) - F(a_1, a_2) - F(b_1, b_2)$
(d) None

Q.68 Let $a_1 < a_2$, $b_1 < b_2$ then we have

(a) $F(a_1, b_2) \geq F(a_1, a_2)$
(b) $F(b_1, a_2) \geq F(a_1, a_2)$
(c) $F(a_1, b_2) \leq F(a_1, a_2)$
(d) Both (a) & (b)

Q.69 For a distribution function F(x,y), we can write -

(a) $F(-\infty, y) = 0$
(b) $F(x, -\infty) = 0$
(c) $F(-\infty, \infty) = 1$
(d) All the three

Q.70 In case of joint p.d.f. f (x, y), the marginal distribution function of X is

(a) $F(x) = \int_{-\infty}^{x} \left\{ \int_{-\infty}^{\infty} (x, y) dy \right\} dx$
(b) $\int_{-\infty}^{\infty} \left\{ \int_{-\infty}^{y} f(x, y) dy \right\} dx$
(c) $\int_{-\infty}^{\infty} \left\{ \int_{-\infty}^{x} f(x, y) dx \right\} dy$
(d) $\int_{-\infty}^{\infty} \left\{ \int_{-\infty}^{\infty} f(x, y) dx \right\} dx$

Q.71 The joint p.d.f. f(x, y) is defined as

(a) $\lim_{\delta x \to \infty, \delta y \to 0} \dfrac{P(x \leq X \leq x + \delta x, y \leq Y \leq y + \delta y)}{\partial x \, \partial y}$

(a) $\lim_{\delta x \to 0, \delta y \to 0} \dfrac{P(x - \delta x \leq X \leq x, y - \delta y \leq Y \leq y)}{\partial x \, \partial y}$

(a) $\lim_{\delta x \to 0, \delta y \to 0} \dfrac{P\left(x - \frac{\delta x}{2} \leq X \leq x + \frac{\delta x}{2}, y - \frac{\delta y}{2} \leq Y \leq y + \frac{\delta y}{2}\right)}{\partial x \, \partial y}$

(d) None

Q.72 The marginal p.d.f. of X for discrete variable is given by

(a) $f(x) = \sum_{y} P(x, Y)$
(b) $f(x) = \sum_{x} P(x, Y)$
(c) $f(x) = \sum_{x,y} P(x, Y)$
(d) None

Q.73 The marginal p.d.f. of x for continuous r.v; x is given by

(a) $f(x) = \int_{-\infty}^{\infty} f(x, Y)\, dx$
(b) $f(x) = \int_{-\infty}^{\infty} f(x, Y)\, dy$
(c) $f(x) = \int_{-\infty}^{y} f(x, Y)\, dy$
(d) $f(x) = \int_{-\infty}^{x} f(x, Y)\, dy$

Q.74 From the marginal distributions f(x) and f(y) of two random variables (X, Y), the joint distribution function (density), f(x, y) can

(a) Not be determined (b) Be determined

(c) Not be said with surity (d) None

Q.75 For the joint p.d.f. f(x, y) and marginal density functions f(x), and f(y), we can define the conditional p.d.f. f(y/x) for given X = x, as follows

(a) $f(y/x)=\frac{f(x,y)}{f(y)}$ (b) $f(y/x)=\frac{f(x,y)}{f(x)}$

(c) $f(y/x)=\frac{f(x)}{f(x,y)}$ (d) $f(y/x)=\frac{f(y)}{f(x,y)}$

Q.76 For given joint p.d.f. f(x, y) and marginal density function f(x) and g(y), the two r.v's (X, Y) are said to be stochastically independent iff -

(a) f (x, y) = f (x)g (y) (b) f (x, y) = g (y) f (x)

(c) f (x, y) = f(x /y)g(y / x) (d) Both (a) & (b)

Q.77 For two r.v's (X, Y) being stochastically independent and for possible pairs of real numbers (a_1, b_1) & (a_2, b_2) where ai < bi the value of $P(a_1 < x < b_1) \cap P(a_2 < y < b_2)$ is equal to

(a) $P(a_1 < x \leq b_1)+P(a_2 < Y \leq b_2)$

(b) $P(a_1 < x \leq b_1).P(a_2 < Y \leq b_2)$

(c) $P(a_1 < x \leq b_1) \cup P(a_2 < Y \leq b_2)$

(d) Both (a) & (c)

Q.78 Let X be a continuous r.v. with p.d.f. f(x) and y = g(x) be a strictly monotonic function of (x). Let g(x) be differentiable V x then p.d.f. of Y is given by

(a) $h(y)=f(x).\left|\frac{dx}{dy}\right|$ (b) $h(y)=f(x)\left|\frac{dy}{dx}\right|$

(c) $h(y)=f^2(x).\left|\frac{dx}{dy}\right|$ (d) $h(y)=f^2(x).\left|\frac{dy}{dx}\right|$

Q.79 Let two r.v' s (X, Y) be transferable to new variables as U = u(x y) and V= u(x,y) where (u, v) are continuously differentiable function. Then Jacobian of transformation | J | is equal to -

(a) $|J| = \begin{vmatrix} \frac{\partial x}{\partial u} & \frac{\partial y}{\partial u} \\ \frac{\partial x}{\partial v} & \frac{\partial y}{\partial v} \end{vmatrix}$

(b) $|J| = \begin{vmatrix} \frac{\partial x}{\partial v} & \frac{\partial y}{\partial v} \\ \frac{\partial x}{\partial u} & \frac{\partial y}{\partial u} \end{vmatrix}$

(c) $|j| = \begin{vmatrix} \frac{\partial x}{\partial u} & \frac{\partial y}{\partial u} \\ \frac{\partial v}{\partial x} & \frac{\partial v}{\partial y} \end{vmatrix}$

(d) $|j| = \begin{vmatrix} \frac{\partial u}{\partial x} & \frac{\partial u}{\partial y} \\ \frac{\partial v}{\partial x} & \frac{\partial v}{\partial y} \end{vmatrix}$

Q.80 Let (X, Y) be two independent continuous r.v's then p.d.f. of U = (X-Y) is given by

(a) $h(u) = \int_{-\infty}^{\infty} f(x,x-u)dx$

(b) $\int_{-\infty}^{\infty} f(x)f(x-u)dx$

(c) $h(u) = \int_{-\infty}^{\infty} \frac{f(x)}{f(x-u)} dx$

(d) Both (a) & (b)

Q.81 Let (X, Y) be two independent continuous r.v's. Then p.d.f. of their product U = xy in given by

(a) $\int_{-\infty}^{\infty} f_{xy}\left(x, \frac{u}{x}\right)\frac{dx}{|x|}$

(b) $\int_{-\infty}^{\infty} f_{xy}\left(\frac{u}{x}, x\right)\frac{dx}{|x|}$

(c) $\int_{-\infty}^{\infty} f_{xy}\left(x, \frac{u}{x}\right)\frac{dy}{|y|}$

(d) $\int_{-\infty}^{\infty} f_{xy}\left(x, \frac{u}{x}\right)\frac{dx}{|x|}$

Key Answers

Section-A : True False

1.	True	2.	True	3.	True
4.	True	5.	True	6.	True
7.	True	8.	True	9.	True
10.	True	11.	True	12.	True
13.	True	14.	True	15.	True
16.	True	17.	True	18.	True

19.	True	20.	True	21.	True
22.	True	23.	True	24.	True
25.	True	26.	True	27.	True
28.	True	29.	True	30.	True
31.	True	32.	True	33.	True
34.	True	35.	True	36.	True
37.	True	38.	True	39.	True
40.	False	41.	True	42.	True
43.	True	44.	True	45.	True
46.	True	47.	True	48.	True
49.	True	50.	True	51.	True
52.	True	53.	True	54.	True
55.	True	56.	True	57.	True
58.	True	59.	True	60.	True
61.	True	62.	True	63.	True
64.	True	65.	True	66.	True
67.	True	68.	True	69.	True
70.	True	71.	True	72.	True
73.	True	74.	True	75.	True
76.	True	77.	True	78.	True
79.	True	80.	True	81.	True
82.	True	83.	True	84.	True
85.	True	86.	True	87.	True
88.	True	89.	True	90.	True
91.	True	92.	True	93.	True
94.	True	95.	True	96.	True
97.	True	98.	True	99.	True
100.	True	101.	True	102.	True
103.	True	104.	True	105.	True
106.	True	107.	True	108.	True
109.	True	110.	True	111.	True

Section-B : Fill in the Blanks

1. Real, Random
2. Random
3. Dimensional
4. Domain, Real
5. Function, $\leq$
6. Measureable, S, B
7. w or e
8. Real number

9. $(-\infty, a)$
10. $\cup$
11. $\cap$
12. Probability Space
13. $\leq$
14. Random variables
15. Random variables
16. Random variables
17. Random variables
18. Distribution Function
19. Cumulative
20. $(-\infty,\infty),[0,1]$
21. F (b) - F (a)
22. F (b) - F (a)
23. Decrasing
24. (0, 1)
25. Random, Stochastic
26. Discrete
27. Probability mass
28. Distribution
29. Spectrum
30. $E \cap S$
31. (x - 1)
32. Limits
33. Positive number
34. (x, x + dx)
35. (x, x + dx)
36. Probability differential
37. Probability curve
38. (x, x + dx)
39. f (x) dx
40. f (x)dx
41. 1.0
42. Infinite
43. Zero
44. x f (x)
45. f(x)
46. f (x) dx
47. x^r
48. $x^2 f(x)$
49. f (x) dx
50. $|x - \bar{x}|$
51. f (x)
52. f (x) dx
53. < 0
54. Distribution function
55. Decreasing
56. (0, 1)
57. 0.0
58. 1.0
59. Continuous
60. Countable
61. two
62. (x (w), y (w))
63. y (w)
64. $P(A \cap B)$
65. Atmost a countable number
66. Space

67. (x, y)
68. (X = xi $\cap$Y = yj)
69. (X = xi, Y = yj)
70. 1.0
71. j = 1, m
72. Marginal, p.m.f
73. (x, y), (y)
74. 1.0
75. (xi), (yi)
76. Real (x, y), (x, y)
77. $(a_1, b_2), (b_1, a_2)$
78. Non-decreasing
79. Zero
80. Zero
81. d^2 F (x, y)
82. y, Y = y
83. x, dx
84. (x, y), dx, dy
85. (x + ox, y + Sy)
86. $\left(x-\frac{dx}{2}, y+\frac{dy}{2}\right)$
87. (x, y)
88. f (x, y)
89. Joint distribution
90. Marginal density
91. X, x
92. (y, x)
93. f (x, y)
94. f (x). f (y)
95. $(a_1 < x < x < b_1), (a_2 < y < b_2)$
96. Continous
97. $\left|\frac{dx}{dy}\right|$
98. $\frac{(x, y)}{(x. v)}$
99. (x, y)
100. (u-v)
101. (y)
102. $\left(\frac{v}{x}\right)$

Section-C : Multiple Choice Questions

1. (a)
2. (d)
3. (c)
4. (a)
5. (b)
6. (a)
7. (b)
8. (a)
9. (b)
10. (b)
11. (a)
12. (a)
13. (a)
14. (a)
15. (d)
16. (d)
17. (d)
18. (d)
19. (d)
20. (c)
21. (b)
22. (b)
23. (d)
24. (d)

25. (a)
26. (b)
27. (d)
28. (d)
29. (d)
30. (a)
31. (b)
32. (d)
33. (d)
34. (b)
35. (c)
36. (a)
37. (d)
38. (a)
39. (d)
40. (a)
41. (b)
42. (d)
43. (b)
44. (a)
45. (a)
46. (a)
47. (a)
48. (b)
49. (d)
50. (d)
51. (a)
52. (d)
53. (d)
54. (a)
55. (b)
56. (b)
57. (a)
58. (a)
59. (b)
60. (b)
61. (d)
62. (a)
63. (a)
64. (a)
65. (a)
66. (a)
67. (a)
68. (d)
69. (d)
70. (a)
71. (a)
72. (a)
73. (b)
74. (a)
75. (b)
74. (d)
77. (b)
78. (a)
79. (a)
80. (d)
81. (a)

Chapter - 7

Mathematical Expectation

Section-A : Write True / False

Q.1 The average value of a random fenamenon is termed as "Mathematical Expectation" or "Expected value" of random fenamenon.

Q.2 The mathematical expression to compute the expected value of a discrete random variable X with its p.m.f. f(x) is defined as

$$E(X) = \sum_{x} xf(x)$$

Q.3 The mathematical expression to compute the expected value of a continuous random variable X with its p.d.f. f(x) is defined as

$E(X) = \int_{-\infty}^{\infty} xf(x)dx$ provided the integral is convergant.

Q.4 E(X) exists if $E|x|$ exists

Q.5 Consider an indicator variable $X = I_A$ so that

$I_A = \begin{cases} 1, \text{if A happens} \\ 0, \text{if } \bar{A} \text{ happens} \end{cases}$ and $E(I_A) = P(A)$.

Q.6 We can write that

$$V(X) = E(X^2) - [E(X)]^2$$

Q.7 $V(I_A) = P(A).P(\overline{A})$

Q.8 Let g(x) be a function of r.v. X and it itself is a r.v. Then, we have

$$E(g(x)) = \int_{-\infty}^{\infty} g(x)f(x)dx \text{ for continuous r.v.}$$

$$\& \; E(g(x)) = \sum_{x} g(x)f(x) \text{ for discrete r.v.}$$

Q.9 If $g(x) = X^r$ then $E(X_r) = \int_{-\infty}^{\infty} x^r f(x)dx$

Q.10 We have $\mu_2 = \sigma_x^2 = \int_{-\infty}^{\infty} (x - \overline{x})^2 f(x)dx$

Q.11 If (X & Y) are two r.v's then

$$E(X + Y) = E(x) + E(y)$$

Q.12 If (Xi, i =1, 2, ..., n) are n r.v.'s then $E\left(\sum_{i=1}^{n} X_i\right) = \sum_{i=1}^{n} E(X_i)$

Q.13 If (X & Y) are two independent r.v's then E(XY) = E(X) E(Y)

Q.14 For n independent random variables (Xi, i = 1, 2, ... n), we can write that

$$E\left(\prod_{i=1}^{n} Xi\right) = \prod_{i=1}^{n} E(Xi)$$

Q.15 If a is a constant then we can write that $E(a\psi(x)) = aE(\psi(x))$

Q.16. We have $E(X - \overline{X}) = 0$

Q.17 In have to report that in general ; $E(X^2) \neq (E(x))^2$

Q.18 For n constants and n random variables (ai, Xi, i = 1, 2, ..., n) respectively we have

$$E\left(\sum_{i=1}^{n} aiXi\right) = \sum_{i}^{n} E(aiXi)$$

Q.19 If two r.v's (X, Y) are such that $Y \leq X$, then $E(Y) \leq E(X)$

Q.20 We have the result that $| E(X) | \leq E | X |$, provided the expectation exists.

Q.21 If μ_r' excists then μ_s' exists for s, where $1 \leq s \leq r$

Q.22 If (X & Y) are two independent r.v's then

$$E [h (x) g (y)] = E [h (x)] E [g (y)]$$

Q.23 For two r.v's (X, Y), we can write cov (X, Y) in mathematical expression form as

$$COV (X, Y) = E(XY) - E(X) E(Y)$$

Q.24 If (X, Y) are independent r.v's then we have $COV (X, Y) = E(XY) - E(X) E(Y)$

Q.25 $COV (ax, by) = a.b.COV (X, Y)$

Q.26 $COV = \left(\frac{X - X}{\sigma_x}, \frac{Y - \bar{Y}}{\sigma_y} \right) = \frac{1}{\sigma_x \sigma_y} COV(X, Y)$

Q.27 Let $\sum^{n} a_i X_i$ be a linear function of n r.v's then we can write

$$V\left(\sum_{i=1}^{n} aiXi \right) = \sum_{i=1}^{n} a_i^2 v(Xi) + 2 \text{ ai aj} \sum_{i \neq j}^{n} COV(Xi, Xj)$$

Q.28 If $\sum_{i=1}^{n} a_i X_i$ is a linear function of n independent random variables Xi, (i = 1, 2, ..., n) then we have

$$V\left(\sum_{i=1}^{n} aiXi \right) = \sum_{i=1}^{n} a_i^2 v(Xi)$$

Q.29 Let (X, Y) be two r.v's taking real values. Then cauchy - Schwartz inequality is given as

$$[E(XY)]^2 < E(X^2) E (Y^2)$$

Q.30 A continuous function g(x) on the interval I is convex if for every x_1 and x_2, $(x_1 + x_2)/2 \in I$, we have

$$\frac{g(x_1 + x_2)}{2} \leq \frac{1}{2} g(x_1) + \frac{1}{2} g(x_2)$$

Q.31 For $(x_1, x_2) \in I$, we have

$g[\lambda x_1 + (1-\lambda)x_2] \leq \lambda g(x_1) + (1-\lambda) g(x_2)$

Q.32 If g is continuous and convex function on the interval I, and X is a r.v. whose values are in I with probability one, then Jonson's inequality is $E[g(x)] \geq g[E(x)]$ provided the expectation exists.

Q.33 A continuous function g is concave on the interval I if (-g) is convex.

Q.34 If g is a continuous and concave function on the interval I and x is a r.v. whose values are in I with probability 1.0, then $E[g(x)] \leq g[E(x)]$ provided the expectation exists.

Q.35 Let f and g be monotone functions on some subsets of the real lineR and X be a r.v. whose range is in the subset almost surely (a.s.). If the expectation exists, then

$E[f(x)\, g(x)] \geq E[f(x)]\, E[g(x)]$

or $E[g(x)\, g(x)] \leq E[f(x)]\, E[g(x)]$

according as f and g are monotone in the same or apposite directions.

Q.36 For $\alpha > 0, \beta > 0$ the $f(x) = x^\alpha$ and $g(x) = x^\beta$ are monotonic in the same direction.

Q.37 From the Question (36), we have

$E(x^\alpha x^\beta) = E(x^{\alpha+\beta}) > E(x^\alpha)\, E(x^\beta)$ for $\alpha > 0, \beta > 0$

Q.38 For $\alpha > 0$, we have

$E(x^\alpha x^{-1}) \leq E(x^\alpha)\, E(x^{-1}) = E(x^\alpha)\, E(\frac{1}{x}) > E(x^{\alpha-1})$

Q.39 We can write that

$E(x^2)\, E(x^2) \geq E(x^2 x^{-2}) = 1 \Longrightarrow E(x^2) \geq \frac{1}{E(x^2)}$

Q.40 Let (X & Y) be two r.v's having finite means. Then

$E[\mathrm{Min}(X, Y)] \leq \mathrm{Min}[E(x), E(y)]$

Q.41 We have for in Q(40), $E[\mathrm{Max}(X,Y)] \geq \mathrm{Max}[E(x), E(y)]$

Q.42 $E[\mathrm{Min}(X, Y) + \mathrm{Max}(X, Y)] = E(X) + E(Y)$

Q.43 For any two r.v's (X & Y)

$\left[E(X+Y)^2\right]^{\frac{1}{2}} \leq \left[E(X^2)\right]^{\frac{1}{2}} + \left[E(Y^2)\right]^{\frac{1}{2}}$

Q.44 For two continuous r.v's (X & Y), we have

$$E[g(X, Y)] = \int_{-\infty}^{\infty} \int_{-\infty}^{\infty} g(x,y).f(x, y)dx.dy$$

Q.45 For two discrete r.v's (X, Y), we have

$$E[g(X, Y)] = \sum_{i} \sum_{j} g(xi,yj)P(X = xi \cap Y = yj)$$

Q.46 The (rth, sth) product moment of two r.v's (X, Y) respectively about origin is defined as

$$\mu'_{rs} = E(X^r Y^s) = \int_{-\infty}^{\infty} \int_{-\infty}^{\infty} X^r Y^s .f(x, y)dxdy$$

Q.47 The (rth, sth) central moment of two r.v's (X, Y) respectively is given by

μ_{rs} $E[(X - \mu_x)^r (Y - \mu_Y)^s]$, for $E(X) = \mu_x$, $E(Y) = \mu_Y$

Q.48 For conditional expectation, we have

$$E[g(X, Y) \mid Y = yj] = \frac{\sum_{i=1}^{\infty} g(xi, yj) P(X = xi \cap Y = yj)}{P(Y = yj)}$$

Q.49 The conditional variance of X given $Y = y_j$ is defined as

$v(X/Y = yj) = E[\{X-E(x \mid Y = y_j)\}^2 / Y = y_j]$

Q.50 For continuous random variables (X, Y), the conditional expectation of g (X, Y) is defined as

$$E[g(X, Y)/Y = y] = \int_{-\infty}^{\infty} g(x,y).f(x / y)dx = \frac{\int_{-\infty}^{\infty} g(x,y).f(x,y)dx}{f(y)}$$

Q.51 The expected value of x is equal to the expectation of the conditional expectation of x given y i.e.

$E[E(X \mid Y=y) = E(X)$

Q.52 The variance of X can be regarded as consisting of two parts, the expectation of the conditional variance and the variance of conditional expectation, i.e.

$V(X) = E[V(X I Y)] + V[E(X \mid Y)]$

Q.53 Let A and B be two mutually exclusive events then

$$E(X \mid AUB) = \frac{P(A)E(X/A)+P(B)E(X/B)}{P(A \cup B)}$$

Q.54 $E(X) = P(A)\ E(X/A)\ P(A)\ E(X/\bar{A})$

Section -B : Fill in the Blanks

Q.1 The average value of a random variable X is known as " .... " of it.

Q.2 The expected value of a random variable X with its p.m.f. f(x) can be worked out from the expression.

$$E(X) = \sum_{x} \ldots\ldots$$

Q.3 The expected value of a continuous r.v. X with its p.d.f. f(x) is calculated from the expression

$$E(X) = \int_{-\infty}^{\infty}$$

Q.4 The expected value of X exists if (................) exists.

Q.5 For an indicator variable $X = I_A$' we write as

$$I_A = \begin{cases} 1, \text{if} \ldots\ldots\ldots \\ 0, \text{if} \ldots\ldots\ldots \end{cases} \quad \text{and } E(I_A) = P(A)$$

Q.6 The variance ex in term of mathematical expectation can be defined as

$$V(X) = E(X^2) - (\ldots\ldots\ldots\ldots)^2$$

Q.7 Variance of an incator function I_A is worked out as $V(I_A)$ = ...

Q.8 Let g(x) be a function of continuous random variable X. with its p.d.f. f(x), then -

$$E(g(x)) = \int_{-\infty}^{\infty} g(x) \ldots\ldots$$

Q.9 Let g(x) be a function of discrete r.v. X with its p.m.f. f(x) then

$$E(g(x)) = \sum_{x} \ldots\ldots$$

Q.10 Let $g(x) = x^r$ be continuous function of X with its p.d.f. f(x) then

$$E(x^r) = \int_{-\infty}^{\infty} (.......)dx$$

Q.11 The population variance $\sigma^2 x$ of a continuous r.v. X is defined as -

$$\sigma_x^2 = \int_{-\infty}^{\infty} (.....)^2 f(x)dx$$

Q.12 Let (X & Y) be two r.v's then, we have E(X + Y) = ...

Q.13 Let (Xi, i = 1, 2, 3, ..., n) be n r.v's then -

$$E\left(\sum_{i}^{n} Xi\right) = ...$$

Q.14 For n independent r.v's say (Xi, i = 1, 2, ..., n), we can write that

$$E\left(\prod_{i=1}^{n} Xi\right) = \prod_{i=1}^{n} ....$$

Q.15 If 'a' is a constant and ψ (x) is a function of x then

$E(a\psi (x))$ a.E (.....)

Q.16 For the situation described in Q (15), we have

$E(\psi(x)+a)= a+$ ....

Q.17 We have the result that

$E(X - \bar{X}) =$ ....

Q.18 We have to report that in general

$E (x^2)$ .... $(E(x))^2$

Q.19 For a linear function of n independent r.v's (Xi, i = 1, 2, ..., n), we have

$$E\left(\sum_{i=1}^{n} aiXi\right) = \sum_{i=1}^{n} ....$$

Q.20 Two random variables (X, Y) are such that $Y \leq X$, then E(Y) .... E(X)

Q.21 We have the result that

$| E(x) |$ .... $E | x |$, provided the expectation exists

Q.22 If μ'_r exists then μ'_s will also exist for (1 .... s..........r).

Q.23 If (X & Y) are two independent r.v's then

$E[h(x).g(y)] = E(\ldots\ldots\ldots\ldots)\ E(\ldots\ldots\ldots\ldots)$

Q.24 For two r.v's (X, Y) the COV (X,Y) can be expressed in terms of mathematical expectation as COV (X, Y) = E(XY) - (.......) (........)

Q.25 If (X, Y) are independent r.v's, then we have COV (X,Y) = E(.........) - E(X) E(Y)

Q.26 COV (ax, by) = ... COV (X, Y)

Q.27 $COV = \left(\frac{X-X}{\sigma_x}, \frac{X-\overline{Y}}{\sigma_y}\right) = \frac{1}{\sigma_x\sigma_y} COV\ (\ldots\ldots\ldots)$

Q.28 Let $\sum_{i=1}^{n}$ aiXi be a linear function of in r.v's then we can write

$$V\left(\sum_{i=1}^{n} aiX_i\right) = \sum_{i=1}^{n} a_i^2 V\ (Xi) + \ldots\ldots\ldots\ldots\ldots\ldots \sum_{i\neq 1}^{n} COV\ (Xi, Yj)$$

Q.29 If $\sum_{i}^{n}$ aiXi is the linear function of n independent r.v's (Xi, i = 1, 2, ..., n) then

$$V\left(\sum_{i}^{n} aiXi\right) = \sum_{i=1}^{n} \ldots\ldots\ldots\ldots$$

Q.30 Let (X, Y) be two r.v's taking real values then cauchy - schwartz in equality is given as

$[E(XY)]^2 \leq \ldots\ldots\ldots$

Q.31 A continuous function g(x) on the interval I, is convex if for every (x_1, & x_2), (x_1, + x_2) ϵ I and we have

$$g\left(\frac{x_1 + x_2}{2}\right) \leq \frac{1}{2} \ldots\ldots + \frac{1}{2} \ldots\ldots\ldots$$

Q.32 For (x_1, x_2) ϵ I, we have

$g[\lambda x_1 + (1-\lambda)x_2] \leq \lambda \ldots\ldots\ldots + (1 - \lambda) \ldots\ldots\ldots$

Q.33 If g is continuous and convex function on the interval I, and X is a r.v. whose values are in I, with probability 1, then Jenson' s Inequality is

E[g(x)] ≥ .... provided expectation exists

Q.34 Let f and g be monotone functions on some subsets of real line R, and X be a r.v. whose range is in the subset almost surely. If g and f are monotone function in the same direction then

E[f(x) g(x)] ≥ E[f(x)] E[g(x)] ....

Q.35 In Q(34), if f and g are monotone functions in oposite directions, then

E[f (...) g(x)] .... E[f(x)] E[g(x)]

Q.36 If $\alpha > 0$, $\beta > 0$ then the functions f (x) = x^α and g(x) = x^β are monotone in the ....

Q.37 Form the Q(36), we have

$E(x^\alpha x^\beta) \geq$ E(..........) E (.............) for $\alpha > 0, \beta > 0$

Q.38 A continuous function g is concave on an interval I if (-g) is

....

Q.39 For $\alpha > 0$, we have

$E(X^\alpha X^{-1}) <$ E(.........) E(...............)

Q.40 We can write that

$E(X^2)\, E(X^{-2})$ .... $E\,(x^2\, x^{-2})$

Q.41 Let (X, Y) be two r.v's having finite means. Then

E[Min. (X, Y)] .... Min [E(x), E(y)]

Q.42 In case of Q (41), we have

E[Max (X, Y)]............ Max [E(X), E(Y)]

Q.43 E[Min(X, Y) + Max (X, Y)] = ...

Q.44 For any two r.v's (X, Y), we have

$[E(X + Y)^2]^2$ .... $[E(x^2)]^{1/2}$ + [................]$^{1/2}$

Q.45 For two continuous r.v's (X & Y) with p.d.f. f(x, y), we have

$E[g(x, y)] = \int_{-\infty}^{\infty} \int_{-\infty}^{\infty} g(x,y)$..........

Q.46 For two discrete r.v's (X, Y), we have

$$E[g(x, y)] = \sum_i \sum_j g\,(xi, yi)P(xi......yj)$$

Q.47 The (rs)th product moment of two r.v's (X, Y), about origin is defined as

$$\mu'_{rs} = E(X^r y^s) = \int_{-\infty}^{\infty} \int_{-\infty}^{\infty} (x^r, y^s).........$$

Q.48 For two r.v's (X, Y), the (rs)th central moment μ_{rs} is defined as

$$\mu'_{rs} = E[(X - \mu_x)^r (......)^s]$$

Q.49 For conditional expectation, we have

$$E\,[g\,(X, Y) / Y = y_j] = \frac{\sum_i^{\infty} g(x_i, y_j)P(x_i......y_j)}{P(Y = y_j)}$$

Q.50 The conditional variance of X given Y = y_j is defined as

$$V\,(X/Y = y_j) = [\{X\text{-}E(X / ....)\}^2 / Y = y_j]$$

Q.51 For continuous r.v's (X, Y), the conditional expectation of g(x, y) is defined as

$$E[g(x, y)/Y = y] = \int_{-\infty}^{\infty} g\,(x, y)f(..........)dx$$

Q.52 For Q(51), we can write it also as

$$E[g(x, y)/Y = y] = \frac{\int_{-\infty}^{\infty} g(x, y)f(.........)dx}{f(y)}$$

Q.53 In case of two r.v's (X, Y), the expected value of X is equal to the expectation of the conditional expectation of X given Y = y i.e. mathematically it is -
E [E(X/Y = ...)]

Q.54 In case of two r.v's (X, Y), the variance of X can be regarded as consisting of two parts, the expectation of conditional variance and the variance of conditional expectation i.e.
V(x) = E (v (X/ ......] + ... [.........(x/y)]

Q.55 Let A and B be two mutually exclusive events then

$$E(x/AUB) = \frac{P(A)E(X/.....)+P(B).E(X/.....)}{P(A\cup B)}$$

Q.56 $E(X) = P(A).\ E(X/A) + P(\overline{A}).\ E(X/\overline{A})$

Section-C: Multiple Choice Questions

Q.1 The average value of a r.v. is known as its -

(a) Arithematic mean (b) Expected value

(c) First moment about origin (d) All the three

Q.2 The average value of a d iscrets r.v .X with its p.m.f. f(x) can be worked out from the expression -

(a) $\sum_x xf(x)$ (b) $\sum_x x^o f(x)$

(c) $\sum_x \sqrt{x}f(x)$ (d) $\int_{-\infty}^{\infty} xf(x)dx$

Q.3 The average value of a continuous r.v. X with it p.d.f. f(x) in the form of mathematical expectation is defined by the expression

(a) $\int_{-\infty}^{\infty} x^2 f(x)dx$ (b) $\int_{-\infty}^{\infty} xf(x)dx$

(c) $\sum_x xf(x)$ (d) Both (b) & (c)

Q.4 Consider an indicator variable $X = I_A$ so that -

(a) $I_A = 1$ if A happens (b) $I_A =$ if $\overline{A}$ happens

(c) $I_A = 0$ if A does not happen (d) Both (a) & (b)

Q.5 We can define $V(x) = \sigma_X^2$ as follows

(a) $\sigma_X^2 = E(X^2) - (E(X))^2$ (b) $(E(X))^2 - E(X^2)$

(c) $-[(E(X))^2 - E(X^2)]$ (d) Both (b) & (c)

Q.6 We can define $V(I_A)$ as follows -

(a) $V(I_A) = P(A)P(\overline{A})$ (b) $V(I_A) = P(A)P(A^{-1})$

(c) $V(I_A) = P(A)P(A^1)$ (d) $V(I_A) = P(A)P(\overline{\overline{A}})$

Q.7 Let g(x) be a continuous function of r.v. X then E(g(x)) can be defined as

(a) $\sum_{X} g(x)f(x)$ (b) $\int_{-\infty}^{\infty} g(x)f(x)dx$

(c) $\int_{-\infty}^{\infty} g(x)(dF(x))dx$ (d) Both (b) & (c)

Q.8 Let $g(x) = x^r$ be a function of continuous r.v. X with its p.d.f. f(x) then $E(X^r)$ can be expressed as

(a) $\int_{-\infty}^{\infty} x^r f(x)dx$ (b) $\int_{-\infty}^{\infty} x^r (dF(x))dx$

(c) $\sum_{X} x^r f(x)$ (d) Both (a) & (b)

Q.9 Let g(x) be a function of discrete random variable X with its p.m.f. as p(x) then E(g(x)) can be worked out by the expression

(a) $\int_{-\infty}^{\infty} g(x)p(x)dx$ (b) $\sum_{X} g(x)P(x)$

(c) $\sum_{X} g(x)p(x)dx$ (d) Both (b) & (c)

Q.10 The variance $\sigma^2 x$ of a random variable X, can be worked out from the expression

(a) $(x - \bar{x})^2 f(x)dx$ (b) $(x - \bar{x})f(x)dx$

(c) $(x - \bar{x})^2 p(x)$ (d) Both (a) & (b)

Q.11 Let Xi (i = 1,2, ..., n) be n r.v's then we can write

(a) $E\left(\sum_{i}^{n} Xi\right) = \sum_{i=1}^{n} E(Xi)$ (b) $E\left(\sum_{i} Xi\right) = \sum_{i}^{n} xi\ p(xi)$

(c) $\int_{1}^{n} xi\ f(xi)dx$ (d) Both (a) & (c)

Q.12 Let (X, Y)be two independent r.v's then we can write -

(a) E(XY) = E(X) E(Y) (b) E(XY) = E(Y) E(X)

(c) $E(X, Y) = [E(x^2)\ E(y^2)]^{1/2}$ (d) Both (a) & (b)

Q.13 For n independent r.v's (Xi, i = 1, 2, ..., n). We can write that

(a) $E\left(\prod_{i}^{n} Xi\right) = \prod_{i=1}^{n} E(Xi)$ (b) $E\left(\prod_{i=1}^{n} Xi\right) = \sum_{i=1}^{n} E(Xi)$

(c) $E\left(\prod_{i=1}^{n} Xi\right) = \prod_{i=1}^{n} E(Xi) - 1$ (d) $E\left(\prod_{i=1}^{n} Xi\right) = \left(1 - \sum_{i}^{n} E(Xi)\right)$

Q.14 If 'a' is a constant then we can write that

(a) $E(a\psi(x)) = aE(\psi(x))$ (b) $E(a\psi(x)) = a\psi(E(x))$

(c) $E(a\psi(x)) = a\psi(\overline{E(X)})$ (d) None

Q.15 We can write that

(a) $E(a + \psi(x)) = a + E(\psi(x))$ (b) $E(a + \psi(x)) = a + E(\psi(x))$

(c) $E(a + \psi(x)) = aE(\psi(x))$ (d) None

Q.16 We have result that

(a) $E(x - \bar{x}) > 0$ (b) $E(x - \bar{x}) \leq 0$

(c) $E(x - \bar{x}) = 0$ (d) $E(x - \bar{x}) = 0$

Q.17 In general, we can report that -

(a) $E(x^2) > (E(x))^2$ (b) $E(x^2) < (E(x))^2$

(a) $E(x^2) \neq (E(x))^2$ (d) None

Q.18 For a linear function $\sum^{n} a_i x_i$ for n random variables, we can write that

(a) $E\left(\sum_{i=1}^{n} aixi\right) = \sum_{i=1}^{n} aiE(xi)$ (b) $E\left(\sum_{i}^{n} aixi\right) = \sum_{i}^{n} E(aixi)$

(c) $E\left(\sum_{i}^{n} aixi\right) = \sum_{i}^{n} ai^2 E(xi)$ (d) Both (a) & (b)

Q.19 For two r.v's (X, Y) such that Y < X, we can write that

(a) $E(y) \geq E(x)$ (b) $E(y) \geq E(x)$

(c) $E(y) < E(x)$ (d) $E(y) > E(x)$

Q.20 If E(X) exists then we can write that

(a) $|E(x)|<E|x|$ (b) $|E(x)|>E|x|$

(c) $|E(x)|\leq E|x|$ (d) None

Q.21 If μ'_r exists then μ'_s exists for

(a) $s > r$ (b) $s \geq r$

(c) $s <r$ (d) $1 \leq s \leq r$

Q.22 If (X, Y) be two independent r.v's then E[g(x) h(y)] is equal to -

(a) E(g(x)) E(h(y)) (b) g(E(x)) h(E(y))

(c) E(h(y)) E(g(x)) (d) Both (a) & (c)

Q.23 For two r.v's (X, Y), the COV (X, Y) can be worked out from the expression

(a) E(X Y) - E(X) E(Y) (b) E(X) E(Y) - E(X Y)

(c) - [E(X) E(Y) - E (X Y) (d) Both (a) & (c)

Q.24 If (X, Y) are two independent r.v's then COV (X, Y) can be calculated from the expression -

(a) E(X Y) - E(X) E(Y) (b) - [E(X) E(Y) - E(X Y)]

(c) E(Y X) - E(Y) E(X) (d) All the three

Q.25 COV (ax, by) can be calculated from the formula -

(a) a.b COV (X, Y) (b) ba. COV (Y, X)

(c) ba COV (X, Y) (d) All the three

Q.26 The COV $= \left(\frac{X-\overline{X}}{\sigma_x}, \frac{Y-\overline{Y}}{\sigma_y}\right)$ is equal to

(a) $\frac{1}{\sigma_x\sigma_y}$ COV (X,Y) (b) $\frac{1}{\sigma_x\sigma_y}$ COV $(X-\overline{A}),(Y-\overline{A})$

(a) $\frac{1}{\sigma_y\sigma_x}$ COV (Y,X) (d) Both (a) & (c)

Q.27 Let $\sum_{i=1}^{n}$ aiXi be a linear function of n r.v's (Xi, i = 1 ...., n), then we can write that $V\left(\sum_{i}^{n} aiXi\right)$ is equal to

(a) $\sum_{i=1}^{n} a_i^2 V(Xi) + 2ai\,aj \sum_{i \neq j}^{n} COV(Xi, Yj)$

(b) $\sum_{i=1}^{n} a_i V(Xi) + ai\,aj \sum_{i \neq j} COV(Xi, Yj)$

(c) $\sum_{i} a_i^2\ v(Xi) - 2ai\ aj \sum^{n} COV(Xi, Yj)$

(d) $\sum_{i=1} a_i^2\ v(X_i^2) + 2ai\ aj \sum_{i \neq j}^{n} COV(Xi, Yj)$

Q.28 Let $\left(\sum_{i=1}^{n} aiXi\right)$ be a linear function of n independent r.v's Xi(i = 1, 2, ..., n) then $V\left(\sum_{i=1}^{n} aiXi\right)$ is equal to -

(a) $\sum_{i}^{n} a_i^2 v(Xi)$

(b) $\sum_{i}^{n} ai\ v(Xi)$

(c) $\sum_{i}^{n} v(Xi) a_i^2$

(d) Both (a) & (c)

Q.29 Let (X, Y) be two r.v's taking real values, then cauchy - Schwartz inequality is given by

(a) $(E(XY))^2 \le E(X^2)E(Y^2)$

(a) $(E(XY))^2 = E(X^2)E(Y^2)$

(c) $(E(XY))^2 \le E(X)E(Y^2)$

(d)$(E(XY))^2 \ge E(X^2)E(Y^2)$

Q.30 A continuous function g(x) on the interval I is convex if for every x_1 and x_2 and ½ $(x_1 + x_2) \in I$, then we can have -

(a) $g\left(\frac{x_1 + x_2}{2}\right) \le \frac{1}{2} g(x_1) + \frac{1}{2} g(x_2)$

(b) $g\left(\frac{x_1 + x_2}{2}\right) \le g\left(\frac{x_1}{2}\right) + g\left(\frac{x_2}{2}\right)$

(c) $g\left(\frac{x_1 + x_2}{2}\right) \ge \frac{1}{2} g(x_1) + \frac{1}{2} g(x_2)$

(d) Both (a) & (b)

Q.31 For (x_1, x_2) e I, we can write -

(a) $g[\lambda x_1+(1-\lambda)x_2] \leq \lambda g(x_1)+(1-\lambda)g(x_2)$

(b) $g[\lambda x_1+(1-\lambda)x_2] \geq \lambda g(x_1)+(1-\lambda)g(x_2)$

(c) $\lambda g(x_1)+(1-\lambda)g(x_2) \geq g[\lambda x_1+(1-\lambda)x_2]$

(d) Both (a) & (c)

Q.32 Let g be a continuous and convex function on interval I and X be a r.v. whose values are in I with probability 1.0 then "Jenson's Inequality" is

(a) $E[g(x)] \geq g[E(x)]$
(b) $E[g(x)] \leq g[E(x)]$

(c) $g[E(x)] \leq E[g(x)]$
(d) Both (a) & (c)

Q.33 Let g be a continuous and concave function on the interval I and x is a r.v. whose values are in I with probability 1.0 then

(a) $E[g(x)] \leq g[E(x)]$
(b) $g[E(x)] \geq E[g(x)]$

(c) Both (a) & (b)
(d) $E[g(x)] \geq g[E(x)]$

Q.34 Let f and g be monotone function on some subsets of real line R in the same direction and X be a r.v. whose range is in the subset almost surely, then

(a) $E[f(x)g(x)] \geq E[f(x)]E[g(x)]$

(b) $E[f(x)]E[g(x)] \leq E[f(x)]E[g(x)]$

(c) $E[f(x)g(x)] = E[f(x)]E[g(x)]$

(d) Both (a) & (b)

Q.35 If in Q(34), f and g are monotone functions in the oposite direction then

(a) $E[g(x)f(x)] \leq E[g(x)]E[f(x)]$

(b) $E[g(x)f(x)] \geq E[g(x)]E[f(x)]$

(c) $E[g(x)]E[f(x)] \geq E[g(x)f(x)]$

(d) Both (a) & (c)

Q.36 For a > 0, $\beta > 0$ then $f(X) = x^{\alpha}$ and $g(X) = x^{\beta}$ are -
(a) Monotonic function
(b) Not monotonic functions
(c) Monotone function in same direction
(d) Both (a) & (c)

Q.37 In Q(36) we have
(a) $E(X^{\alpha}X^{\beta})=E(X^{\alpha+\beta})$
(b) $E(X^{\alpha}X^{\beta})\geq E(X^{\alpha}).E(X^{\beta})$
(c) $E(X^{\alpha}).E(X^{\beta})\leq E(X^{\alpha}X^{\beta})$
(d) All the three

Q.38 For $\alpha > 0$, we have
(a) $E(X^{\alpha}X^{-1}) \leq E(X^{\alpha})E(X^{-1})$
(b) $E(X^{\alpha}X^{-1}) = E(X^{\alpha})(X^{-1})$
(c) $E(X^{\alpha})E(X^{-1}) \leq E(X^{\alpha}X^{-1})$
(d) Both (a) & (c)

Q.39 We have the result that
(a) $E(X^{2})E(X^{-2}) \geq E(X^{2}X^{-2})$
(b) $E(X^{2}.\overline{X}^{-1})=E(X^{\alpha})(X^{-1})$
(c) Both (a) & (b)
(d) $E(X^{2}.\overline{X}^{-2})E(X^{2})E(\overline{X}^{-2})$

Q.40 Let (X, Y) be two r.v's having finite means then,
(a) E[Min X, Y)] ≤ Min [E(X), E(Y)]
(b) Min [E(X), E(Y)] ≥ E[Min (X, Y)]
(c) Min [E(X), E(Y)] ≥ E (Min (X, Y)]
(d) E[Min (X, Y)] ≥ Min [E(X), E(Y)]

Q.41 For in Q(40), we have
(a) E[Max (X, Y)] ≥ Max [E(X), E(Y)]
(b) Max [E(X), E(Y)] ≤ E[Max (X, Y)]
(c) Both (a) & (b)
(d) Max [E(X), E(Y)] = E[Max (X, Y)]

Q.42 We have the result that
(a) E[Min (X, Y) + Max (X, Y)] = E(X) + E(Y)
(b) E[Min (X, Y) + max (X, Y)] ≥ E(X) + E(Y)
(c) E[Min (X, Y) + Max (X, Y)] ≤ E(X) + E(Y)
(d) None

Q.43 For any two r.v's (X, Y), we have

(a) $\left(E(x+y)^2\right)^{\frac{1}{2}} \leq \left(E(X^2)\right)^{\frac{1}{2}} + \left(E(Y^2)\right)^{\frac{1}{2}}$

(b) $\left(E(x^2)\right)^{\frac{1}{2}} + \left(E(y^2)\right)^{\frac{1}{2}} \geq \left(E(X+Y)^2\right)^{\frac{1}{2}}$

(c) $\left(E(x+y)^2\right)^{\frac{1}{2}} = \left(E(X^2)\right)^{\frac{1}{2}} + \left(E(Y^2)\right)^{\frac{1}{2}}$

(d) Both (a) & (b)

Q.44 The (r,s)th product moment of two r.v's (X, Y) about origin is defined as

(a) $\mu'_{rs} = E(X^r Y^s)$

(b) $\mu'_{rs} = \int\int X^r Y^s f(x,y)\,dxdy$

(c) $\mu'_{rs} = \sum_i \sum_j (x^r y^s) P(X = x_i, Y = y_j)$

(d) All the three

Q.45 The (rs)th central product moment of two r.v's (X, Y) is defined as

(a) $\mu'_{rs} = E\left[(X-\bar{X})^r (Y-\bar{Y})^s\right]$

(b) $\mu'_{rs} = \int_{-\infty}^{\infty}\int_{-\infty}^{\infty} (X-\bar{X})^r (Y-\bar{Y})^s f(x,y)\,dx\,dy$

(c) $\mu'_{rs} = \sum_i \sum_j (X-\bar{X})^r (Y-\bar{Y})^s P(X = xi, Y = yj)$

(d) All the three

Q.46 For continuous r.v's (X, Y), the conditional expectation of g (X, Y) for Y = y is defined as E [g(x, y)/Y = y] and is given as

(a) $\int_{-\infty}^{\infty} g(x, y) f(x / y)\,dx$

(b) $\dfrac{\int_{-\infty}^{\infty} g(x,y) f(x,y)\,dx}{f(y)}$

(c) $\int_{-\infty}^{\infty} g(x, y) f(y / x)\,dy$

(d) Both (a) & (b)

Q.47 The expected value of X is equal to the expectation of the conditional expectation of x given Y = y. It is given mathematically as

(a) E (X) = E[E (X / Y = y)] (b) (X) = E[E (X / Y^2 =y)]

(c) E(X)= $\int_{-\infty}^{\infty} Xf(x / Y = y)dx$ (d) Both (a) & (c)

Key Answers

Section-A : True False

1. True	2. True	3. True
4. True	5. True	6. True
7. True	8. True	9. True
10. True	11. True	12. True
13. True	14. True	15. True
16. True	17. True	18. True
19. True	20. True	21. True
22. True	23. True	24. True
25. True	26. True	27. True
28. True	29. True	30. True
31. True	32. True	33. True
34. True	35. True	36. True
37. True	38. True	39. True
40. True	41. 'True	42. True
43. True	44. True	45. True
46. True	47. True	48. True
49. True	50. True	51. True
52. True	53. True	54. True

Section-B Fill in the Blanks

1. Expected value	2. x f(x)
3. x f(x)dx	4. E |X|
5. A happens, $\overline{A}$ happens	6. E(x)
7. P(A) P($\overline{A}$)	8. f(x)dx
9. g(x) f(x)	10. X^r f(x)
11. $(X- \overline{X})^2$	12. E(X) + E(Y)
13. $\sum_i E(Xi)$	14. E (Xi)

15. $\psi(x)$ 16. $E(\psi(x))$
17. 0.0 18. $\neq$
19. ai E(Xi) 20. $\leq$
21. $\leq$ 22. $\leq, \leq$
23. h(x), g(x) 24. E(X), E(Y)
25. XY 26. ab
27. (X, Y) 28. 2ai aj
29. a_i^2 V(Xi) 30. E (X^2) E (Y^2)
31. g (x_1), g (x_2) 32. g (x_1), g (x_2)
33. E (g (x)) 34. E (g(x))
35. $\leq$ 36. Same direction
37. X^α, X^β 38. Convex
39. X^α, X^{-1} 40. $\geq$
41. $\leq$ 42. $\geq$
43. E(X) + E(Y) 44. $\leq$, $E(Y^2)$
45. f (x, y) dx dy 46. $\cap$
47. f(x, y) dx dy 48. $(Y - \overline{Y})$
49. $\cap$ 50. $Y = y_j$
51. (X/Y = y) 52. (x y)
53. y 54 Y, v [E]
55. A, B 56. $\overline{A}, \overline{A}$

Section-C : Multiple Choice Questions

1. (d) 2. (a) 3. (b)
4. (d) 5. (d) 6. (a)
7. (d) 8. (d) 9. (b)
10. (d) 11. (d) 12. (d)
13. (a) 14. (a) 15. (a)
16. (c) 17. (c) 18. (d)
19. (b) 20. (a) 21. (d)
22. (d) 23. (d) 24. (d)
25. (d) 26. (d) 27. (a)
28. (d) 29. (a) 30. (d)

31. (d)	32. (d)	33. (c)
34. (d)	35. (d)	36. (d)
37. (d)	38. (d)	39. (c)
40. (c)	41. (c)	42. (a)
43. (d)	44. (d)	45. (d)
46. (d)	47. (a)	

Chapter - 8

Generating Functions Law of Large Numbers and Central Limit Theorems

Section-A : Write True / False

Q.1 The moment generating function (m.g.f) of a random variable X (about origin having the probability function f (x) is defined as

$$M_x(t) = E(e^{tx}) = \int_{-\infty}^{\infty} e^{tx} f(x)dx$$ for continous probability distribution.

Q.2 The m.g.f. for discrete probability function f (x), is given as

$$M_x(t) = E(e^{tx}) = \sum_x e^{tx} f(x)$$

Q.3 In Mx (t), t is the real parameter and it is assumed that the fuction E (etx) is absolutely convergent for some positive number h such that (-h < t < h).

Q.4 We write M_x (t) as

$M_x(t) = \sum_{r=0}^{\infty} \left| \frac{t^r}{\underline{|r}} \right| \mu'_r$ where μ_r = E (X^r) is the rth moment about origin.

Q.5 In the m.g.f i.e. M_x (t), the coefficient of $\frac{t^r}{r!}$ gives the rth moment of X about origin.

Q.6 The M_x (t) generates moment of random variable X, hence, it is known as "Moment generating function".

Q.7 In m.g.f. we can obtain rth moment about origin (say μ_r) from equation

$$\mu_r = \left| \frac{d^r}{dt^r} M_x(t) \right|_{t=0}$$

Q.8 In general the m.g.f of a r.v.X about a point (x = a) is defined as M_x (t) (about x = a) = E ($e^{t(x-a)}$)

Q.9 A random variable X may have no moments although its m.g.f exists.

Q.10 For a discrete r.v with probability function

$$f(x) = \begin{cases} \frac{1}{x(x+1)}, & x = 1, 2, \ldots\ldots \\ 0, & otherwise \end{cases}$$

no moment exists but M_x (t) exist for $t \leq 0$.

Q.11 A random variable X can have m.g.f. but this m.g.f. does not generate moments.

Q.12 A discrete r.v X with probability function P (x = 2^x) = $\frac{e^{-1}}{x!}$, x = 0, 1, 2 .... have m.g.f. but it does not generate moments.

Q.13 A r.v. X can have all or some moments but rn.g.f. does not exist except perhaps at one point.

Q.14 For student's t distribution and snedecor's F distribution, the moments of all orders exist but m.g.f for these two distributions do not exist.

Q.15 The m.g.f. is a poor tool in comparison with characteristic function because the domain of the dummy parameter 't' of m.g.f depends on the distribution of r.v. under consideration while c.f. exist for all real t, $(-\infty < t < \infty)$.

Q.16 Let c be a constant then $M_{ex}(t) = M_x(ct)$.

Q.17 The m.g.f. of the sum of a number of independent random variables, is equal to the product of their respective m.g.f.

Q.18 The m.g.f. of a r.v. depends on its change of origin and scale both.

Q.19 The m.g.f. of a distribution, if it exists, uniquelly determines the distribution.

Q.20 For a given probability function, there exists only one m.g.f. (if it exists) and vice-versa.

Q.21 Cummulant generating function $K_x(t)$ of a r.v. X is defined as $K_x(t) = \log_e(M_x(t))$

Q.22 K_r = coefficient of $\frac{t^r}{r!}$ in $K_x(t)$ is known as rth cumulant.

Q.23 We have the following results from cumulant generating function.

$K_1 = \mu_1$ = Mean, $K_2 = (\mu_2 - \mu_1'^2) = s^2x$, $K_3 = \mu_3$ and $\mu_4 = K_4 + 3K_2^2$

Q.24 We can get the rth cumulant from cumulant generating function as follows

$$K_r = \left|\frac{d^r}{dt^r} K_x(t)\right|_{t=0}$$

Q.25 The rth cumulant of the sum of n independent r.v.'s is equal to thesum of the rth cumulant of the individual r.v's i.e.

$$K_r\left(\sum_{i=1}^{n} Xi\right) = \sum_{i=1}^{n} K_r(Xi)$$

Q.26 Except first cumulant, all the cumulants are invariant (independent) of change of origin.

Q.27 All the cumulants are not invariant of change of scale.

Q.28 The characteristic function (c.f.) is not more serviceable than m.g.f. of a r.v. X.

Q.29 The c.f. of a continous random variable X usually denoted by ϕ_x (t) is defined as

$$\phi_x(t) = E(e^{itx}) = \int_{-\infty}^{\infty} e^{itx} f(x)\, dx$$

Q.30 The c.f. of a discrete r.v. X usually denoted by f_x (t) is defined as

$$\phi_x(t) = E(e^{itx}) = \sum_x e^{itx} f(x)$$

Q.31 If F. (x) is the distribution function of a continous r.v. X then

$$\phi_x(t) = \int_{-\infty}^{\infty} e^{itx}\, dF(x)$$

Q.32 The c.f. of a r.v. X is given in an expanded form as f. (t) = 1 + it

$$\mu_1 + \frac{(it)^2}{2!}\mu_2 + \ldots\ldots\ldots\ldots + \frac{(it)^r}{r!}\mu_r + \ldots\ldots\ldots$$

Q.33 The c.f. of a r.v. X can be rewritten as

$$\phi_x(t) = \sum_{r=0}^{\infty} \frac{(it)^r}{r!}\mu'_r$$

Q.34 The cumulant generating function K. (t) in terms of f_x (t) is given by

$$K(t) = \sum_{r=1}^{\infty} \frac{(it)^r}{r!} K_r \text{ where } K_r = \text{coefficient of } \frac{(it)^r}{r!} \text{ in } K_x(t)$$

Q.35 An advantage of c.f. is that it uniquely determines the distribution function through a theorem known as "Uniqueness theorem of c.f.".

Q.36 For all real t, we have $\phi(0) = 1$.

Q.37 We have the result that $|\phi(t)| \le 1$

Q.38 ϕ (t) is continous function of t in $(-\infty < t < \infty)$.

Q.39 $\phi_x(-t)$ and $\phi_x(t)$ are conjugate functions i.e. $\phi_x(-t) = \overline{\phi x(t)}$ where $\overline{\phi x(t)}$ is the complex conjugate of $\phi_x(t)$.

Q.40 If the distribution function of a r.v. X is symmeterical about zero i.e. if f (-x)=f(x) then ϕ_x (t) is real valued and even function of it.

Q.41 If X is some r.v.with characteristic function ϕ_x (t) and if μ_r = E (x^r) exist then

$$\mu_r = (-i)^r \left| \frac{\partial^r}{\partial t^r} \phi_x \ (t) \right| t = 0$$

Q.42 ϕ_x (t) can be expressed in the form as

$$\phi_x\ (t) = 1 + it\ \mu_1 + \frac{(it)^2}{2!}\ \mu_2 + \ldots\ldots. + \frac{(it)^r}{r!}\ \mu_2 + \ldots\ldots\ldots\ldots\ldots\ldots.$$

Q.43 ϕ_{cx} (t) = ϕ_x (ct) where c is a constant.

Q.44 If (X_1, X_2) are two independent r.v's then ϕ_{x1+x2} (t) = ϕ_{x1}(t) ϕ_{x2}(t)

Q.45 If (Xi, i = 1, 2, ..., n) are n independent r.v's we have

$$\phi_{\sum_i^n Xi}(t) = \sum_{i=1}^{n} \phi_{xi}(t)$$

Q.46 The result ϕ_{x1+x2} (t) = ϕ_{x1}(t) ϕ_{x2}(t) does not imply that X_1 and X_2 are independent.

Q.47 The joint characteristic function ϕ_{x1+x2} (t_1, t_2) is equal to ϕ_{x1}(t_1). ϕ_{x2} (t_2).

Q.48 If U = (x - a)/h, where a & h being constants, we can write ϕ_μ

$$(t) = \phi_u e^{\frac{-iat}{h}} . \phi_x \left(\frac{t}{h}\right)$$

Q.49 The ch. function of a standard normal variate Z = (x - M)/σ is given as

$$\phi_z(t) = e^{\frac{-i\mu t}{\sigma}} . \phi_x\left(\frac{t}{\sigma}\right)$$

Q.50 The following is a set of sufficient but not necessary conditions, due to Polya for a function (I) (t) to be the ch. function.

(1) ϕ (0) = 1 (2) ϕ (t) = ϕ (-t)

(3) ϕ (t) is continous

(4) ϕ (t) is convex for t > 0 i.e. for t_1, t_2 > 0, we have

$$2\phi\left[\frac{1}{2}(t_1+t_2)\right] \leq \phi(t_1)\phi(t_2)$$

(5) $\lim_{t\to\infty}\phi(t)=0$

Q.51 (Due to Cramer) : A given, bounded and continous function ϕ should be the ch. function of a disttribution, the n & s condition is that ϕ (t) = 1 and that the function ϕ (x, A) = $\int_0^A\int_0^A \phi(t-u)e^{(t-u)ix}\, dt\, du$ is real and non negative $\forall$ x and $\forall$ A>0

Q.52 The central limit theorem states that "If Xi (i = 1, 2, ..., n) be independent r.v's with E (Xi) = μ_1 and V (Xi) = σ^2_i , then under certain very general conditions, the r.v. $S = \sum_i^n Xi$ is assympotically normal with mean $\mu = \sum_i^n \mu_i$ and Variance $\sigma^2 = \sum_i^n \sigma_i^2$".

Q.53 The central limit theorem was first time stated by "Laplace in 1812".

Q.54 The regorous proof of C.L.T. was gien by "Liapounoff" in 1901.

Q.55 De-Moivre's Laplace theorem states that if.

$$Xi = \begin{cases} 1\ with\ probability\ p \\ 0\ with\ probability\ q \end{cases}$$

then r.v. $S_n = \sum_i^n Xi$ is assympototically nomral as n —> ∞ where Xi's are independent.

Q.56 Under C.L.T, the binomial distribution tands to normal distribution as n —> ∞.

Q.57 Lindeberg-Levy Theorem states that
If Xi (i = 1, 2, ..., n) are iid random variables with E (Xi) = μ and V (Xi) = σ^2 $\forall$ i = 1, 2, ..., n.

then the sum $S_n = \sum_{i}^{n} Xi$ is assymptotially normal with mean = $n\mu$ and variance = $n.\sigma^2$.

Q. 58 Liapounoff's central limit theorem is applied inzeneralised case when the variables are not iid

Q.59 Both the central limit theorem (C.L.T.) and weak law of large numbers (W.L.L.N) hold for a sequence $\{X_n\}$ of iid. r.v. with finite mean μ and variance σ^2.

Q.60 Let X be a r.v. following a distribution with a single parameter θ which is itself a random variable. Then we say that the r.v. X has a compound or composed distribution".

Q.61 Compound Bionimial Distribution

Let X be a binomial variate with parameter (n,p) and following p.m.f. given as

$P(X = x)\ n_{cx} p_x q^{n-x}$, x = 0, 1, 2, ..., n

Let n instead of being a fixed constant, is also a r.v. following Poisson law with parameter λ. Then

$$P(n = k) = \frac{e^{-\lambda}\lambda^k}{K!}, K = 0, 1, 2, ...$$

In such case X is said to have "Compound binomial distribution".

Q.62 The joint p.m.f. of X and n is given by

$$P(X = r\ n = K) = P(n = K)\ P(X = r/n = k) = \frac{e^{-\lambda}\lambda^k}{K!}\binom{K}{r}p^r q^{K-r}$$

Q.63 $P(X = r) = \sum_{K=r}^{\infty} P(X = r \cap n = k) = \frac{e^{-\lambda p}(\lambda p)^r}{r!}$

which is the probability function of a Poisson variate with parameter " λp".

Q.64 In Q (63), the mean and variance of compound binomial distribution are given as

Mean (Compound binomial distn) = λp

Variance (Compound binomial distn) = λp

Q.65 The probability of an insect laying n eggs is given by the Poisson distribution $e^{-\lambda}. \lambda^n/n!$ and the probability of an egg developing is "p". Assuming natural independence of eggs, the probability of a total of K survivors is given by the Poisson distribution with parameter 'λp'.

Q.66 The probability that a radioactive substance gives off n Beta particles in a unit of time is P (λ), (n = 0, 1, 2, ...). The probability that a given particle will strike a counter and be registered as 'p'. Then probability of registering n Beta particles in a unit of time is P (λp).

Q.67 If the probability of number of hits by lighting during any time interval 't' is P (λt) and the probability of its hitting and damaging an individual is 'p' then (assuming atochastic independence) of total damage during time 't' is P (λtp).

Q.68 **Truncated Distribution :** Let X be a r.v. with p.d.f. (or p.m.f.) f (x). the distribution of X is said to be trumcated at the point X = a if all the values of X ≤ *a* are discarded. Hence the p.d.f. (or p.m.f) of the distribution, truncated at X = a is given by

$$g(x) = \frac{f(x)}{P(X > a)}, x > a \Rightarrow$$

$$g(x) = \begin{cases} \dfrac{f(x)}{\sum f(x)}, & x > a \ (\textit{for discrete r.v.}) \\ x > a & \\ \dfrac{f(x)}{\int_a^{\infty} f(x)dx}, & x > a \ (\textit{for continous r.v.}) \end{cases}$$

Q.69 Let X ~ B(n,p) . Then truncated distribution at x = 0, is given by

$$g(x) = \frac{f(x)}{P(X > 0)} = \frac{f(x)}{1 - f(0)} = \frac{1}{1 - q^n} \cdot {}^nC_x p^x q^{n-x}, \ x = 1,2,3....$$

Section-B : Fill in the Blanks

Q.1 The M.G.F. of a r.v. X (about origin) having the probability function f(x), is defined as

$$M_x(t) = E(e^{tx}) = \int_{-\infty}^{\infty} \ldots\ldots\ldots\ldots dx \text{ for continous p.d.f.}$$

Q.2 The M.G.F. for discrete probability function f (x) is defined as

$$M_x(t) = E(e^{tx}) = \sum_x \ldots\ldots\ldots$$

Q.3 In $M_x(t)$, t is real parameter and it is assumed hat the function $E(e^{tx})$ is absolutely .... for some positive number 'h' such that (-...... < t < ....).

Q.4 We write $M_X(t)$ as

$$M_X(t) = \sum_{r=0}^{\infty} \frac{t^r}{r!} \cdot \mu_r \text{ where } \mu'_r = \ldots\ldots \text{ is the rth moment about origin.}$$

Q.5 In the function $M_x(t)$, the coefficient of $\frac{t^r}{r!}$ give the .... of X about origin.

Q.6 The function $M_x(t)$, generates moments of random variable X, hence it is known as " .... ".

Q.7 From a M.G.F. $M_X(t)$, we can obtain rth moment about origin (say μ_r) from the equation

$$\mu_r = \left| \frac{d^r}{dt^r} \ldots\ldots\ldots\ldots\ldots\ldots \right|_{t=0}$$

Q.8 In general, the M.G.F. of a r.v. X about a point 'a' is defined as $M_x(t) = E(e^{\ldots\ldots\ldots\ldots})$.

Q.9 A random variable X may have no moments although its .... exists.

Q.10 A r.v. X can have M.G.F. but this M.G.F does not generates

Q.11 For student's t and snedecor's F distributions, the moments of all orders exist but .... of these two distributions do not exist.

Q.12 Let "c" be a constant then $M_{ct}(x) = M_X(\ldots\ldots\ldots)$.

Q.13 The M.G.F. of sum of a number of indeperident random variables is equal to the .... of their respective.

Q.14 The M.G.F. of a r.v. depends on its change of .... and .... both.

Q.15 The M.G.F of a distribution, if it exists, uinquely determines the ....

Q.16 For a given probability function, there exists only one .... (if it exists) and vice-versa.

Q.17 Kumulant generating function of a r.v. X, is defined as $K_x(t) = \log_e$ (..........), provided the R.H.S. can be expanded as a .... series in power of t.

Q.18 K_r = coefficient of .... in $K_x(t)$ is known as the rth kumulant.

Q.19 From kumulant generating function $K_x(t)$, we have $K_1 = \mu_1 =$ ....

Q.20 From kumulant generating function $K_x(t)$, we have $K_2 =$ ....

Q.21 From kumulant generating function $K_x(t)$, we have $\mu_4 = K_4 +$ ....

Q.22 We can get Kth cumulant from the equation given as

$K_r = |$ .... $K_x(t) |_{t=0}$

Q.23 Let (Xi, i = 1, 2, ...n) be n independent r.v's then we have

$$K_r\left(\sum_{i=1}^{n} Xi\right) = \text{..................}$$

Q.24 Except first cumulant, all the Kumulants are invariant (independent) of change of ....

Q.25 All cumulants are not invariant of change of ....

Q.26 The characteristic function $\phi_x(t)$ is more serviceable then .... generating function $M_x(t)$.

Q.27 The characteristic function $\phi_x(t)$ of a continous r.v. X is defined

as $f_x(t) = E(e^{itx}) = \int_{-\infty}^{\infty}$ ....dt

Q.28 The c.f. of a discrete r.v. usually denoted by ϕ_x (t) is defined as

$$\phi_x(t) = \sum_x \text{...........}$$

Q.29 If F_x (x) is the distribution function of continuous r.v. X then

$$\phi_x(t) = \int_{-\infty}^{\infty} \text{..........}$$

Q.30 The c.f. of a r.v. X is given in an expanded form as ϕ_x (t) =

$$1 + it\mu_1 + \frac{(it)^2}{2!}\mu_2 + \text{...........}$$

Q.31 The c.f. of a r.v. X can be rewritten as

$$\phi_x(t) = \sum_{r=0}^{\infty} \text{..........}\mu'_r$$

Q.32 The Kumulant generatig function of a r.v. X in the expanded form can be written as

$$K(t) = \sum_{r=1} \text{..........}K_r$$

Q.33 The advantage of c.f. is that it uniquely determines the " ...."through a theorem known as " ...." of characteristic function.

Q.34 For all real t, we have ϕ (0) = ...

Q.35 We have the result that $|\phi_x(t)| \leq$ ....

Q.36 ϕ (t) is a .... function of t in $(-\infty < t < \infty)$

Q.37 ϕ (-t) and ϕ (t) are .... functions i.e. $\phi_x(-t) = \overline{\phi_x(t)}$ where $\overline{\phi_x(t)}$ is the complex conjugate of ϕ_x (t).

Q.38 If the distribution function of a r.v. X is .... about Zero i.e. f (-x) = f (x), then ϕ_x (t) is real valued function of (t).

Q.39 If ϕ_x (t) is a characteristic function of a r.v. X and if $\mu_r = E(X^r)$ then

$$\mu_r = (-i)^r \left|\text{............}\right|_{t=0}$$

Q.40 Let c be a constant then ϕ_{cx} (t) = ...

Q.41 Let (Xi, i = 1, 2, ..., k) be k independent r.v.'s then we have

$$\phi_{\sum_{i}^{\mu} Xi}(t) = \text{....................}$$

Q.42 Let X_1 & X_2 be two independent r.v's then ϕx_1+x_2 (t) = ...

Q.43 The result ϕx_1+x_2 (t) = ϕx_1 (t) ϕx_2 (t) does not imply that X_1 and X_2 are ....

Q.44 The joint characteristic function $\phi x_1 x_2 (t_1 t_2)$ = (.........)(..........)

Q.45 Let U= $\frac{(x-a)}{h}$, where a and h are constants. We can write

$$\phi_u(t) = e^{\text{..................}} \phi_x\left(\frac{t}{h}\right)$$

Q.46 The ch. function of a stanudard normal variate z defined as

$$Z = \frac{x-\mu}{\sigma}, \text{ is written as } \phi_Z(t) = e^{\frac{i\mu t}{\sigma}} .\phi_x(.......)$$

Q.47 For a characteristic function ϕ_x (t), we have
$\lim_{t\to\infty}$ it ϕ (t) = ...

Q.48 Let (Xi, i = 1, 2, 3, ...,n) be independent r.v's with E (Xi) = μ_i and v (Xi) = σ_i^2, then "Central limit theorem" states that

$$U = \sum_{i=1}^{n} Xi \text{ is assumptotically ...."}$$

Q.49 In Q (45), the mean $\bar{x}$ and variance σ^2 are given as $\bar{x}$ = ..., σ^2 = ...

Q.50 The Central limit theorem was first time stated by .... in ....

Q.51 The rigorous proof of central limit theorem (C.L.T.) was given by" ....".

Q.52 Let (Xi, i = 1, 2, ..., n) be independent random variables. Then De-Moivre's Laplace theorem states that if

$$Xi = \begin{cases} 1 \text{ with probability } p \\ 0 \text{ with probability } q \end{cases}$$

then r.v. $S_n = \sum_{i}^{n} Xi$ is "......................" as $n \to \infty$.

Q.53 Under C.L.T, the binomial distribution tends to .... as $n \to \infty$.

Q.54 Let (Xi, i = 1, 2, ..., n) are iid as N (μ, σ^2). Then Lindberg-Levy Theorem states that

$$S_n = \sum_{i}^{n} Xi \sim N(n\mu, n\sigma^2)$$

Q.55 Liapouneff's C.L.T is also applicable in a very general case when variables are ....

Q.56 Both C.L.T and W.L.L.N hold for a sequence $\{X_n\}$ of .... random variables with finite mean .... and variance

Q.57 Let X be a r.v. followig a distribution with a single parameter θ which itself is a random variable. Then r.v. X is said to have a " ...." or composed distribution.

Q.58 Let X be a r.v following b (n, p). Let p be a r.v. following P (λ) given as

$$P(n=k) = \frac{e^{-\lambda}\lambda^k}{K!}, \; K = 0, 1, 2, \ldots\ldots\ldots\ldots$$

Then in such case the r.v. X is said to posses a " ...." distribution.

Q.59 In case of compound binomial distribution, the joint p.m.f. of X and n is given as

P (X = r, n = k) = P (n = k). P (X = r/n = k)

$$= \frac{\ldots\ldots\ldots\ldots}{K!}\binom{K}{r}\ldots\ldots\ldots$$

Q.60 In Q (59), the compound binomial p.d.f. is given as P (X = r)

$$= \frac{e^{-\lambda p}(\ldots\ldots)^r}{r!}$$

Q.61 In compound binomial distribution given in Q (60), the mean and variance are given as $\overline{X}$ = ..., σ^2 = ...

Q.62 The probability of an insect laying n egg. is given by " ...." distribution with its p.m.f. given as $P(X = n) = \frac{(..........)}{n!}$

Q.63 Let X be a r.v. with p.d.f. given as f (x). Then if all values of $X \leq a$ are discarded then the distributed of X is said to be "..................".

Section -C : Multiple Choice Questions

Q.1 The M.G.F. of a r.v. X about origin with its probability density function f (x) is defined as

(a) $M_x(t) = \int_{-\infty}^{\infty} e^{tx} f(x) dx$

(b) $M_x(t) = \int_{-\infty}^{\infty} e^{-tx} f(x) dx$

(c) $M_x(t) = \int_{-\infty}^{\infty} e^{itx} f(x) dx$

(d) $M_x(t) = \int_{-\infty}^{\infty} e^{-itx} f(x) dx$

Q.2 The M.G.F. for discrete probability function f (x) is given as

(a) $M_x(t) = \sum_x e^{-tx} f(x)$

(b) $M_x(t) = \sum_x e^{tx} f(x)$

(c) $M_x(t) = \sum_x e^{itx} f(x)$

(d) $M_x(t) = \sum_x e^{-itx} f(x)$

Q.3 In defining M.G.F. of r.v. X, it is assumed that the function $E(e^{tx})$ is absolutely

(a) Convergent for some positive h such that (-h < t < h)

(b) Divergent for some positive h such that (-h < t < h)

(c) Convergent for some negative h such that (-h < t < h)

(d) Divergent for some negative h such that (-h < t < h)

Q.4 With $\mu_r = E(X^r)$ the $M_X(t)$ can be written in expanded form as

(a) $M_x(t) = \sum_{r=0}^{\infty} \frac{t^{-r}}{r!} \mu_r'$

(b) $M_x(t) = \sum_{r=0}^{\infty} \frac{t^{r}}{r} \mu_r'$

(c) $M_x(t) = \sum_{r=0}^{\infty} \frac{t^{r}}{r!} \mu_r'$

(d) $M_x(t) = \sum_{r=0}^{\infty} \frac{t^{r}}{2r} \mu_r'$

Q.5 In M_x (t), the rth moment of X about origin is the coefficient of

(a) $\frac{t^r}{r!}$ (b) $\frac{t^{-r}}{r!}$

(c) $\frac{t^r}{r}$ (d) $\frac{t^{-r}}{r}$

Q.6 In M.G.F. say M_x (t), the rth moment μ_r about origin can be ontained from the equation

(a) $\mu_r^{'} = \left|\frac{d}{dt} M_x(t)\right|_{t=0}$ (b) $\mu_r^{'} = \left|\frac{dr}{dt^r} M_x(t)\right|_{t=0}$

(c) $\mu_r^{'} = \left|\frac{d}{dt} M_x(t)\right|_{t=1}$ (d) $\mu_r^{'} = \left|\frac{dr}{dt^r} M_x(t)\right|_{t=1}$

Q.7 In general the M.G.F. say M_x (t) of a r.v. X about a point x = a, is written as

(a) $M_x(t) = E(e^{t(x+a)})$ (b) $M_x(t) = E(e^{t(x-a)})$

(c) $M_x(t) = E(e^{-t(x+a)})$ (d) $M_x(t) = E(e^{-(x-a)})$

Q.8 For student's t distribution, the
(a) Moments of all orders exist
(b) Moments of all orders do not exist
(c) M.G.F does not exist
(d) Both (a) & (c)

Q.9 For Snedecor's F distribution, the
(a) Moments of all orders exist
(b) Moments of all orders do not exist
(c) M.G.F does not exist
(d) Both (a) & (c)

Q.10 Let c be a constant then M_{cx} (t) is equal to

(a) $M_{cx}(t) = M_x(-ct)$ (d) $M_{cx}(t) = M_x(ct)$

(c) $M_{cx}(t) = M_x(c^2t)$ (b) $M_{cx}(t) = M_x(-c^2t)$

Q.11 Let (Xi, i = 1, 2, ..., n) be independent random variables then $M_{\Sigma xi}{}^{(t)}$ is equal to

(a) $M_{\sum X_i}(t) = \prod_{i=1}^{n} M_{X_i}(t)$ (b) $M_{\sum_i X_i}(t) = \sum_{i=1} M_{X_i}(t)$

(c) $M_{\sum_i X_i}(t) = \prod_{i=1} \frac{1}{M_{X_i}(t)}$ (d) $M_{\sum X_i}(t) = \sum_{i=1}^{n} \frac{1}{M_{X_i}(t)}$

Q.12 The M.G.F. of a r.v. X is dependent on its change of
(a) Origin (b) Scale
(c) Origin and Scale both (d) None

Q.13 The M.G.F. fo a distribution if exists, uniquely determines the
(a) Distribution
(b) Probability density function
(c) Probability function
(d) All the three

Q.14 Cumulant generating function of a random variable X is defined as
(a) $K_x(t) = \log_e(M_x(t))$ (b) $K_x(t) = \log_{10}(M_x(t))$
(c) $K_x(t) = \log_e(M_x(t!))$ (d) $K_x(t) = \log_e(M_x(t^2))$

Q.15 The rth cumulant Kr in cumulant generating function is the

(a) Coefficient of $\frac{t^r}{r!}$ (b) Coefficient of $\frac{t^r}{r}$

(b) Coefficient of $\frac{t^r}{r^2}$ (d) Coefficient of $\frac{t^r}{2r}$

Q.16 From cumulant generating function, we have
(a) $\mu_4 = K_4 + 3\,K^2_2$ (b) $\mu_4 = K_4 + 3\,K_2$
(c) $\mu_4 = K_4 + K^2_2$ (d) $\mu_4 = K_4 + K^3_2$

Q.17 The rth cumulant Kr from cumulant generatig function can be obtained from the

(a) $K_r = \left|\frac{d^r}{dt^r} K_x(t)\right|_{t=0}$ (b) $K_r = \left|\frac{d^r}{dt^r} K_x(t)\right|_{t=1}$

(c) $K_r = \left|\frac{d}{dt} K_x(t)\right|_{t=0}$ (d) $K_r = \left|\frac{d}{dt} K_x(t)\right|_{t=1}$

Q.18 Let (Xi, i = 1, 2, ..., n) be n independent random variables, then rth cumulant Kr of $\sum_{i=1}^{n}(xi)$ is given as

(a) $K_r = \prod_{i=1}^{n} Kr(Xi)$ (b) $K_r = \sum_{i=1}^{n} K_r(Xi)$

(c) $K_r = \sum_{i=1}^{n} \frac{1}{K_r(Xi)}$ (d) $K_r = \prod_{i=1} \frac{1}{Kr(Xi)}$

Q.19 All cumulants except first one are invariant (independent) of
(a) Change of origin (b) Change of scale
(c) Change of origin & scale (d) None

Q.20 The characteristic function ϕ_x (t) of a r.v. X is
(a) More serviceable than m.g.f.
(b) Less serviceable than m.g.f.
(c) Equally serviceable as m.g.f.
(d) Not comparable

Q.21 The characteristic function ϕ_x (t) of a continous r.v. x is defined as

(a) $\phi_x(t) = \int_{-\infty}^{\infty} e^{-itx} f(x)\,dx$ (b) $\phi_x(t) = \int_{-\infty}^{\infty} e^{itx} f(x)\,dx$

(c) $\phi_x(t) = \int_{-\infty}^{\infty} e^{tx} f(x)\,dx$ (d) $\phi_x(t) = \int_{-\infty}^{\infty} e^{-tx} f(x)\,dx$

Q.22 The ch. f. of a discrete r.v. X usually denoted by ϕ_x (t) is defined as

(a) $\phi_x(t) = \sum_x e^{itx} p(x)$ (b) $\phi_x(t) = \sum_x e^{-itx} p(x)$

(c) $\phi_x(t) = \sum_x e^{tx} p(x)$ (d) $\phi_x(t) = \sum_x e^{-tx} p(x)$

Q.23 If F (x) is the distribution function of a continous r.v. X then its ch. function is defined as

(a) $\phi_x(t) = \int_{-\infty}^{\infty} e^{-itx} dF(x)\,dx$ (b) $\phi_x(t) = \int_{-\infty}^{\infty} e^{itx} dF(x)\,dx$

(c) $\phi_x(t) = \int_{-\infty}^{\infty} e^{tx} dF(x)\,dx$ (d) $\phi_x(t) = \int_{-\infty}^{\infty} e^{-tx} dF(x)\,dx$

Q.24 The ch. f. of a r.v. X in expanded form is written as

(a) $\phi_x(t) = \sum_{r=0}^{\infty} \frac{(it)^r}{r!} \mu_r'$ (b) $\phi_x(t) = \sum_{r=0}^{\infty} \frac{(it)^r}{r} \mu_r'$

(c) $\phi_x(t) = \sum_{r=0}^{\infty} \frac{(-it)^r}{r!} \mu_r'$ (d) $\phi_x(t) = \sum_{r=0}^{\infty} \frac{(-it)^r}{r} \mu_r'$

Q.25 The cumulant generating function K x (t) in expanded form of is written as

(a) $K_x(t) = \sum_{r=1} \frac{(it)^r}{r!} K_r$ (b) $K_x(t) = \sum_{r=1} \frac{(-it)^r}{r!} K_r$

(c) $K_x(t) = \sum_{r=1} \frac{(it)^r}{r} K_r$ (d) $K_x(t) = \sum_{r=1} \frac{(-it)^r}{r} K_r$

Q.26 The advantage of ch. function is that it uniquelly determines the

(a) Distribution function   (b) Probability function
(c) p.d.f. of continuous r.v.   (d) All the three

Q.27 For all real t, we have the result that

(a) ϕ (0) = 1   (b) ϕ (0) = 0
(c) ϕ' (0) = 1   (d) ϕ' (0) = 0

Q.28 We have the result that

(a) $|\phi (x)| \geq 1$ (b) $|\phi (x)| \leq 1$
(c) $|\phi' (x)| \geq 1$ (d) $|\phi' (x)| \leq 1$

Q.29 The ch. function ϕ_x (t) is
(a) Continous function of t in ($-\infty < t < \infty$)
(b) Discrete function of t in ($-\infty < t < \infty$)
(c) Continous function of t in ($0 < t < \infty$)
(d) Continous function of t in ($-\infty < t < 0$)

Q.30 If p.d.f. of a r.v. X is symmetrical about zero i.e. if f (-x) = f (x), then ϕ_x (t) is real valued and
(a) Symmetrical, even function of t
(b) Symmetrical, odd function of t
(c) Asymmetrical, even function of t
(d) Asymmetrical, odd function of t

Q.31 Let ϕ_x (t) be a ch. function of r.v. X and μ_r = E (X^r), then

(a) $\mu_r' = (-i)^r \left| \frac{d^r}{dt^r} \phi_x(t) \right|_{t=0}$ (b) $\mu_r' = (-i)^r \left| \frac{d^r}{dt^r} \phi_x(t) \right|_{t=1}$

(c) $\mu_r' = (-i)^r \left| \frac{d}{dt} \phi_x(t) \right|_{t=0}$ (d) $\mu_r' = (-i)^r \left| \frac{d}{dt} \phi_x(t) \right|_{t=1}$

Q.32 The ch. function ϕ_x (t) can be written in expanded form as

(a) $\phi_x(t) = \sum_{r=0} \frac{(it)^r}{r!} \mu_r'$ (b) $\phi_x(t) = \sum_{r=0} \frac{(it)^r}{r} \mu_r'$

(c) $\phi_x(t) = \sum_{r=0} \frac{(it)^{-r}}{r!} \mu_r'$ (d) $\phi_x(t) = \sum_{r=0} \frac{(it)^{-r}}{r} \mu_r'$.

Q.33 For a constant 'c' we can write

(a) $\phi_{cx}(t) = \phi_x(ct)$ (b) $\phi_{cx}(t) = \phi_x\left(\frac{t}{c}\right)$

(c) $\phi_{cx}(t) = \phi_x\left(\frac{t}{c^2}\right)$ (d) $\phi_{cx}(t) = \phi_x\left(\frac{t}{2c}\right)$

Q.34 If (X_1, X_2) are two independent r.v.'s then, we can write

(a) $\phi_{x_1+x_2}(t) = \phi_{x_1}(t) + \phi_{x_2}(t)$ (b) $\phi_{x_1+x_2}(t) = \phi_{x_1}(t)\phi_{x_2}(t)$

(c) $\phi_{x_1+x_2}(t) = \phi_{x_2}(t).\phi_{x_1}(t)$ (d) Both (b) & (c)

Q.35 If (Xi, i = 1, 2, ..., n) are n independent r.v.'s then we can write

(a) $\phi_{\sum_i^n Xi}(t) = \prod_{i=1}^{n} \phi_{x_i}(t)$ (b) $\phi_{\sum_{i=1} Xi}(t) = \sum_{i=1}^{n} \phi_{x_i}(t)$

(c) $\phi_{\sum Xi}(t) = \sum_{i=1}^{n} \frac{1}{\phi_{x_i}(t)}$ (d) $\prod_{i=1} \frac{1}{\phi_{Xi}(t)}$

Q.36 The result $\phi x_1 + x_2$ (t) = ϕx_1 (t) ϕx_2 (t)

(a) Does not imply that X_1 & X_2 are independent
(b) Imply that X_1 & X_2 are independent
(c) Imply that X_1 and X_2 are not dependent
(d) Both (b) & (c)

Q.37 The joint characteristic function $\phi x_1 x_2\ (t_1, t_2)$ is equal to

(a) $\phi_{x_1x_2}(t_1,t_2)=\phi_{x_1}(t_1)\phi_{x_2}(t_2)$

(b) $\phi_{x_1x_2}(t_1,t_2)=\sum_{i=1}^{2}\phi_{x_1}(t_i)$

(c) $\phi_{x_1x_2}(t_1,t_2)=\phi_{X_1}(t_1)/\phi_{X_2}^{-1}(t_2)$ (d) Both (a) & (c)

Q.38 If U = $\frac{(x-a)}{h}$,where a & h are contants, we can write

(a) $\phi_u(t)=e^{\frac{-iat}{h}}.\phi_x\left(\frac{t}{h}\right)$ (b) $\phi_u(t)=e^{\frac{iat}{h}}.\phi_x\left(\frac{t}{h}\right)$

(c) $\phi_u(t)=e^{iat}\phi_x\left(\frac{t}{h^2}\right)$ (d) $\phi_u(t)=e^{-iat}\phi_x\left(\frac{t}{h}\right)$

Q.39 The ch. function of a standard normal variate Z = $\frac{(x-\mu)}{\sigma}$ is given as

(a) $\phi_z(t)=e^{\frac{i\mu t}{\sigma}}.\phi_x\left(\frac{t}{\sigma}\right)$ (b) $\phi_z(t)=e^{\frac{-it\mu}{\sigma}}.\phi_x\left(\frac{t}{\sigma}\right)$

(c) $\phi_z(t)=e^{i\mu t}.\phi_x\left(\frac{t}{\sigma}\right)$ (d) $\phi_z(t)=e^{-it\mu}.\phi_x\left(\frac{t}{\sigma}\right)$

Q.40 If (Xi, i = 1, 2, ..., n) are independent r.v.'s with E (Xi) = μ_i, V (Xi) = σ^2_i, then "C.L.T" states that under certain general conditions, the r.v. S = $\sum_{i=1}^{n}$ Xi is

(a) Normally distributed (b) Assymptotically normal
(c) Not normally distributed
(d) Not assymptotically normal

Q.41 In Q40, the r.v. S is assymptotically normally distributed as

(a) $\mu=\sum_{i}^{n}\mu_i,\ \sigma^2=\sum_{i=1}^{n}\sigma_i^2$ (b) $\mu\neq\sum_{i=1}^{n}\mu_i,\ \sigma^2=\sum_{i=1}^{n}\sigma_i^2$

(c) $\mu=\sum_{i=1}^{n}\mu_i,\ \sigma^2\neq\sum_{i=1}^{n}\sigma_i^2$ (d) $\mu\neq\sum_{i=1}^{n}\mu_i,\ \sigma^2\neq\sum_{i=1}^{n}\sigma_i^2$

Q.42 The central limit theorem was first time reported by
(a) Lindeberge levy (b) Liapouneff
(c) D-Moivefe laplace (d) Karl Pearson

Q.43 The rigorous proof of "C.L.T." was given by
(a) Liapouneff (b) Lindeberge levey
(c) Karl Pearson (d) De-Moivor laplace

Q.44 Let {Xi, i = 1, 2, ..., n} be sequenec of independent r.v.'s. Then De-moivre laplace theorem states that if

$Xi = \begin{cases} 1\ with\ probability\ p \\ 0\ with\ probability\ q \end{cases}$ the r.v. $S_n = \sum_{i=1}^{n} Xi$ ..... is

(a) Assymptotically normal as $n \to \infty$
(b) Not assymptotically normal as $n \to \infty$
(c) Assymptotically Poisson as $n \to \infty$
(d) Not assymptotically Poisson as $n \to \infty$

Q.45 Under C.L.T., the binomial distribution tends to
(a) Normal distribibition as $n \to \infty$
(b) Poisson distribution as $n \to \infty$
(c) Negative binomial distribution as $n \to \infty$
(d) None of the above

Q.46 Let (Xi, i = 1, 2, ..., n) be i.i.d random variables with E (Xi) = μ and V (Xi) = σ^2 $\forall$ i = 1, 2, ..., n. The sum $S_n = \sum_{i}^{n} Xi$ is distributed as

(a) Assymptotically normal with mean = $n\mu$, variance = σ^2
(b) Assymptotically normal with mean = $n\mu$, variance = $n\,\sigma^2$
(c) Assymptotically normal with mean = μ, variance = $n\,\sigma^2$
(d) Assymptotically normal with $n\,\mu$ and variance = σ^2

Q.47 In Q (46) the distribution of S_n has been proved by the theorem reported by :
(a) Lindeberg levy (b) De-moivre laplace
(c) Karl Pearson (d) Liapnouneff

Q.48 Both "Central limit theore" (C.L.T) and "Weak law of large numbers" (W.L.L.N.) hold for a sequence $\{X_n\}$ of

(a) i.i.d r.v.'s with mean μ and variance σ^2

(b) i.i.d. r.v.'s with mean μ_i and σ^2

(c) i.i.d. r.v.'s with mean $n\mu$, variance $n\,\sigma^2$

(d) i.i.d. r.v.'s with mean $\sum_{i=1}^{n} \mu_i$ and variance $\sum_{i}^{n} \sigma_i^2$

Q.49 Let X be a r.v. following a distribution with parameter θ which is itself a r.v. Then distirbution of X is known as

(a) Compound distribution (b) Composed distribution

(c) Both (a) & (b) (d) Simple distribution

Q.50 Let X be a r.v. following b (n, p). Suppose n is itself a r.v. following Poisson distribution with parametr k. Then X is said to have

(a) Compound binomial distribution

(b) Compound poisson distribution

(c) Compound normal distribution as $n \to \infty$

(d) Compound poisson distribution as $n \to \infty$

Q.51 In Q (50), the mean and variance of compound binomial distribution are given as

(a) $\bar{x} = \lambda p$ and $\sigma^2 = \lambda p$ (b) $\bar{x} = np$ and $\sigma^2 = npq$

(c) $\bar{x} = p$ and $\sigma^2 = pq$ (d) None

Q.52 The probability of an insect laying n eggs is given by Poisson distribution as P (X) and probability of developing an egg is 'p'. Assuming natural independence of egg, the probability of a total of "k" survivors is given by

(a) Poisson distribution as P (λp)

(b) Poisson distribution as P (λ)

(c) Poisson distribution as P (K.p)

(d) Poisson distribution as P $(k\lambda)$

Q.53 Let X be a r.v. with p.d.f (or p.m.f) as f (x). If all values of $x \leq a$, are discarded from distribution, then the distribution of X is said to be

(a) Truncated at the point $x = a$

(b) Not truncated at the point $x = a$

(c) Compoound at the point $x = a$

t(d) Identified at the point $x = a$

Key Answers

Section-A : True False

1. True	2. True	3. True
4. True	5. True	6. True
7. True	8. True	9. True
10. True	11. True	12. True
13. True	14. True	15. True
16. True	17. True	18. True
19. True	20. True	21. True
22. True	23. True	24. True
25. True	26. True	27. True
28. False	29. True	30. True
31. True	32. True	33. True
34. True	35. True	36. True
37. True	38. True	39. True
40. True	41. True	42. True
43. True	44. True	45. True
46. True	47. True	48. True
49. True	50. True	51. True
52. True	53. True	54. True
55. True	56. True	57. True
58. True	59. True	60. True
61. True	62. True	63. True
64. True	65. True	66. True
67. True	68. True	69. True

Section-B : Fill in the Blanks

1. $e^{tx} f(x)$	2. $e^{tx} f(x)$
3. Convergent, $(-h < t < h)$	4. $E(X^r)$
5. rth moment	6. m.g.f
7. $M_x(t)$	8. $t(x - a)$
9. m.g.f.	10. Moments
11. m.g.f.	12. ct

13. Product, m.g.f.
14. Origin, scale
15. Distribution
16. m.g.f.
17. M_x (t), convergent
18. t^r |r|
19. Mean
20. Variance
21. $3K_2^2$
22. $\frac{d^r}{dt^r}$
23. $\sum_{i=1}(Xi)$
24. Origin
25. Scale
26. Moment
27. e^{itx} f (x)
28. e^{itx} f (x)
29. e^{itx} d f (x)
30. $\sum_{i=1} \frac{(it)^r}{r!} \mu'_r$
31. $\frac{(it)^r}{r!}$
32. $\frac{(it)^r}{r!}$
33. Distribution function, Uniqueness theorem
34. =1
35. ≤ 1
36. Continuous
37. Conjugated
38. Symmetrical, even
39. $\left| \frac{d^r}{dt^r} \phi_x (t) \right|$
40. ϕ_x (t)
41. $\prod_{i=1} \phi_{xi}$ (t)
42. ϕ_{x1} (t) ϕ_{x2} (t)
43. Independent
44. ϕ_{x1} (t_1) ϕ_{x2} (t_2)
45. iat / h
46. t / σ
47. 0
48. Normal
49. $\sum_i \mu_i, \sum_i \sigma_i^2$
50. Laplace in 1812
51. Liapouneff
52. Assymptotically normal
53. Normal distribution
54. nμ, n σ^2
55. Not i.i.d.
56. i.i.d., μ, σ^2
57. Compound
58. Compound binomial

59 . $e^{-\lambda} \lambda^k, p^r \ q^{k-r}$

60. λp

61. λ p

62. $e^{-\lambda} . \lambda^n$

63. Truncated at the pont x = a

Section C : Multiple Choice Questions

1. (a)	2. (b)	3. (a)
4. (c)	5. (a)	6. (b)
7. (b)	8. (d)	9. (d)
10. (b)	11. (a)	12. (c)
13. (d)	14. (a)	15. (a)
16. (a)	17. (a)	18. (b)
19. (a)	20. (a)	21. (b)
22. (a)	23. (b)	24. (a)
25. (a)	26. (d)	27. (a)
28. (b)	29. (a)	30. (a)
31. (a)	32. (a)	33. (a)
34. (d)	35. (a)	36. (d)
37. (d)	38. (a)	39. (b)
40. (b)	41. (a)	42. (c)
43. (a)	44. (a)	45. (a)
46. (b)	47. (a)	48. (a)
49. (c)	50. (a)	51. (a)
52. (a)	53. (a)	

❐❐❐

Chapter - 9

Discrete Distributions

9.1 Section-A : Write True / False

Q.1 The probabilistic laws for different values of a random variable are made on the basis of given conditions. These conditions are known as "prior considerations".

Q.2 If the probabilistic laws are made for different values of a random variable on the basis of results obtained from an experiment, then such results are known as "a posterior inferences".

9.1.1 Discrete Uniform Distribution

Q.3 A random variable X is said to have a discrete uniform distribution over the range [1, n] if its p.m.f is written as

$$P(X = x) = \begin{cases} \frac{1}{n} & \text{for } x = 1, 2, \ldots, n \\ 0 & \text{otherwise} \end{cases}$$

Q.4 In discrete uniform distribution, the n is known as the "parameter of the distribution, which takes values of the positive integers.

Q.5 Discrete uniform distribution is also known as "discrete rectangular distribution".

Q.6 For a die experiment and for an experiment with a deck of cards, the discrete uniform distribution is appropriate.

Q.7 The expected value of discrete uniform r.v. (Mean of discrete uniform distribution) is given as $E(X) = \frac{n+1}{2}$

Q.8 The expected value of square of r.v. X for discrete uniform distribution, is given as $E(X^2) = \frac{(n+1)(2n+1)}{6}$

Q.9 The variance of discrete uniform distribution is given as

$$V(X) = \frac{(n-1)(n+1)}{12}$$

Q.10 The m.g.f. of X is given as

$$M_x(t) = \frac{e^t(1-e^{nt})}{e(1-e^t)}$$

9.1.2 Bernouille Distribution

Q.11 A r.v. Xis said to have a Bernouille distribution with parameter p if its p.m.f. is given by

$$P(X = x) = \begin{cases} p^x (1-p)^{1-x} & \text{for } x=0,1 \\ 0 & \text{otherwise} \end{cases}$$

Q.12 A random experiment whose outcomes are success "S" and failure "F" with probabilities p and q respectively is called a Bernouille trial.

Q.13 Mean and variance of Bernouille distribution are given as Mean $(\bar{x}) = p$, $v(x) = pq$

Q.14 For Bernoulli r.v. X, we have $E(X^2) = p$.

Q.15 The m.g.f. of Bernouille variate X is $M_x(t) = (q + pe^t)$

Q.16 If a degenerate r.v. X degenerates at the point $x = c$, then $v(x) = 0$

Q.17 The m.g.f. of a degeneratve r.v. X at c is given as $M_x(t) = e^{ct}$.

9.1.3 Binomial Distribution

Q.18 Binomial distribution was discovered by James Bernouille.

Q.19 The probability of x success and consequently (n - x) failure in n independent trials is given as
$P\ (x\ \text{success}) = {}^nC_x\, p^x\, q^{n-x}$

Q.20 A random variable X is said to follow binomial distribution if assumes only non-negative values and p.m.f. is given by

$$p\,(x) = \begin{cases} {}^nC_x\, p^x\, q^{n-x}, & x=0,1,2,\ldots\ldots,n \\ 0, & \text{otherwise} \end{cases}$$

Q.21 Binomial distribution is a discrete distribution. It is denoted as $X \sim B\,(n, p)$

Q.22 Let us suppose that n trials constitute an experiment. If this experiment is repeated N times, the frequency function of binomial distribution is given as $f\,(x) = N\,p\,(x)$

Q.23 In binomial experiment, trials are independent of each other

Q.24 In binomial distribition, the probability of success 'p' remains constant from trial to trial.

Q.25 The problems of tossing a die or drawing cards from a pack of cards with replacement leads to binomial distribution.

Q.26 Mean of $B\,(n, p) = np$

Q.27 Variance of $B\,(n, p) = npq$

Q.28 For B (n, p), skewness $\beta_1 = \dfrac{(1-2p)^2}{npq}$

Q.29 For B (n, p), the kurtosis $\beta_2 = 3 + \dfrac{1-6pq}{npq}$

Q.30 For binomial distribution B (n, p), we have Mean < Variance

Q.31 For binomial distribution, we have $\mu_3 = npq\,(q - p)$

Q.32 Let $(n + 1)\,p$ be not an integer. Let $(n + 1)\,p = m + f$ where m is an integer part. Then m is an unique model value of B (n, p).

Q.33 Let in B (n, p), the $(n + 1)\,p = m$ (an integer). Then two model values are m and (m - 1).

Q.34 The moment generating function of B (n, p) is given $M_x(t) = (q + p\,e^t)^n$

Q.35 Binomial distribution does not possers the additive or reproductive property

Q.36 If Xi (i = 1, 2, ..., k) are independent binomial variates with parameters (p, ni, i = 1, 2, k). Then $\sum_{i=1}^{n} Xi \sim B\left(\sum_{i=1}^{n} ni, p\right)$

Q.37 The characteristic function $\phi_x(t)$ of B (n, p) is given as $\phi_x(t) = (q + pe^{it})^n$

Q.38 The cumulant generating function of B (n, p) is given as $K_x(t) = n \log_e (q + pe^t)$

Q.39 Recurrence relation for the probabilities of binomial distribution is given as

$$p(x+1) = \left(\frac{n-x}{x+1}\right)\frac{p}{q}.p(x)$$

9.1.4 Poisson Distribution

Q.40 Poisson distribution was discovered by the French mathematician and physicist Simen Denis Poisson.

Q.41 Poisson distribution is the limiting case of the bimonial distribution under the conditions $n \to \infty$, $p \to 0$, $np = \lambda$ (finite).

Q.42 A r.v. X is said to follow Poisson distribution if it assumes only non-negative values and its p.m.f is given by

$$p(x, \lambda) = \begin{cases} \dfrac{e^{-\lambda}.\lambda^x}{x!}, & x = 0,1,2,....,\lambda > 0 \\ 0, & \text{otherwise} \end{cases}$$

Q.43 Poisson distribution is denoted as $X \sim P(\lambda)$ where λ is the parameter of the distribution.

Q.44 The distribution function F (x) of P (λ) is given as

$$F(x) = e^{-\lambda}\sum_{r=0}^{x}\frac{\lambda^r}{r!}, \; x = 0, 1, 2, ...$$

Q.45 Poisson distribution occurs when there are events which do not occur as the outcomes of a definite number of trials.

Q.46 The number of deaths from a disease such as heart attack, or Cancer, or due to snake bite follow the Poisson distribution.

Q.47 The number of suicides reported in a particular city follow Poisson distribution.

Q.48 The number of defective materials in a packing manufactured by a good concern follow Poission distribution.

Q.49 The number of fragements received by a surface area 'A' from a fragement of atom bomb follow Poisson distribution.

Q.50 The emission of radio active particle follow Poisson distribution.

Q.51 The mean of Poisson distribution is equal to λ.

Q.52 The variance of Poisson distribution = λ.

Q.53 For P (λ), the $\mu_3 = \lambda$

Q.54 For P (λ), the $\beta_1 = \frac{1}{\lambda}$

Q.55 For P (λ), the $\beta_2 = 3 + \frac{1}{\lambda}$

Q.56 Poisson distribution is always skewed distribution.

Q.57 Let λ be not an integer, Let $\lambda = m + f$ where m is the integer part of λ. Then m is the unimodel of Poisson distribution.

Q.58 Let $\lambda = k$ in P (λ), then distribution is biomodel and modes are (K - 1, K)

Q.59 The m.g.f. of P (λ) is given as $M_x(t) = e^{\lambda}(e^t-1)$

Q.60 The characteristic function of P (λ) distribution is given as $\phi_x(t) = e^{\lambda}(e^{it}-1)$

Q.61 Kumulant generating function of Poissoin distribution $K_x(t) = \lambda(e^t - 1)$

Q.62 All kumulants of P (λ) are same and are equal to λ.

Q.63 For Poisson distribution, we have

$$\text{Skewness} = \beta_1 = \frac{1}{\lambda}, \text{ Kurtosis} = \beta_2 = 3 + \frac{1}{\lambda}$$

Q.64 If Xi (i = 1, 2, ..., k) are independent Poisson variates with parameters (λi, i = 1, 2, ..., K). Then $U = \sum_{i=1}^{k} Xi$ is a Poisson variate with parameter $\sum_{i=1}^{k} \lambda i$.

Q.65 For a Poisson distribution with parameter λ, the recurrence formula for probabilities is given by $p(x + 1) = \left(\frac{\lambda}{x+1}\right) \cdot p(x)$

9.1.5 Negative Binomial Distribution

Q.66 Let f (x, r, p) denote the probability that there are x failures preceeding to the rth success in (r + x) trials. then x is said to have a negative bimonial distribution with parameters (r, p) if its p.m.f. is given by

$$p(x) = \begin{cases} \binom{x+r-1}{r-1} p^r q^x, & x = 0,1,2,\ldots\ldots\ldots \\ 0, & \text{otherwise} \end{cases}$$

Q.67 The negative binomial distribution may be regarded as the generalization of geometric distribution.

Q.68 If in negative binomial distribution, we take r = 1, then p (x) reduces to geometric distribution.

Q.69 The negative binomial distribution when regarded as one having two parameters (p, r) is known as Pascal's distribution.

Q.70 If $p = \frac{1}{Q}$ and $q = \frac{P}{Q}$, then m.g.f. of negative binomial distribution is given as $M_x(t) = (Q - P e^t)^{-r}$

Q.71 The mean of negative binomial distribution is given as Mean = r.P.

Q.72 Variance of negative binomial distribution is given as $\mu_2 = rPQ$

Q.73 As Q > 1, rp < rPQ i.e. Mean < Variance

Q.74 Kumulant generating function of negative binomial distribution is given by $K_x(t) = -r \log e (Q - P e^t)$

Q.75 In terms of (p, q), we have Mean $= \frac{rq}{p}$, $\mu_2 = \frac{rq}{p^2}$

Q.76 Negative binomial distribution reduces to Poisson distribution as $P \to 0, r \to \infty$ such that $rp = \lambda$ (finite).

9.1.6 Geometric Distribution

Q.77 A r.v. X is said to have a geometric distribution if it assumes only non-negative values and its p.m.f is given by

$$p(x) = \begin{cases} q^x p, & x = 0,1,2,\ldots\ldots\ldots 0<p<1,\ q=1-p \\ 0, & \text{otherwise} \end{cases}$$

Q.78 Let there be n independent trials, then the probability that there are x failures preceeding to the first success is given by geometric distribution.

Q.79 The mean of geometric distribution is given by $\frac{q}{p}$.

Q.80 The variance of geometric distribution is given by

$$\mu_2 = \frac{q}{p^2}$$

Q.81 The m.g.f of geometric distribution is given as

$$M_x(t) = \frac{q}{(1-qe^t)}$$

Q.82 For geometric distribution we have Mean < Variance

Q.83 Multinomial distribution can be regarded as the generalisation of binomial distribution.

Q.84 When there are more than two mutually exclusive outcomes of a trial, then such observatins lead to the multinomial distribution.

Q.85 The probability that the event (Ei occurs xi times, i = 1, 2, ..., K) in n independent observations, is given by

$$p(x_1, x_2, \ldots\ldots, x_k) = cp_1^{x_1}\ p_2^{x_2}\ \ldots\ldots\ldots\ldots p_k^{x_k}$$

9.2 Section-B : Fill in the Blanks

Q.1 The probabilistic laws for different values of a r.v. are made on the basis of some given conditions. These conditions are known as " .... " considerations.

Q.2 If the probabilistic laws are made for the different values of a random variable on the basis of results obtained from an experiment, then such results are known as "a ...., ... inference.

9.2.1 Discrete Uniform Distribution

Q,3 A r.v. X is said to have a discrete uniform distribution over the range [1, n] if its p.m.f. is given as

$$P\,(X = x) = \begin{cases} \text{....................}, & x = 0,1,2,....,n \\ o, & \text{otherwise} \end{cases}$$

Q.4 In discrete uniform distribution, n is known as .... of the distribution which takes values of all the positive integers.

Q.5 Discrete uniform distribution is also known as " .... " .

Q.6 For a die experiment and for an experiment with a deck of cards, the appropriate distribution to be applicable is the " .... " .

Q.7 The expected value of a discrete uniform r.v. variable is given as E (x) = ...

Q.8 The expected value of square of r.v. X for discrete uniform distribution is given as E (x^2) = ...

Q.9 The variance of discrete uniform distribution is given as v (x) = ...

Q.10 The m.g.f. of discrete uniform distribution is given as

$$M_x\,(t) = \frac{e^t(1-\text{...........})}{n(1-\text{...........})}$$

9.2.2 Bernoulli Distribution

Q.11 A r.v. X is said to have Bernouille distribution with parameter

'p' if its p.m.f. is given as $\begin{cases} \text{....................} & \text{for } x = 0,1 \\ 0, & \text{otherwise} \end{cases}$

Q.12 A random experiment with outcomes as "success S" and "failure F" with parameters "p, q" is known as " ...." trial.

Q.13 For a Bernouille distribution with parameters (p, q), we have Mean $(\overline{x})$ = ..., v (x) = ...

Q.14 For Bernoulli r.v. X, we have E (x^2) = ...

Q.15 The m.g.f. of a Bernouille r.v. is given as M_x (t) = ...

Q.16 If a degenerate r.v. X degenerates at the point c, then v (c) = ...

Q.17 The m.g.f. of a degenerate r.v. X is given as M_x (t) = ...

9.2.3 Binomial Distribution

Q.18 Binomial distribution was discovered by James ....

Q.19 The probability of x success and consequently (n - x) failures in n independent trials is given by

P (x) = ..., $0 < p < 1, 0 < q < 1$

$p + q = 1$

Q.20 A r.v. X is said to follow binomial distribution if it assumes only non-negative values and its p.m.f. is given by

$$p(x) = \begin{cases} \text{..............}, & x = 0, 1, 2, \ldots\ldots, n \\ 0, & \text{otherwise} \end{cases}$$

Q.21 Binomial distribution is a " .... " distribution and it is denoted as X ~ B(n, ...).

Q.22 Let us consider an experiment following B (x, n, p). If this experiment is repeated N times, the frequency function of binomial distribution is given as f (x) = ...

Q.23 In binomial experiment, the trials are .... of each other.

Q.24 In B (x, n, p) distribution, the probability of success "p" remains .... from trial to trial.

Q.25 The problem of tossing a coin or a die or drawing a card from a pack of cards with replacement leads to a " ...." distribution.

Q.26 Mean of B (n, p) = ...

Q.27 The variance of B (n, p) = ...

Q.28 For B (n, p) the skewness $\beta_1 = \frac{(1-2p)^2}{..........}$

Q.29 For B (n, p), the kurtosis $\beta_2 = 3 + \frac{(1-6pq)}{..........}$

Q.30 For binomial distribution we have relation between its mean and variance as Mean .... Variance

Q.31 For B (n, p), we have μ_3 = ... (q - p)

Q.32 Let (n + 1)p = (m + f) where m is an integer part and f is a fraction. Then an unique model value of B (n, p) = ...

Q.33 Let (n + 1p) = m (an integre). Then the two model values are equal to .... and ....

Q.34 The m.g.f. of B (n, p), are given by M_x (t) = (q + ...)^n

Q.35 Binomial distribution does not possess the additive or .... property.

Q.36 Let (Xi, i = 1, 2, ..., n) be n independent binomial variates with parameters (p, ni, i = 1, 2, ..., n). Then

$$U = \sum_{i=1}^{n} Xi \sim B(............,...........)$$

Q.37 The chracaters function fx (t) of B (n, p) is given as

ϕ_x (t) = (q + ...)^n

Q.38 The cumulant generating function of B (n, p) is given as

K_x (t) = n $\log_e$ (q + ...)

Q.39 The recurrence relation for the probabilities of binomial

distribution B (p, n) is given as p (x + 1) = (.......). $\frac{q}{p}$ P(x)

9.2.4 Poisson Distribution

Q.40 Poisson distribution was discorvered by French mathematician and physicist - Simean ....

Q.41 Poisson distribution is the limiting case of B (n, p) in the conditions $n \to \infty$, $p \to$ .... and $np = \lambda$(finite)

Q.42 A r.v. X is said to follow Poisson distribution if it assumes only .... values and its p.m.f. is given by

$$p(x, \lambda) = \begin{cases} \dfrac{e^{-\lambda}\ldots\ldots}{\ldots\ldots\ldots}, & x = 0,1,2,\ldots\ldots \\ 0, & \text{otherwise} \end{cases}$$

Q.43 Poisson distribution is denoted by X ~ ....where λ is the .... of the distribution.

Q.44 The distribution function F (x) of P (λ) is defined as

$F(x) = e^{-\lambda}$ (.....), x - 0, 1, 2,.....

Q.45 The number of deaths from a disease such as heart attack, or cancer or due to snake bite follow the " .... " distribution.

Q.46 The number of suicides reported in a particular city follows " .... " distribution.

Q.47 The number of defective materials in a packing manufactured by a good concern follow the " .... " distribution.

Q.48 The number of fragements received by a surface area "A" from a fragement of atom bomb follow .... distribution.

Q.49 The emission of radio-active particles follow " .... " distribution.

Q.50 The mean of Poisson distribution is equal to ....

Q.51 The varianec of Poisson distribution is equal to ....

Q.52 For the distribution P (λ), we have μ_3 = ...

Q.53 For the distribution P (λ), we have β_1 = ...

Q.54 For the distribution P (λ), we have $\beta_2 = 3 +$ ....

Q.55 Poisson distribution is always a .... distribution.

Q.56 Let λ be not an integer. Let $\lambda = m + f$ where m is the integer part of λ. Then unimodel of Poisson distribution is equal to .....

Q.57 Let $\lambda = k$, in P (λ) then Poisson distribution is bimodel and modes are equal to (..........,..........).

Q.58 The m.g.f. of Poisson distribution is given as $M_x(t) = e^{\lambda}$ (.......).

Q.59 The characteristic function of P (λ) distribution is given by $\phi_x(t) = e^{\lambda}$ (..........)

Q.60 the kumulant generating function of P (λ) distribution functin is given by $K_x(t) = \lambda$ (..............)

Q.61 All kumulants of P (λ) are .... and are equal to ....

Q.62 For Poisson distribution, we have β_1 = ..., β_2 = ...

Q.63 Let (Xi, i = 1, 2, ..., k) be k independent Poisson variates with parameters (λi, i =1, 2, ..., k). Then $U = \sum_{i=1}^{k} X_i$ is also a " .... " variate with parameter λ = ...

Q.64 For a Poisson varaite with parameter λ, the recurrence formula for probability is given as p (x + 1) = (.............) p (x).

9.2.5 Negative Binomial Distribution

Q.65 A random variable X is said to have a negative binomial distribution with parameters (r, p) is given as

$$p(x) = \begin{cases} (\text{.............})\, p^r q^x, & x = 0, 1, 2, \text{........} \\ 0, & \text{otherwise} \end{cases}$$

Q.66 The negative binomial distribution may be regarded as the generalisation of ....

Q.67 If in negative binomial distribution, we take r = 1, then p (x) reduces to ....

Q.68 If $p = \frac{1}{Q}$ and $q = \frac{P}{Q}$, then m.g.f. of negative binomial distribution is given by $M_x(t) = (\text{............})^{-r}$

Q.69 The mean of negative binomial distribution is given as Mean = ...

Q.70 The variance of negative binoMial distributeon is equal to μ_2 = ...

Q.71 In case of negative binomial distribution, its mean and variance has relation given as Mean .... Variance

Q.72 Kumulant generating function of negative binomial distribution is given by $K_x(t) = -r \log_e (..........)$

Q.73 In terms of (p, q), for negative binomial distribution, we have

$$\text{mean} = \frac{rq}{\ldots\ldots}, \mu_2 = \frac{rq}{\ldots\ldots}$$

Q.74 Negative binomial distribution reduces to Poisson distribution as $p \to$ ...., $r \to$ .... such that $rp = \lambda$ (finite).

9.2.6 Geometric Distribution

Q.75 Let there ben independent trials, then the probability that there are x failures preceeding to first success is given by " .... distribution".

Q.76 A r.v. Xis said to follow the geometric distribution if it assumes only non-negative values and its p.m.f. is given by

$$p(x) = \begin{cases} \ldots\ldots\ldots\ldots, x = 0,1,2,\ldots\ldots, & 0 < p < 1, q = 1-p \\ 0, & \text{otherwise} \end{cases}$$

Q.77 The mean of geometric distribution is given by Mean = ...

Q.78 The variance of geometric distribution is given by

$$\mu_2 = \frac{q}{\ldots}$$

Q.79 The m.g.f. of geometric distribution is given by $M_x(t) = \frac{p}{(\ldots)}$

Q.80 In case of geometric distribution, its mean and variance has the relation given by Mean .... Variance

Q.81 Multinomial distribution can be regarded as the generalisation of " .... " distribution.

Q.82 When there are more than two mutually exclusive outcomes of a trial them such observations lead to the " .... "

Q.83 The probability that the event (Ei, occurs xi times, i = 1, 2, ..., k) in n independent trials, is given by $p(x_1, x_2, \ldots\ldots\ldots, x_k) = c$ ....

9.3 Section-C : Multiple Choice Questions

Q.1 If the probabilistic laws are made for different values of a random variable on the basis of results obtained from an experiment then such results are known as

(a) Posterior inference (b) Prior inference
(c) Latter inference (d) Earlier inference

Q.2 A r.v. X is said to have a discrete uniform distribution over the range [0, n] if its p.m.f is written as

(a) $p(x) = \frac{1}{x}$, x = 0, 1, n (b) $p(x) = \frac{1}{2n}$, x = 0. 1, ..., n

(c) $p(x) = \frac{1}{n^2}$, x = 0, 1, n, ..., n (d) $p(x) = \frac{1}{(n+1)}$ x= 0, 1, ..., n

9.3.1 Discrete Uniform Distribution

Q.3 Discrete uniform distribution is also known as

(a) Discrete rectungular distribution
(b) Discrete geometric distribution
(c) Negative binomial distribution
(d) Hyper geometric distribution

Q.4 For a die experiment as well as for an experiment with a dack of cards the appropriate distribution to be followed is the

(a) Uniform discrete distribution
(b) Discrete geometric distribution
(c) Discrete rectangular distribution
(d) Both (a) & (c)

Q.5 For a discrete uniform distribution over a range [0, 1], the mean is equal to

(a) $\frac{(2n+1)}{1}$ (b) $\frac{n+1}{2}$

(c) $\frac{n(n+1)}{2}$ (d) $\frac{2n+1}{2}$

Q.6 For a discrete uniform distribution over [0, 1], the expected value of square of r.v. X is equal to

(a) $\frac{n(n+1)(2n+1)}{6}$ (b) $\frac{n(n+1)(2n+1)}{12}$

(c) $\frac{(n+1)(2n+1)}{6}$ (d) $\frac{n(n+1)(2n+1)}{8}$

Q.7 The variance of a discrete uniform distribution over [0, 1], is equal to

(a) $\frac{(n-1)(n+1)}{12}$ (b) $\frac{n(n^2-1)}{6}$

(a) $\frac{(n^2-1)}{12}$ (d) Both (a) & (c)

Q.8 The m.g.f. of a discrete uniform distribution over a range [0,1] is given as

(a) $M_x(t) = \frac{e^t(1-e^{nt})}{n(1-e^t)}$ (b) $M_x(t) = \frac{n(1-e^t)}{e^t(1-e^{nt})}$

(c) $M_x(t) = \frac{e^t(1-e^{-nt})}{n(1-e^t)}$ (d) $M_x(t) = \frac{e^t(1-e^{nt})}{n(1-e^{-t})}$

9.3.2 Bernoulli Distribution

Q.9 The p.m.f. of a Bernouille distribution with parameter "p" is given as

(a) $p(x) = \begin{cases} p^x q^{1-x}, & \text{for } x = 0,1,\ p+q=1 \\ 0, & \text{otherwise} \end{cases}$

(b) $p(x) = \begin{cases} q^x p^{1-x}, & \text{for } x = 0,1 \\ 0, & \text{otherwise} \end{cases}$

(c) $p(x) = \begin{cases} {}^nC_x\ p^x q^{1-x}, & \text{for } x = 0,1,\ldots,n \\ 0, & \text{otherwise} \end{cases}$

(d) $p(x) = \begin{cases} {}^nC_x\ q^x p^{n-x}, & x = 0,1,\ldots,n \\ 0, & \text{otherwise} \end{cases}$

Q.10 A random experiment whose outcomes are (S, F) as success and failure respectively, is known as

(a) Binomial trial (b) Bernoulli trial
(c) Negative binomial trial (d) None of these

Q.11 The mean of Bernouille distribution with parameter "p" is equal to

(a) q (b) p
(c) np (d) nq

Q.12 The variance of Bernouille distribution is equal to

(a) npq (b) pq
(c) np (d) $\sqrt{npq}$

Q.13 In case of Bernouille distribution the expected value of square of r.v. X is equal to

(a) $E(x^2) = p$ (b) $E(x^2) = pq$
(c) $E(x^2) = npq$ (d) $E(x^2) = p^2 q^2$

Q.14 The m.g.f. of a Bernouille variate X is written as

(a) $M_x(t) = (q + p e^t)$ (b) $M_x(t) = (p + q e^t)$
(c) $M_x(t) = (q + p e^{-t})$ (d) $M_x(t) = (p + q e^{-t})$

Q.15 The variance of a degenerate random variate at X = c is equal to

(a) C (b) C^2
(b) Zero (d) 1.0

Q.16 The m.g.f. of a degenerate r.v. X at 'c' is equal to

(a) $e^{c^2 t}$ (b) e^{ct}
(c) e^{-ct} (d) e^{ct^2}

9.3.3 Binomial Distribution

Q.17 Binomial distribution was discovered by :

(a) R.A. Fisher (b) James Bernouille
(c) D. Moiver (d) Karl Pearson

Q.18 The probability of x success and consequently (n - x) failures in n independent trial is given as

(a) $p(x) = {}^nC_x q^x (1-q)^{n-x}$ (b) $p(x) = {}^nC_x p^x q^{n-x}$
(c) $p(x) = {}^nC_x p^x (1-p)^{n-x}$ (d) Both (b) & (c)

$(x = 0, 1, 2, \ldots\ldots\ldots, n \quad p + q = 1)$

Q.19 The p.m.f. of B (x, n, p) is written as

(a) $p(x) = {}^nC_x\, p^x q^{n-x}$
(b) $p(x) = {}^nC_x\, q^x p^{n-x}$
(c) $p(x) = {}^nC_x\, p^x (1-p)^{n-x}$
(d) Both (a) & (c)

$\left.\right\}$ $x = 0, 1, 2, ..., n$; $p + q = 1$

Q.20 Let us suppose that n trials constitute an experiment. If this experiment is repeated N times then frequency function of B (x, n, p) is given as

(a) f (x) = N p (x) (b) f (x) = n. p (x)
(c) $f(x) = N^2 p(x)$ (d) (n + 1) p (x)

Q.21 In binomial experiment, trials are

(a) Independent of each other (b) Dependent on each other
(c) Mutually exclusive (d) Both (a) & (c)

Q.22 The problem of tossing a coin or die or drawing cards from a pack of cards with replacement, leads to the

(a) Bernouille distribution (b) Binomial distribution
(c) Negative binomial distribution (d) Both (a) & (b)

Q.23 The mean of binomial distribution B (n, p) is equal to

(a) nq (b) np
(c) n (1 - q) (d) Both (b) & (c)

Q.24 The variance B (n, p) distribution is equal to

(a) npq (b) $\sqrt{npq}$
(c) np (1 - p) (d) Both (a) & (c)

Q.25 For B (n, p) distribution, the skewness 13 1 is equal to

(a) $\beta_1 = \frac{(1-2p)^2}{npq}$ (b) $\beta_1 = \frac{(1-3p)^2}{npq}$

(c) $\beta_1 = \frac{(1-2p)^2}{np(1-p)}$ (d) Both (a) & (c)

Q.26 For B (n, p) distribution, the kurtosis (32 is equal to

(a) $\beta_2 = 3 + \frac{1-6pq}{npq}$ (b) $\beta_2 = 3 + \frac{1-6pq}{np(1-p)}$

(c) Both (a) & (b) (d) $\beta_2 = 3 + \frac{1-3pq}{npq}$

Q.27 For B (n, p) distribution, we have

(a) Mean > Variance
(b) Mean < Variance
(c) Variance < Mean
(d) Both (a) & (c)

Q.28 For B (n, p) distribution, we have

(a) μ_3 = npq (q - p)
(b) μ_3 = -npq (p - q)
(c) μ_3 = np(1 - p) (q - p)
(d) All the three

Q.29 Let (n + 1)p = m + f, where m is an integer part and f is a fraction < 1. Then unique mode of B (n, p) will be equal to

(a) m
(b) m + 1
(c) m^2
(d) $\frac{2m + 1}{2}$

Q.30 In B (n, p), let (n + 1)p = m (an integer). Then two model values of the distribution are equal to

(a) m, m + 1
(b) m, m -1
(c) (m - 1), m
(d) Both (b) & (c)

Q.31 The m.g.f. of B (n, p) is given as

(a) $M_x(t) = (q + p e^t)^n$
(b) $M_x(t) = (q + (1 - q) e^t)^n$
(c) $M_x(t) = (p + q e^t)^n$
(d) Both (a) & (b)

Q.32 Let (Xi, i = 1, 2, ..., k) are independent binomial variates with parameters (p, ni, i = 1, 2, ..., k). Then the variate U = $\sum_{i=1}^{k} Xi$ follows the distribution as

(a) $B\left(\sum_{i=1}^{k} ni, p\right)$
(b) $B\left(\sum_{i=1}^{k} ni, (1-q)\right)$
(c) $B\left(p, \sum_{i=1}^{k} ni\right)$
(d) All the three

Q.33 The characteristic function 4). (t) of B (n, p) is written as

(a) $\phi_x(t) = (q + p e^{it})^n$
(b) $\phi_x(t) = (q + (1-q) e^{it})^n$
(c) $\phi_x(t) = (q + p e^{it})^n$
(d) Both (a) & (b)

Q.34 The cumulative generating function of B (n, p) distribution is given as

(a) $K_x(t) = n \log(q + p e^t)$ (b) $K_x(t) = n \log(p + q e^t)$

(c) $K_x(t) = n \log(q + (1-q) e^t)$ (d) Both (a) & (c)

Q.35 The recurrence relation for the probabilities of B (n, p) distribution is given as

(a) $p(x+1) = \left(\frac{n-x}{x+1}\right) \cdot \frac{p}{q}\, p(x)$

(b) $p(x+1) = \left(\frac{n-x}{x}\right) \cdot \frac{p}{q}\, p(x)$

(c) $p(x+1) = \left(\frac{n-x}{x+1}\right) \cdot \left(\frac{q}{p}\right)^{-1} p(x)$

(d) Both (a) & (c)

9.3.4 Poisson Distribution

Q.36 Poisson distribution was discovered by French mathematician and physicist

(a) Simeon Denis Poisson (b) Simean Deming Poisson

(c) Simeon Deying Poisson (d) Simean Gomeg Poisson

Q.37 Poisson distribution is the limiting case of B (n, p) distribution under the following limiting case :

(a) $n \to \infty, p \to 1, np \to \lambda$ (finite)

(b) $n \to \infty, p \to 0, np \to \lambda$ (finite)

(c) $n \to \infty, q \to 1, np \to \lambda$ (finite)

(d) Both (b) & (c)

Q.38 The p.m.f. of Poisson distribution is written as

(a) $p(x, \lambda) = \begin{cases} \frac{e^{-\lambda}.\lambda^x}{x!}, & x = 0,1,2,\ldots\ldots, \lambda > 0 \\ 0 & \text{,otherwise} \end{cases}$

(b) $p(x, \lambda) = \begin{cases} \frac{e^{-x}.\lambda^\lambda}{x!}, & x = 0,1,2,\ldots\ldots, \lambda > 0 \\ 0 & \text{,otherwise} \end{cases}$

(c) $p(x, \lambda) = \begin{cases} \dfrac{\lambda^x e^{-\lambda}}{x!}, & x = 0,1,2,\ldots\ldots, \lambda > 0 \\ 0 & ,\text{otherwise} \end{cases}$

(d) Both (a) & (c)

Q.39 The distribution function F (x) of P (X) is written as

(a) $F(x) = e^{-\lambda} \sum_{r=0}^{x} \left(\frac{\lambda^r}{r!}\right)$, x = 0, 1, 2, ...

(b) $F(x) = e^{-\lambda} \sum_{r=0}^{x} \left(\frac{e^{-\lambda}}{r!}\right)$, x = 0, 1, 2, ...

(c) $F(x) = e^{-\lambda} \sum_{r=0}^{x} \left(\frac{r!}{\lambda^r}\right)^{-1}$, x = 0, 1, 2, ...

(d) Both (a) & (c)

Q.40 Poisson distribution occur when events are
(a) From an infinite number of trials
(b) From a definite number of trials
(c) Not from a definite number of trials
(d) From a 100 trials

Q.41 The number of death from the following disease follows : Poisson distribution
(a) Heart attack
(b) Cancer
(c) Snake bite
(d) All the three

Q.42 The number of fragements received by a surface area "A" from a fragement of atom bomb follows
(a) Binomial distribution
(b) Poisson distribution
(c) Hyper geometric distribution
(d) Normal distribution

Q.43 The number of suicides reported in a particular city follows
(a) Normal distribution
(b) Binomial distribution
(c) Negative binomial distribution
(d) Poisson distribution

Q.44 Let {Xi, i = 1, 2, ..., n} be sequenec of independent r.v.'s. Then De-moivre laplace theorem states that if

$$Xi = \begin{cases} 1, \text{ with probability } p \\ 0 \text{ with probability } q \end{cases} \text{ the r.v. } S_n = \sum_{i=1}^{n} Xi \text{ ..... is}$$

(a) Assymptotically normal as $n \to \infty$
(b) Not assymptotically normal as $n \to \infty$
(c) Assymptotically Poisson as $n \to \infty$
(d) Not assymptotically Poisson as $n \to \infty$

Q.45 Under C.L.T., the binomial distribution tends to
(a) Normal distribition as $n \to \infty$
(b) Poisson distribution as $n \to \infty$
(c) Negative binomial distribution as $n \to \infty$
(d) None of the above

Q.46 Mean of Poisson distribution is equal to

(a) 2λ (b) λ
(c) $\frac{\lambda}{2}$ (d) None

Q.47 The variance of Poisson distribution P (X) is equal to
(a) λ (b) $\lambda^2 -1$
(c) $\lambda^2 -3$ (d) $\lambda(\lambda - 1)$

Q.48 For Poisson distribution P (λ), the 3rd central moment μ_3 is equal to

(a) λ (b) 2λ
(c) $3\lambda + 1$ (d) $\lambda + 3$

Q.49 For Poisson distribution, the skewness 13 1 is equal to

(a) $\beta_1 = \frac{2}{\lambda}$ (b) $\beta_1 = \frac{\lambda}{3}$
(c) $\beta_1 = \frac{3}{\lambda}$ (d) None

Q.50 For Poisson distribution, we have

(a) $\beta_2 = 3 + \frac{1}{\lambda}$ (b) $\beta_2 = 3 + \frac{2}{\lambda}$
(c) $\beta_2 = 3 + \frac{2}{3\lambda}$ (d) None

Q.51 Let λ be not an integer and $\lambda = m + f$ where m is the integer part and f is the fractional part. Then unimodel value of this distribution is given as

(a) Mode = 2m (b) Mode = $\frac{3}{2}$ m

(c) Mode = $\frac{2}{3}$ m (d) None

Q.52 Let λ = k an integer value. The P (λ) is a bimodel and modes are equal to

(a) 2 k - 1, k (b) K - 1, k

(c) K, K -1 (d) Both '(b) & (c)

Q.53 The m.g.f of Poisson distribution is given as

(a) $M_x(t) = e^{\lambda(e^t-1)}$ (b) $M_x(t) = e^{(e^t-1)\lambda}$

(c) $M_x(t) = e^{-\lambda(e^t-1)}$ (d) Both (a) & (b)

Q.54 The characteristic function of P (λ) is written as

(a) $\phi_x(t) = e^{\lambda(e^{it}-1)}$ (b) $\phi_x(t) = e^{(e^{it}-1)\lambda}$

(c) $\phi_x(t) = e^{-\lambda(e^{it}-1)}$ (d) Both (a) & (b)

Q.55 Kumulant generating function of P (λ) is written as

(a) $K_x(t) = \lambda(e^t - 1)$ (b) $K_x(t) = \lambda(e^{-t} - 1)$

(c) $K_x(t) = \lambda(e^t - 1)\lambda$ (d) Both (a) & (c)

Q.56 All kumulant of P (λ) distribution are equal to

(a) λ (b) 2λ

(c) $\frac{2}{3}\lambda$ (d) $\frac{3}{2}\lambda$

Q.57 Let (Xi, i = 1, 2, ..., k) be k independent Poisson variates with parameters (Xi, i = 1, 2, , k). Then $U = \sum_{i}^{k} Xi$ is

(a) Poisson variate with parameter equal to $2\sum_{i=1}^{k} Xi$

(b) Poisson variate with paramete $\sum_{i=1}^{k} \lambda i$

(c) Not Poisson variate with parameter $\sum_{i=1}^{k} \lambda i$

(d) None of these

Q.58 For Poisson variate X, with parameter λ, the recurrence formula for probabilities is given by

(a) $p(x+1) = \left(\frac{\lambda}{x+1}\right)p(x)$ (b) $p(x+1) = \left(\frac{x+1}{\lambda}\right)^{-1} p(x)$

(c) $p(x+1) = \left(\frac{x+1}{\lambda}\right)p(x)$ (d) Both (a) & (b)

9.3.5 Negative Binomial Distribution

Q.59 Let there be x failures preceeding to rth success in (r + x) trials. Then p.m.f. of negative binomial distribution is given by

(a) $p(x) = \begin{cases} \binom{x+r-1}{r-1} p^r q^x, & x = 0,1,2,\ldots\ldots. \\ 0 & ,\text{otherwise} \end{cases}$

(b) $p(x) = \begin{cases} \binom{x+r-1}{r-1} q^x p^r, & x = 0,1,2,\ldots\ldots. \\ 0 & ,\text{otherwise} \end{cases}$

(c) $p(x) = \begin{cases} \binom{x+r-1}{r-1} p^r (1-p)^x, & x = 0,1,2,\ldots\ldots. \\ 0 & ,\text{otherwise} \end{cases}$

(d) Both (a) & (c)

Q.60 The negative binomial distribution is the generalization of
(a) Binomial distribution (b) Geometric distribution
(c) Bernoulli distribution (d) Uniform distribution

Q.61 Negative binomial distribution having two parameters (p, r) is known as
(a) Pascal's distribution (b) Pearsonian distribution
(c) Demoiver distribution (d) Geometric distribution

Q.62 The mean of negative binomial distribution with parameter (p, r) is equal to

(a) Mean = rP (b) Mean = rQ

(c) Mean = r (1 - Q) (d) Both (a) & (c)

where $p = \frac{1}{Q}$ and $q = \frac{P}{Q}$

Q.63 The variance g of negative binomial distribution is equal to

(a) $\mu_2 = r\,P\,Q$ (b) $\mu_2 = r\,(1 - P)\,(1 - Q)$

(c) $\mu_2 = r\,P^2\,Q^2$ (d) Both (a) & (c)

where $p = \frac{1}{Q}$ and $q = \frac{P}{Q}$

Q.64 In case of negative binomial distribution, the man and variance of it has the following relation

(a) Mean < Variance (b) Mean > Variance

(c) Variance > Mean (d) Both (a) & (c)

Q.65 The mean of negative binomial distribution with parameters (p, r) is equal to

(a) Mean = $\frac{rq}{p}$ (b) Mean = $\frac{rq}{q}$

(c) Mean = $\frac{rq}{r}$ (d) None

Q.66 The variance μ_2 of negative binomial distribution with parameters (p, r) is equal to

(a) $\mu_2 = \frac{rq}{p^2}$ (b) $\mu_2 = \frac{r(1-p)}{p^2}$

(c) $\mu_2 = \frac{rq}{q^2}$ (d) Both (a) & (b)

Q.67 Negative binomial distribution reduces to Poisson distribution under the condition

(a) $p \to 0, r \to \infty$, such that $rp = \lambda$ (finite)

(b) $p \to 0, r \to N$, such that $rp = \lambda$ (finite)

(c) $p \to 1, r \to \infty$, such that $rp = \lambda$ (finite)

(d) $q \to 0, r \to \infty$, such that $rp = \lambda$ (finite)

9.3.6 Geometric Distribution

Q.68 The p.m.f of geometric distribution is given by the function written as

(a) $p(x)=\begin{cases} q^x p, & x=0,1,2,\ldots\ldots\ldots, \quad 0<p<1 \\ 0 \quad , & \text{otherwise} \qquad p+q=1 \end{cases}$

(b) $p(x)=\begin{cases} p^x q, & x=0,1,2,\ldots\ldots\ldots, \quad 0<p<1 \\ 0 \quad , & \text{otherwise} \end{cases}$

(c) $p(x)=\begin{cases} q^x(1-q), & x=0,1,2,\ldots\ldots\ldots, \quad 0<p<1 \\ 0 \quad , & \text{otherwise} \end{cases}$

(d) Both (a) & (c)

Q.69 The probability that there are x failures preceeding to first success in n independent trials, is given by

(a) Bernouille distribution (b) Geometric distribution

(c) Negative binomial distribution (d) None of these

Q.70 The mean of geometric distribution is equal to

(a) Mean $= \dfrac{q}{p}$ (b) Mean $= \dfrac{p}{q}$

(c) Mean $= \dfrac{(1-p)}{(1-q)^2}$ (d) All the three

Q.71 The variance of geometric distribution is equal to

(a) $\mu = \dfrac{q}{p^2}$ (b) $\mu_2 = \dfrac{q}{(1-q)^2}$

(c) $\mu_2 = \dfrac{(1-p)}{(1-q)^2}$ (d) All the three

Q.72 The m.g.f. of geometric distribution is written as

(a) $M_x(t) = \dfrac{p}{(1-qe^t)}$ (b) $M_x(t) = \dfrac{q}{(1-qe^t)}$

(c) $M_x(t) = \dfrac{p}{(1-qe^{-t})}$ (d) None

Q.73 For geometric distribution, the mean and variance has relation given as
(a) Mean < Variance (b) Mean > Variance
(c) Variance > Mean (d) Both (a) & (c)

Q.74 Multinational distribution is the generalisation of
(a) Binomial distribution (b) Geometric distribution
(c) Negative binomial distribution (d) None

Q.75 When there are more than two mutually exclusive outcomes of a trial, then such observations lead to the
(a) Multinomial distribution (b) Binomial distribution
(c) Mutlivariate normal distribution (d) None

Key Answers

Section-A : True False

1. True	2. True	3. True
4. True	5. True	6. True
7. True	8. True	9. True
10. True	11. True	12. True
13. True	14. True	15. True
16. True	17. True	18. True
19. True	20. True	21. True
22. True	23. True	24. True
25. True	26. True	27. True
28. False	29. True	30. True
31. True	32. True	33. True
34. True	35. True	36. True
37. True	38. True	39. True
40. True	41. True	42. True
43. True	44. True	45. True
46. True	47. True	48. True
49. True	50. True	51. True
52. True	53. True	54. True
55. True	56. True	57. True

58. True
59. True
60. True
61. True
62. True
63. True
64. True
65. True
66. True
67. True
68. True
69. True
70. True
71. True
72. True
73. True
74. True
75. True
76. True
77. True
78. True
79. True
80. True
81. True
82. True
83. True
84. True
85. True

Section-B : Fill in the Blanks

1. a prior
2. Posterior
3. $\frac{1}{n}$
4. Parameter
5. Discrete rectangular
6. Discrete uniform distribution
7. $\frac{n+1}{2}$
8. $\frac{(n+1)(2n+1)}{2}$
9. $\frac{(n-1)(n+1)}{12}$
10. $(1 - e^{nt}) / n (1 - e^{t})$
11. $p^{x} q^{(1-x)}$
12. Bernouille
13. p, pq
14. p
15. (q + pet)
16. Zero
17. e^{ct}
8. Bernouille
19. $^{n}C_{x} p^{x} q^{n-x}$
20. $^{n}C_{x} p^{x} q^{n-x}$
21. Discrete, (p)
22. N p (x)
23. Independent
24. Constant
25. Binomial
26. np
27. npq
28. npq
29. npq
30. >
31. npq
32. m
33. (m - 1), m
34. pe^{t}

35. Reproductive

36. $\left(\sum_{i=1}^{k} ni, p\right)$

37. pe^{it}

38. n log $(q + p.e^{t})$

39. $\left(\frac{n-x}{x+1}\right)$

40. Denis Poisson

41. $n \to \infty$, p->0

42. Non negative,

43. P (λ), Parameter

44. $\left(\frac{\lambda^{r}}{r!}\right)$

45. Poisson

46. Poisson

47. Poisson

48. Poisson

49. Poisson

50. λ

51. λ

52. λ

53. $\frac{1}{\lambda}$

54. $\frac{1}{\lambda}$

55. Skewed

56. m

57. (K - 1), k

58. $(e^{t} - 1)$

59. $(e^{it} - 1)$

60. $\lambda (e^{t} - 1)$

61. Same, λ

62. $\frac{1}{\lambda}, 3 + \frac{1}{\lambda}$

63. Poisson, $\left(\sum_{i=1}^{k} \lambda i\right)$

64. $\left(\frac{\lambda}{x+1}\right)$

65. $\binom{x+r-1}{r-1}$

66. Geometric distribution

67. Geometric distribution

68. $(Q - Pe^{t})$

69. rP

70. rPQ

71. Mean < Variance

72. $(Q - Pe^{t})$

73 . p, p^2

74. 0

75. Geometric distribution

76. $q^{x} p$

77. q/p

78. q/ p^2

79. $(1 - qe^{t})$

80. $<$

81. Binormial

82. Multinomial

83. $p_{1}^{X_{1}}, p_{2}^{X_{1}}, \ldots\ldots, p_{k}^{k}$

Section C : Multiple Choice Questions

1. (a)	2. (a)	3. (a)
4. (d)	5. (b)	6. (c)
7. (d)	8. (a)	9. (a)
10. (b)	11. (b)	12. (b)
13. (a)	14. (a)	15. (c)
16. (b)	17. (b)	18. (d)
19. (d)	20. (a)	21. (d)
22. (b)	23. (d)	24. (d)
25. (d)	26. (c)	27. (d)
28. (d)	29. (a)	30. (d)
31. (d)	32. (d)	33. (d)
34. (d)	35. (d)	36. (a)
37. (d)	38. (d)	39. (d)
40. (a)	41. (d)	42. (b)
43. (d)	44. (a)	45. (a)
46. (b)	47. (a)	48. (a)
49. (d)	50. (a)	51. (d)
52. (d)	53. (d)	54. (d)
55. (d)	56. (a)	57. (b)
58. (d)	59. (d)	60. (b)
61. (a)	62. (d)	63. (d)
64. (d)	65. (a)	66. (d)
67. (a)	68. (d)	69. (b)
70. (d)	71. (d)	72. (a)
73. (d)	74. (a)	75. (a)

❑❑❑

Chapter - 10

Continuous Distributions

Section-A : Write True / False

10.1.1 Normal Distribution

Q.1 A r.v. X is said to have a normal distribution with parameters (μ, σ^2) if its p.d.f. is given by the probability law

$$f(x,\mu,\sigma^2)=\frac{1}{\sqrt{2\pi\sigma}}e^{-\frac{1}{2}\left(\frac{x-\mu}{\sigma}\right)^2} \qquad \begin{array}{l}-\infty < x < \infty \\ -\infty < \mu < \infty\end{array} \qquad \sigma > 0$$

Q.2 When X is distributed as N(μ, σ^2) then it is expressed as X ~ N(μ,σ^2).

Q.3 Normal distribution is the limiting case of binomial distribution under the following case -
(i) $n \to \infty$
(ii) Neither p nor q is very small

Q.4 Mean, Median and mode of normal distribution are equal to μ.

Q.5 The mean μ and variance σ^2 are the two parameters of normal distribution.

Q.6 As x increases numerically the (x) decreases rapidally.

Q.7 The maximum probability of N (μ, σ^2) occurs at the point x =μ.

Q.8 For N (μ, σ^2), we have

$$\text{Max } f(x) = \frac{1}{\sqrt{2\pi\sigma}}$$

Q.9 For N (μ, σ^2), we have $\beta_1 = 0$ and $\beta_2 = 3$

Q.10 The add order moments of N (μ, σ^2) are equal to zero.

Q.11 The even order moments of N (μ, σ^2) are given by the expression

$\mu_{2r} = 1.3.5 \text{ .... } (2r-1)\sigma^2, (r = 0, 1, 2, \text{ ....})$

Q.12 Linear combination of independent normal variates is also normal variate.

Q.13 X axis is an asymptotic to the normal curve.

Q.14 The points of inflextion of the curve are given as

$$x = \mu \pm \sigma, \ f(x) = \frac{1}{\sqrt{2\pi\sigma}}, \ e^{-\frac{1}{2}}$$

Q.15 Mean deviation about mean of the N (μ, σ^2) is equal to

$$\sqrt{\frac{2}{\pi}}\sigma = \frac{4}{5}\sigma$$

Q.16. Quartile deviation of normal distribution is equal to

$$\text{Q.D.} = \frac{(Q_3 - Q_1)}{2} = \frac{2}{3}\sigma$$

Q.17 We have the relation for normal distribution N(μ, σ^2). QD : M.D. : S.D.: as 10 : 12 : 15

Q.18 We have for N (μ, σ^2) $P(\mu - \sigma < x < \mu + \sigma) = 0.6826$

Q.19 $P(\mu - 2\sigma < x < \mu + 2\sigma) = 0.9544$

Q.20 $P(\mu - 3\sigma < x < \mu + 3\sigma) = 0.9944$

Q.21 If X and Y are independent standard normal variate then the variates U = X + Y and V = X - Y are independently distributed as U ~ (0, 2) and V ~ N(0, 2).

Q.22 If X1, X2, ..., Xn are iid. r.v's with finite variance then

$$\overline{x}_n = \frac{1}{n}\sum_{i=1}^{n} xi \text{ and } s^2 = \frac{1}{(n-1)}\sum_{i}^{n}(xi - \overline{x}n)^2$$

are independent.

Q.23 The m.g.f. of N(μ, σ_2) is given by the formula -

$$M_x(t) = e^{\mu t + \frac{t^2\sigma^2}{2}}$$

Q.24 M.G.F. of standard normal distribution, N(0,1) is written as

$$M_z(t) = e^{\left(\frac{t^2}{2}\right)}$$

Q.25 Kumulant generating function of N(μ, σ_2) is given as

$$Kx(t) = \mu t + \frac{1}{2} t^2\sigma^2$$

Q.26 Let Xi (I =1, 2, ..., n) be n independent variates distributed as N (μ_i, σ_i^2, = 1, 2, ..., n). Then the linear function

$$U = \sum_{i}^{n} aixi \sim N\left(\sum_{i=1}^{n} ai\mu i, \sum_{i=1}^{n} a_i^2\sigma_i^2\right)$$

Q.27 The entire theory of small sample tests viz, t, F, X^2 etc. is based on the fundamental assumption that the parent populations from which samples are drawn, follow normal distribution.

Q.28 Normal distribution finds large applications for statistical quality control in industry for setting control limits.

10.1.2 Rectangular (or Uniform) Distribution

Q.29 A random variable X is said to have a continuous rectangular (Uniform) distribution over an interval (a, b) i.e. (-∞ < a, b < ∞) if its p.d.f. is given as

$$f(x, a, b) = \begin{cases} \frac{1}{(b-a)}, & \text{if } a<b \\ 0 & \text{, otherwise} \end{cases}$$

Q.30 A uniform or rectungular distribution on the interval (a, b) is written as x ~ U [a, b] or x ~ R[a, b].

Q.31 In rectangular distribution R[a, b], the two parameters are "a" and "b".

Q.32 The uniform distribution U[a, b] is also known as rectangular distribution, since the curve y = f(x), makes a rectangle over the x axis and between the ordinates at x = a and x = b.

Q.33 The mean and variance of U[a, b] distribution are given as

$$\text{Mean }(x) = \frac{a+b}{2}, \; V(X) = \frac{(b-a)^2}{12}$$

Q.34 M.G.F. of rectangular distribution is given as

$$Mx(t) = \frac{e^{bt-at}}{t(b-a)}, \quad t \neq 0$$

Q.35 The characteristic function of rectangular distribution is given as

$$\phi_x(t) = \frac{e^{ibt-iat}}{it(b-a)}, \; t \neq 0$$

Q.36 Mean deviation about mean is equal to $\left(\frac{b-a}{4}\right)$

1 0.1 .3 Gamma Distribution

Q.37 A.r.v. X is said to have a gamma distribution with parameter λ > 0, if its p.d.f. is given by

$$f(x) = \begin{cases} \dfrac{e^{-x}x^{\lambda-1}}{\lceil\lambda}, \lambda > 0, \; 0 < x < \infty \\ 0 \qquad , \text{otherwise} \end{cases}$$

Q.38 A continuous r.v. X having the following p.d.f.

$$f(x) = \begin{cases} \dfrac{a^{\lambda}e^{-ax}x^{\lambda-1}}{\lceil\lambda}, a > 0, \lambda > 0, \; 0 < x < \infty \\ 0 \qquad , \text{otherwise} \end{cases}$$

is said to have a gamma distribution with two parameters (a, λ).

Q.39 The M.G.F. of gamma distribution about origin is given by $M_x(t) = (1\text{-}t)^{-\lambda}, |t| < 1$

Q.40 Kumutand generating function of gamma distribution r (λ) is given as

$k_x(t) = -\lambda \log_e (1 - t), |t| < 1$

Q.41 Mean and variance of gamma distribution r (λ) is given as Mean = Variance = λ

Q.42 The standard gamma variate Z is defined as

$$Z = \frac{x - \lambda}{\sqrt{\lambda}}$$

Q.43 The standard gamma variate tends to standard normal variate if $\lambda \to \infty$.

Q.44 For the two parameters gamma distribution, the m.g.f. is given as

$$Mx(t) = \left(1 - \frac{t}{a}\right)^{-\lambda}, t < a$$

Q.45 For two parameters gamma distribution, γ (a, λ), the kumulant g.f. is given as

$$K_x(t) = -\lambda \log_e \left(1 - \frac{t}{a}\right)$$

Q.46 If (Xi, i = 1, 2, ..., k) are independent gamma variates with parameters (λi, i = 1, 2, ..., k) respectively then

$$U = \sum_{i}^{n} Xi \sim \gamma\left(\sum_{i=1}^{k} \lambda i\right)$$

Q.47 In general if Xi ~ γ (a, λi), i = 1, 2, ..., k are independent, r.v's then

$$V = \sum_{i}^{n} Xi \sim \gamma\left(a, \sum_{i=1}^{k} \lambda i\right)$$

10.1.4 Beta Distribution of First Kind

Q.48 A r.v. X is said to have a beta distribution of first kind with parameters (μ, v), μ > 0, v > 0, if its p.d.f. is given by

$$f(x) = \begin{cases} \frac{1}{\beta(\mu, v)} x^{\mu-1}(1-x)^{v-1}, & (\mu, v) > 0, 0 < x < 1 \\ 0 & \text{, otherwise} \end{cases}$$

Q.49 The beta distribution of first kind is denated by β_1 (μ, v).

Q.50 In case of first kind of beta distribution b1 (μ, v), for μ = v = 1, it reduces to uniform distribution on [0, 1].

Q.51 The mean of beta distribution b1 (μ, v) is equal to Mean = μ/ (μ + v).

10.1.5 Beta Distribution of Second Kind

Q.52 A r.v. X is said to have a beta distribution of second kind with parameters (μ,v), μ > 0, v > 0, if its p.d.f. is given by

$$f(x) = \begin{cases} \frac{1}{\beta(\mu, v)} \frac{x^{\mu-1}}{(1+x)^{\mu+v}}, & (\mu, v) > 0, 0 < x < \infty \\ 0 & \text{, otherwise} \end{cases}$$

Q.53 The beta distribution of second kind is denoted by β_2 (μ, v).

Q.54 If X ~ b2 (μ, v), then $Y = \frac{1}{1+X} \sim \beta_1 (\mu, v)$

Q.55 The mean of β_2 (μ, v) is given as

$$\text{Mean} = \frac{\mu}{v-1}, v > 1$$

10.1.6 Exponental Distribution

Q.56 A r.v. X is said to have an exponental distribution with parameter θ > 0, if its p.d.f. is given by

$$f(x, \theta) = \begin{cases} \theta . e^{-\theta x}, & x \geq 0 \\ 0 & \text{, otherwise} \end{cases}$$

Q.57 The mean and variance of exponential distribution are given as

$$\text{Mean} = \frac{1}{\theta}, \text{Variance} = \frac{1}{\theta^2}$$

Q.58 Let (Xi, i = 1, 2, ..., n) be n independent exponential variate with parameter (θi, i = 1, 2, ..., n) then $Z = \text{Min}(X_1, X_2, \ldots, X_n)$ has an exponential distribution

with parameter $\left(\sum_{i}^{n} \theta i\right)$.

10.1.7 Cauchy Distribution

Q.59 A r.v. X is said to have a standard cauchy distribution if its p.d.f. is given by

$$f(x) = \frac{1}{\pi(1+x^2)}, \quad -\infty < x < \infty$$

Q.60 In standard cauchy distribution ie in x ~ C (1, 0), X is known as standard cauchy variate.

Q.61 The characteristic function of standard cauchy distribution is given as $\phi_x(t) = x\, e^{-|t|}$

Q.62 If (Xi, i = 1, 2) are independent cauchy variates with parameters (Xi, μi, i = 1, 2) respectively then $U = (X_1 + X_2)$ is a cauchy variate with parameters $\left(\sum_{i}^{2} \lambda i, \sum_{i=1}^{2} \mu i\right)$

Q.63 In general sense, the mean of cauchy distribution does not exist.

10.2 Section - B : Fill in the Blanks

10.2.1 Normal Distribution

Q.1 The p.d.f. of a normal distribution $N(\mu, \sigma^2)$ is written as $f(x, \mu, \sigma^2) = \frac{1}{\sqrt{2\pi}\sigma} e^{-\frac{1}{2}(.....)^2}$ $\begin{matrix} -\infty < x < \infty \\ -\infty < \mu < \infty \end{matrix}$ $0 < \sigma^2 < \infty$

Q.2 A r.v. X following normal distribution with its mean μ and variance σ^2 is expressed as

X ~ N (...........)

Q.3 Normal distribution is the limiting case of binomial distribution under the conditions -

(i) n →

(ii) Neither 'p' nor 'q' is very ....

Q.4 Mean, median and mode of normal distribution $N(\mu, \sigma^2)$ are equal to

Q.5 Two parameters of normal distribution are denoted by .... and .... respectively.

Q.6 A s X increases numerically, f(x) .... rapidly.

Q.7 The maximum probability of $N(X, \mu, \sigma^2)$ occurs at the point x = ...

Q.8 For $N(\mu, \sigma^2)$ distribution, we have Max f (x) = ...

Q.9 For normal distribution $N(\mu, \sigma^2)$, we have shewness β_1 = ..., kurtosis β_2 = ...

Q.10 The odd order moments of $N(\mu, \sigma^2)$ are equal to ....

Q.11 The even order moments μ_{2r} of $N(\mu, \sigma^2)$ distribution are expressed as

μ_{2r} = 1.3.5 .... (........) σ^2, r = 0, 1, 2, ...

Q.12 The linear combination of independent normal variates is also ....

Q.13 X axis is .... to the normal curve.

Q.14 The points of inflexion of normal curve are given as

$$x = \text{.............},\ f(x) = \frac{1}{\sqrt{2\pi\sigma}}\ \text{.............}$$

Q.15 The mean deviation of normal distribution $N(\mu, \sigma^2)$ about mean is equal to ....

Q.16 Quartile deviation of normal distribution $N(\mu, \sigma^2)$ is equal to ....

Q.17 In case of normal distribution $N(\mu, \sigma^2)$, the quartile deviation, mean deviation and standard deviation have values in proportional as

Q.D. : M.D. : S.D. : are : ....

Q.18 According to area property of normal distribution,

$P(\mu - \sigma < x < \mu + \sigma)$ = ...

Q.19 $P(\mu - 2\sigma^2 < x < \mu + 2\sigma)$ = ...

Q.20 $P(\mu - 3\sigma < x < \mu + 3\sigma)$ = ...

Q.21 Let (X, Y) be independent standard normal variates then U = (X + Y), and V = (X - Y) are .... distributed as

U = N(.................) and V ~ N (...................)

Q.22 Let (Xi, i = 1, 2, ... Xn) be i.i.d. r.v's with finite variance then -

$$\bar{x} = \frac{1}{n}\sum_{i}^{n} xi \text{ and } s^2 = \frac{1}{(n-1)}\sum_{i}^{n}(xi - \bar{x})^2 \text{ are ....}$$

Q.23 The M.G.F. of normal distribution N (μ, σ^2) is written as Mx(t) = ...

Q.24 The M.G.F. of standard normal variate N (0, 1) is given as Mz(t) = ...

Q.25 Kumulant generating function of normal distribution is written as Kx(t) = ...

Q.26 Let (Xi, i = 1, 2, ... K) be k independent r.v's distributed as N(μi, σ_i^2, i = 1, 2, ..., k). Then the linear function.

$$U = \sum_{i=1}^{k} aiXi \sim N\ (...........)$$

Q.27 The entire theory of small sample test viz t, F, X^2, etc. is based on the fundamental assumption that the parent population from which the samples are drawn, follows ....

Q.28 Normal distribution finds large applications in statistical .... in industry for setting control limits.

10.2.2 Rectangular Distribution

Q.29 The p.d.f. of a rectangular distribution over an interval (a, b) is written as

$$f(x, a, b) = \begin{cases} ...., \text{if } a < b \\ 0 \quad , \text{otherwise} \end{cases}$$

Q.30 A uniform or rectangular .... over an interval (a, b) is expressed as -
X ~ R [...............]

Q.31 In rectangular distribution R[a, b], the parameters are given as .... and ....

Q.32 The uniform distribution U[a, b] is also known as ...., since the curve y = f(x) makes the rectangle over x axis and between the ordinates at x = ... and ....

Q.33 The mean and variance of U[a, b] distribution are given as

Mean = ..., V(x) = ...

Q.34 The M.G.F. of rectangular distribution is given as

Mx (t) = ,t # 0

Q.35 The characteristic function of rectangular distribution is given as

$$\phi_x(t) = \frac{\text{............}}{it(b-a)}, t \neq 0$$

Q.36 Mean deviation of U[a, b] is equal to ....

10.2.3 Gamma Distribution

Q.37 The p.d.f. of gamma distribution with parameter $\lambda > 0$, is given as

$$f(x) = \begin{cases} \text{..........}, \lambda > 0,\ 0 < x < \infty,\ a > 0 \\ 0 \quad\quad ,\text{otherwise} \end{cases}$$

Q.38 A gamma distribution with two parameters (a, λ) has the p.d.f. given as

$$f(x, a, \lambda) = \begin{cases} \text{........}, \quad a > 0, \lambda > 0, 0 < x < \infty \\ 0 \quad\quad ,\text{otherwise} \end{cases}$$

Q.39 The m.g.f. of gamma distribution about origin is given by - Mx(t) = ..., $|t| < 1$

Q.40 Kumulant generating function of gamma distribution $\gamma(\lambda)$ is written as

Kx(t) = - λ ...., $|t| < 1$

Q.41 Mean and variance of gamma distribution $\gamma(\lambda)$ is given as

Mean = variance = ...

Q.42 The standard gamma variate z is defined as Z = ...

Q.43 The standard gamma variate tends to standard normal variate if $\lambda \rightarrow$ ....

Q.44 For two parameter gamma distribution $\gamma(\lambda, a)$, the m.g.f. is given as Mx (t) = (............), ...

Q.45 For two parameters gamma distribution γ (a, λ), the kumulant generating function is given as:
$Kx(t) = -\lambda \log_e (\ldots\ldots\ldots\ldots)$

Q.46 Let (Xi, i = 1, 2, ..., k) be k independent gamma variates with parameters (λi, i = 1, 2, ..., k) then

$$U = \sum_{i}^{k} Xi \sim \gamma (\ldots\ldots)$$

Q.47 In general if Xi ~ y (a, λi), i = 1, 2, ..., k, are independent r.v's then

$$V = \sum_{i=1}^{k} Xi \sim \gamma (\ldots\ldots)$$

10.2.4 Beta Distribution of First Kind

Q.48 The p.d.f. of beta distribution of first kind with parameters

$$(\mu, v), \mu > 0 \text{ is given as } f(x, \mu, v) = \begin{cases} \dfrac{1}{\beta(\mu, v)} \ldots\ldots\ldots, (\mu, v) > 0 \quad 0 < x < 1 \\ 0 \qquad , \text{otherwise} \end{cases}$$

Q.49 Beta distribution of first kind with parameter (μ, v) is donated as : β_1 (..............)

Q.50 In case of first kind of beta distribution β_1 (μ, v), for μ = v =1, it reduces to on [0, 1]

Q.51 The mean of beta distribution $\beta_1(\mu, v)$ is equal to ....

10.2.5 Beta Distribution of Second Kind

Q.52 The p.d.f. of beta distribution of second kind with parameters (μ, v), μ > 0, v > 0 is given as

$$f(X, \mu, v) = \begin{cases} \dfrac{1}{\beta(\mu, v)} \ldots\ldots\ldots, (\mu, v) > 0 \quad 0 < x < \infty \\ 0 \qquad , \text{otherwise} \end{cases}$$

Q.53 The beta distribution of second kind is denoted by β_2 (.............)

Q.54 If X ~ $\beta^2(\mu, v)$, then $Y = \dfrac{1}{1 + x} \sim$ ....

Q.55 The mean of β_2 (μ, v) is equal to

10.2.6 Exponential Distribution

Q.56 The p.d.f. of an exponential distribution with parameter $\theta > 0$, is written as

$$f(X, \theta) = \begin{cases} \text{........}, x \geq 0 \\ 0 \quad , \text{otherwise} \end{cases}$$

Q.57 The mean and variance of exponential distribution with parameter θ are respectively equal to

Mean = ..., Variance = ...

Q.58 Let (Xi, i = 1, 2, ..., n) be n independent exponential with parameters (0i, i = 1, 2, ..., n) then Z = Min (X_1, X_2, ..., X_n), has an .... with parameter $\sum_{i}^{n} \theta_1 = \theta$

10.2.7 Cauchy Distribution

Q.59 The p.d.f. of a standard cauchy distribution is given as f(x) = ..., $-\infty < x < \infty$

Q.60 In standard cauchy distribution say X : ~ c(1, 0), the variate X is known as " .... " variate.

Q.61 The characteristic function of standard cauchy distribution is given as $\phi_x(t)$ = ...

Q.62 If (Xi, i = 1,2) are independently distributed as cauchy distributions with parameters (λi, μi, i = 1,2) respectively then U = ($X_1 + X_2$) is a cauchy variate with parameters

$$\left(\sum_{i=1}^{2} \lambda i, \sum_{i=1}^{2} \mu i\right)$$

Q.63 In general, the mean of .... distribution does not exist.

10.3 Section-C : Multiple Choice Questions

10.3.1 Normal Distribution

Q.1 The p.d.f. of a normal distribution N(μ, σ^2) is written as

(a) $f(x, \mu, \sigma^2) = \frac{1}{\sqrt{2\pi\sigma}} e^{-\frac{1}{2}\left(\frac{x-\mu}{\sigma}\right)^2}$

(b) $f(x,\mu,\sigma^2)=\frac{1}{\sqrt{2\pi\sigma^2}}e^{-\frac{1}{2}\frac{(x-\mu)^2}{\sigma^2}}$

(c) $f(x,\mu,\sigma^2)=\frac{1}{\sqrt{2\pi\sigma^2}}e^{\frac{1}{2}\left(\frac{x-\mu}{\sigma}\right)^2}$

(d) Both (a) & (b)

Q.2 With a normal variable X, the normal distribution with mean μ and variance σ^2, it is denoted as

(a) $X \sim N(\sigma^2, \mu)$
(b) $X \sim N(\mu, \sigma^2)$
(c) $X \sim N(\sigma, \mu)$
(b) $X \sim N(\mu^2, \sigma^2)$

Q.3 Mean, median and mode of normal distribution $N(\mu, \sigma^2)$ is equal to

(a) μ^2
(b) μ
(c) σ
(d) $\sqrt{\mu}$

Q.4 Parameters of normal distribution are donated by

(a) (μ^2, σ^2)
(b) (μ, σ^2)
(a) (μ, σ)
(a) $(\sqrt{\mu}, \sigma^2)$

Q.5 As x increases numerically, f(x) -

(a) Decreases rapidly
(b) Increases rapidly
(c) Remains constant
(d) Nothing can be said

Q.6 Maximum probability of $N(x, \mu, \sigma^2)$ occurs at the point

(a) x = Mean
(b) x = Median
(c) x = Mode
(d) All the three

Q.7 For normal distribution $N(\mu, \sigma^2)$. We have

(a) Max $f(x) = \frac{1}{\sqrt{2\pi\sigma}}$
(b) Max $f(x) = \frac{1}{\sqrt{2\pi\sigma^2}}$
(c) Max $f(x) = \frac{1}{2\pi\sigma^2}$
(d) Both (a) & (b)

Q.8 For normal distribution $N(\mu, \sigma^2)$, the skewness β_1 is given as -

(a) β_1
(b) $\beta_1 = 1$
(c) $\beta_1 = \sqrt{1}$
(c) $\beta_1 = \sqrt{2}$

Q.9 For normal distribution, the kurtosio β_2 is given as

(a) $\beta_2 = \sqrt{3}$
(b) $\beta_2 = 3$
(c) $\beta_2 = 2 + \sigma$
(d) $\beta_2 = 1 + \sigma$

Q.10 The odd orderm om eniEofN (μ, σ^2) are equal to

(a) 1
(b) $\sqrt{3}$
(c) 0.0
(d) $\sqrt{2}$

Q.11 The even order moments of $N(\mu, \sigma^2)$, are equal to

(a) $\mu_{2r} = 2.4.6.........,2r\sigma^2$
(b) $\mu_{2r} = 1.3.5........, 2r\sigma^2$
(c) $\mu_{2r} = 1.3.5.........,(2r-1)\sigma^2$
(d) $\mu_{2r} = 2.4.6........,2r\sigma^r$

Q.12 The linear combination of independent normal variates follows

(a) Normal distribution
(b) Uniform distribution
(c) Rectangular distribution
(d) Both (b) & (c)

Q.13 X axis is .... to normal curve

(a) Perpendicular
(b) Parallel
(c) Asymptotic
(d) None

Q.14 The point of inflextion of the normal curve are given at the points

(a) $x = \mu \pm \sigma$
(b) $x = \mu \pm 2\sigma$
(c) $x = \mu \pm \sqrt{2}\sigma$
(d) $x = \mu \pm \sqrt{3}\sigma$

Q.15 Mean deviation of normal distribution about mean is equal to

(a) $\frac{2}{3}\sigma$
(b) $\frac{4}{5}\sigma$
(c) $\frac{4}{\sqrt{5}}\sigma$
(d) $\frac{3}{5}\sigma$

Q.16 Quartile deviation of normal distribution $N(\mu, \sigma^2)$ is equal to

(a) $\frac{2}{3}\sigma$
(a) $\frac{3}{2}\sigma$
(c) $\frac{\sqrt{3}}{2}\sigma$
(d) $\frac{\sqrt{2}}{3}\sigma$

Q.17 For normal distribution, the following relation exists
(a) Q.D : M.D: S.D: as 10 : 12 : 15
(b) Q.D: M.D: S.D: as 15, 12, 10
(c) Q.D: M.D: S.D: as 12 : 15 : 17
(d) Q.D: M.D: S.:D as 10 : 11 : 12

Q.18 From normal probability curve, we have
(a) $P(\mu - \sigma < x < \mu + \sigma) = 0.6826$
(b) $P(\mu - \sigma < x < \mu + \sigma) = 0.4826$
(c) $P(\mu - \sigma < x < \mu + \sigma) = 0.5826$
(d) $P(\mu - \sigma < x < \mu + \sigma) = 0.7826$

Q.19 From the probability table of normal distribution, we have
(a) $P(\mu - 2\sigma < x < \mu + 2\sigma) = 0.8544$
(b) $P(\mu - 2\sigma < x < \mu + 2\sigma) = 0.9544$
(c) $P(\mu - 2\sigma < x < \mu + 2\sigma) = 0.9944$
(d) $P(\mu - 2\sigma < x < \mu + 2\sigma) = 0.8626$

Q.20 From area property of normal distribution, we have
(a) $P(\mu - 3\sigma < x < \mu + 3\sigma) = 0.9944$
(b) $P(\mu - 3\sigma < x < \mu + 3\sigma) = 0.8844$
(c) $P(\mu - 3\sigma < x < \mu + 3\sigma) = 0.7744$
(d) $P(\mu - 3\sigma < x < \mu + 3\sigma) = 0.6826$

Q.21 If x and y are independent standard normal variates, then $U = (X + Y)$ is distributed as
(a) N (0, 2) (b) N (2, 0)
(c) N (2, 4) (d) (1, 2)

Q.22 If X and Y are independent standard normal variates then $V = (X - Y)$ is distributed as
(a) N (2, 4) (b) N (0, 2)
(c) N (1, 2) (d) N (2, 0)

Q.23 The M.G.F. of normal distribution $N(\mu, \sigma^2)$ is written as

(a) $M_x(t) = e^{\mu t + \frac{t^2\sigma^2}{2}}$ (b) $M_x(t) = e^{\mu t^2 + \frac{t\sigma^2}{2}}$

(c) $M_x(t) = e^{\frac{\mu t + t^2\sigma^2}{2}}$ (d) $M_x(t) = e^{t\mu^2 + \frac{t^2\sigma^2}{2}}$

Q.24 M.G.F. of standard normal variate is written as

(a) $M_z(t) = e^{\frac{t^2}{2}}$ (b) $M_z(t) = \frac{1}{\sqrt{2\pi}} e^{\frac{t^2}{2}}$

(c) $M_z(t) = \frac{1}{\sqrt{2\pi}} e^{-\frac{t^2}{2}}$ (d) $M_z(t) = \frac{1}{2\pi} e^{\frac{t^2}{2}}$

Q.25 Kumutant generating function of N(μ, σ^2) is given as

(a) $K_x(t) = \mu t + \frac{1}{2} t^2 \sigma$ (b) $K_x(t) = \mu t + \frac{t^2\sigma^2}{2}$

(c) $K_x(t) = \log\left(\mu t + \frac{t^2\sigma^2}{2}\right)$ (d) $K_x(t) = \log\left(\mu t + \frac{t\sigma}{2}\right)$

Q.26 Let (Xi, i = 1, 2, ..., k) be k independent r.v's distributed as N (μi, σ_i^2, i = 1, 2, ..., k) respectively. Then the

linear function U = aiXi is distributed as

(a) $U \sim N\left(\sum_{i=1}^{K} ai\mu i, \sum_{i=1}^{K} a_i^2\sigma_i^2\right)$ (b) $U \sim N\left(\sum_{i=1}^{K} a_i^2\mu i, \sum_{i=1}^{K} a_i^2\sigma_i^2\right)$

(c) $U \sim N\left(\sum_{i=1}^{K} a_i^2\mu i, \sum_{i=1}^{K} ai\sigma_i^2\right)$ (d) $U \sim N\left(\sum_{i=1}^{K} ai\mu i, \sum_{i=1}^{K} ai\sigma_i^2\right)$

Q.27 The entire theory of small sample test viz t, F, χ^2 etc. is based on the fundamental assumption that the parent population from which the samples are drawn, follows

(a) Normal distribution (b) Uniform distribution

(c) Gauss distribution (d) Both (a) & (c) -

Q.28 Normal distribution finds large application for statistical quality control in industry for setting

(a) Point estimation (b) Confidence limits

(c) Control limits (d) Limit of point estimate

10.3.2 Rectangular Distribution

Q.29 The p.d.f. of rectangular distribution over an interval (a, b) is given by

(a) $f(x) = \frac{1}{(a+b)}$ (b) $f(x) = \frac{1}{(b-a)}$

(c) $f(x) = \frac{1}{(a-b)}$ (d) $f(x) = \frac{1}{(b^2-a^2)}$

Q.30 A rectangular distribution on an interval (a, b) is written as

(a) $X \sim R[b, a]$ (b) $X \sim R[a^2, b^2]$

(c) $X \sim R[a, b]$ (d) $X \sim R[b^2, a^2]$

Q.31 In rectangular distribution R[a, b], the two parameters are

(a) b & a (b) a & b

(c) a^2 & b^2 (d) b^2 & a^2

Q.32 An uniform distribution U[a, b], is also known as

(a) Rectangular distribution (b) Geometric distribution

(c) Normal distribution (d) Misokartic distribution

Q.33 The mean of uniform distribution U[a, b] is equal to

(a) $\bar{x} = \frac{a^2 + b^2}{2}$ (b) $\bar{x} = \frac{a + b}{2}$

(c) $\bar{x} = \frac{b + a}{2}$ (d) Both (b) & (c)

Q.34 The variance of uniform distribution U[a, b] is equal to

(a) $\sigma^2 = \frac{b^2 + a^2}{12}$ (b) $\sigma^2 = \frac{b - a}{12}$

(c) $\sigma^2 = \frac{(b - a)^2}{12}$ (d) $\sigma^2 = \frac{b^2 - a^2}{10}$

Q.35 The M.G.F. of uniform distribution U(a, b] is wirtten as
ebt-at

(a) $M_x(t) = \frac{e^{bt-at}}{t(b-a)}$ (b) $M_x(t) = \frac{e^t(b-a)}{t(b-a)}$

(c) $M_x(t) = \frac{e^{t(b-a)}}{t(b^2-a^2)}$ (d) Both (a) & (b)

Q.36 The characteristic function of uniform distribution U(a, b) is written as

(a) $\phi_x(t) = \frac{e^{it(b-at)}}{t(b-a)}$ (b) $\phi_x(t) = \frac{e^{i(tb-ta)}}{i(tb-ta)}$

(c) Both (a) & (b) (d) None

Q.37 Mean deviation of uniform distribution U[a, b] about mean is equal to

(a) $\frac{(b-a)}{4}$ (b) $\frac{(b-a)^2}{4}$

(c) $\frac{(b^2-a^2)}{4}$ (d) $\frac{(b+a)(b-a)}{4}$

10.3.3 Gamma Distribution

Q.38 The p .d .f. of gamma distribution with parameter λ > 0, is given as

(a) $f(x)=\frac{e^{-x}x^{\lambda-1}}{\lceil\lambda}$ ...., $\lambda>0,\ 0<x<\infty$

(b) $f(x)=\frac{e^{-\lambda}x^{\lambda-1}}{\lceil\lambda}$ ...., $\lambda>0,\ 0<x<\infty$

(c) $f(x)=\frac{e^{-\lambda}\lambda^{x-1}}{\lceil\lambda}$ ...., $\lambda>0,\ 0<x<\infty$

(d) $f(x)=\frac{e^{-\lambda}\lambda^{x+1}}{\lceil\lambda}$ ...., $\lambda>0,\ 0<x<\infty$

Q.39 The p.d.f. of gamma distribution with parameters (a, λ) is written as

(a) $f(x)=\frac{a^{\lambda}e^{-ax}x^{\lambda+1}}{\lceil\lambda}, a>0,\lambda>0, 0<x<\infty$

(b) $f(x)=\frac{a^{\lambda}e^{-ax}x^{\lambda-1}}{\lceil\lambda}, a>0,\lambda>0, 0<x<\infty$

(c) $f(x)=\frac{a^{\lambda}e^{ax}x^{\lambda-1}}{\lceil\lambda}, a>0,\lambda>0, 0<x<\infty$

(d) None

Q.40 The M.G.F. of gamma distribution about origin with parameter λ, is given as

(a) $Mx(t) = (1 - t)^{-\lambda}, |t| < 1$ (b) $Mx(t) = (1 - t)^{\lambda}, |t| < 1$

(c) $Mx(t) = (1 + t)^{-\lambda}, |t| < 1$ (d) $Mx(t) = (1 + t)^{-\lambda}, |t| < t$

Q.41 Kumulant generating function of gamma distribution γ (λ) is given as

(a) $Kx(t) = \lambda\ (1 - t)$ (b) $Kx(t) = -\lambda\ (1 - t)$

(c) $Kx(t) = -\lambda\ (1 + t)$ (d) $Kx(t) = \lambda\ (1 + t)$

Q.42 M ean and variance of γ (λ) distribution are given as

(a) $\bar{x} = \sigma^2 = 2\lambda$ (b) $\bar{x} = \sigma^2 = \lambda^2$

(c) $\bar{x} = \sigma^2 = \lambda$ (d) $\bar{x} = \sigma^2 = 1 + \lambda$

Q.43 The standard gamma variate z is defined as

(a) $Z = \frac{X-\lambda}{\lambda}$ (b) $Z = \frac{X-\lambda}{\sqrt{\lambda}}$

(c) $Z = \frac{X-\lambda}{\sqrt{\lambda^2}}$ (d) $Z = \frac{X-\lambda}{\lambda\sqrt{2}}$

Q.44 The standard gamma variate tends to standard normal variate if

(a) $\lambda \to 0$ (b) $\lambda \to 1$

(c) $\lambda \to \frac{1}{2}$ (d) $\lambda \to \infty$

Q.45 For two parameters (a, λ) gamma distribution, the m.g.f. is given as

(a) $M_x(t) = \left(1 - \frac{t}{a}\right)^{-\lambda}$ (b) $M_x(t) = \left(1 - \frac{t}{a}\right)^{\lambda}$

(c) $M_x(t) = \left(1 + \frac{t}{a}\right)^{-\lambda}$ (d) $M_x(t) = \left(1 + \frac{t}{a}\right)^{\lambda}$

Q.46 For two parameters (a, λ) gamma distribution, the kumulant generating function is given as

(a) $K_x(t) = -\lambda \log_e \left(1 - \frac{t}{a}\right)$ (b) $K_x(t) = -\lambda \log_e \left(1 + \frac{t}{a}\right)$

(c) $K_x(t) = \lambda \log_e \left(1 - \frac{t}{a}\right)$ (d) $M_x(t) = \lambda \log_e \left(1 + \frac{t}{a}\right)$

Q.47 Let (Xi, i = 1, 2, ..., k) be k independent gamma variates with parameters (Xi, i = 1, 2, ..., k), then the linear function

$U = \sum_{i=1}^{k} Xi$ follow the

(a) Gamma distribution with parameter $\sum_{i=1}^{k} \lambda i$

(b) Gamma distribution with parameter $\sum_{i=1}^{k} \frac{1}{\lambda i}$

(c) Gamma distribution with parameter $\sum_{i}^{k} 2\lambda i$

(d) None

Q.48 In general it x ~ λ (a, λi, i = 1, 2, ..., k) are independent r.v's then, we have

(a) $\sum_{i}^{k} Xi \sim \gamma\left(a, \sum_{i=1}^{k} \lambda i\right)$

(b) $\sum_{i=1}^{k} Xi \sim \gamma\left(a, \sum_{i}^{k} \lambda_i^2\right)$

(c) $\sum_{i=1}^{k} Xi \sim \gamma\left(a, 2\sum_{i=1}^{k} Xi\right)$

(d) None

10.3.4 Beta Distribution of First kind

Q.49 Beta distribution of first kind with parameters (μ, v), μ > 0, v >0, has its p.d.f. written as

(a) $f(x) = \frac{1}{\beta(\mu, v)} . x^{\mu-1} (1-x)^{v-1}$

(b) $f(x) = \frac{1}{\beta(\mu, v)} . x^{(1+\mu)} (1-x)^{(1+v)}$

(c) $f(x) = \frac{1}{\beta(\mu, v)} . x^{(1-\mu)} (1-x)^{(1-v)}$

(d) None

Q.50 In case of β_1 (μ, v) distribution, for μ = v = 1, it reduces to
(a) Uniform distribution on [0, 1]
(b) Rectangular distribution on [0, 1]
(c) Both (a) & (b)
(d) None

Q.51 The mean of β_1 (μ, v) is equal to

(a) $\bar{x} = \frac{\mu}{\mu + v}$

(b) $\bar{x} = \frac{v}{\mu + v}$

(c) $\bar{x} = \left(\frac{\mu}{\mu + v}\right)^{-1}$

(d) $\bar{x} = \frac{\mu}{\mu + v}$

10.3.5 Beta Distribution of Second Kind

Q.52 The beta distribution of second kind β_2 (μ, v), $\mu > 0$, v > 0, has its p.d.f. written as

(a) $f(x)=\frac{1}{\beta(\mu,v)}\frac{x^{\mu-1}}{(1+x)^{\mu+v}}$, $0 < x < \infty$

(b) $f(x)=\frac{1}{\beta(\mu,v)}\frac{x^{1+\mu}}{(1+x)^{\mu+v}}$,

(c) $f(x)=\frac{1}{\beta(\mu,v)}\frac{x^{1+\mu}}{(1+x)^{\mu-v}}$, (d) None

Q.53 If X ~ β_2 (μ, v) then Y = $\frac{1}{(1+x)}$, follows

(a) β_1 (μ, v) distribution (b) β_2 (μ, v) distribution
(c) γ (μ, v) distribution (d) None

Q.54 The mean $\overline{X}$ of β_2 (μ, v) is equal to

(a) $\overline{X} = \frac{\mu}{v+1}$ (b) $\overline{X} = \frac{\mu}{v-1}$

(c) $\overline{X} = \frac{\mu+1}{v+1}$ (d) $\overline{X} = \frac{\mu+1}{v}$

10.3.6 Exponential Distribution

Q.55 An exponential distribution with parameter $\theta > 0$, has its p.d.f. given as

(a) $f(x, \theta) = \theta. e^{-\theta x}, x \geq 0$ (b) $f(x, \theta) = \frac{1}{\theta} e^{-\theta x}, x \geq 0$

(c) $f(x, \theta) = \frac{1}{\theta.e^{-\theta x}}$ (d) None

Q.56 The mean of exponential distribution e (x, θ), has its mean $\overline{X}$ equal to

(a) $\overline{X} = \theta$ (b) $\overline{X} = \theta^{-1}$

(c) $\overline{X} = \frac{1}{\theta}$ (d) Both (b) & (c)

Q.57 The variancre of exponential distribution e (X, 9) is equal to

(a) $\sigma^2 = \frac{1}{\theta^2}$ (b) $\sigma^2 = \theta^{-2}$

(c) $\sigma^2 = \theta^2$ (d) Both (a) & (b)

Q.58 Let (Xi, i = 1, 2, ..., k) be k independent exponential variates with parameters (θi, i = 1, 2, ..., k) then Z = Min (X_1, X_2, ..., X_k) has an

(a) Exponential distribution with parameter $\left(\sum_{i}^{k} \theta i\right)$

(b) Exponential distribution with parameter $\left(\sum_{i}^{k} \frac{1}{\theta i}\right)$

(c) Exponential distribution with parameter $\left(\sum_{i}^{k} \theta_i^2\right)$

(d) None of the above

10.3.7 Cauchy Distribution

Q.59 The standard cauchy distribution has its p.d.f. given as

(a) $\frac{1}{\pi(1+x^2)}, -\infty < x < \infty$ (b) $\frac{1}{\pi(1+x)}, -\infty < x < \infty$

(c) $\frac{1}{2\pi(1+x^2)}, -\infty < x < \infty$ (d) $\frac{1}{2\pi(1+x)}, -\infty < x < \infty$

Q.60 The characteristic function of standard cauchy distribution is written as

(a) $\phi_x(t) = e^{|t|}$ (b) $\phi_x(t) = e^{-|t|}$

(c) $\phi_x(t) = e^{-t^2}$ (d) $\phi_x(t) = e^{\frac{1}{t}}$

Q.61 Let (Xi, i =1, 2) be independent cauchy variates with parameters (λi, μi, i = 1, 2) respectively then U = (X_1 + X_2) has

(a) Cauchy distribution with parameters $\left(\sum_{i}^{2} \lambda i, \sum_{i}^{2} \mu_i^2\right)$

(b) Cauchy distribution with parameters $\left(\sum_{i}^{2} \lambda_i^2, \sum_{i}^{2} \mu_i^2\right)$

(c) Beta distribution with parameters $\left(\sum_{i}^{2} \lambda i, \sum_{i=1}^{2} \mu_i\right)$

(d) None of the above

Key Answers

Section-A : True False

1. True	2. True	3. True
4. True	5. True	6. True
7. True	8. True	9. True
10. True	11. True	12. True
13. True	14. True	15. True
16. True	17. True	18. True
19. True	20. True	21. True
22. True	23. True	24. True
25. True	26. True	27. True
28. True	29. True	30. True
31. True	32. True	33. True
34. True	35. True	36. True
37. True	38. True	39. True
40. True	41. True	42. True
43. True	44. True	45. True
46. True	47. True	48. True
49. True	50. True	51. True
52. True	53. True	54. True
55. True	56. True	57. True
58. True	59. True	60. True
61. True	62. True	63. True

Section-B : Fill in the Blanks

1. $\left(\frac{x-\mu}{\sigma}\right)$	2. (μ, σ^2)
3. 0, Small	4. μ
5. μ, σ^2	6. Decreases
7. μ	8. $\frac{1}{\sqrt{2\pi\sigma}}$
9. 0, 3	10. Zero
11. (2r - 1)	12. Normal variate
13. Asymptotic	14. $x = \mu \pm \sigma, e^{-\frac{1}{2}}$
15. $\frac{4}{5}\sigma$	16. $\frac{2}{3}\sigma$

17. 10 : 12 : 15

18. 0.6826

19. 0.9544

20. 0.9944

21. (0, 2), (0, 2)

22. Independent

23. $e^{\mu t + \frac{t^2\sigma^2}{2}}$

24. $e \frac{t^2}{2}$

25. $\mu t + \frac{1}{2} t^2\sigma^2$

26. $\left(\sum_{i}^{k} ai\mu i, \sum_{i=1}^{k} a_i^2\sigma_i^2\right)$

27. Normal distribution

28. Quality control

29. $\frac{1}{b—a}$

30. Distribution, [a, b]

31. a, b

32. Rectangular, a, b

33. $\frac{a+b}{2}, \frac{b^2 - a^2}{12}$

34. $\frac{e^{bt—at}}{t(b—a)}$

35. eit (b - a)

36. $\frac{b—a}{4}$

37. $\frac{e^{-x}x^{\lambda-1}}{\lceil\lambda}$

38. $\frac{a^{\lambda}e^{-ax}x^{\lambda-1}}{\lceil\lambda}$

39. $(t — t)^{\lambda}$

40. log (1 - t)

41. λ

42. $\frac{x—\lambda}{\sqrt{\lambda}}$

43. ∞

44. $\left(1-\frac{t}{a}\right)^{-\lambda}$, $t < a$

45. $\left(1-\frac{t}{a}\right)$

46. $\left(\sum_{i=1}^{k} \lambda i\right)$

47. $\left(a, \sum_{i=1}^{k} \lambda i\right)$

48. $x^{\mu-1}(1 - x)^{v-1}$

49. (μ, v)

50. Uniform distribution

51 $\frac{\mu}{\mu+v}$

52. $\frac{x^{\mu-1}}{(1+x)^{\mu+v}}$

53. (μ, v)

54. $\beta_1 (\mu, v)$

55. $\frac{\mu}{v—1}$

56. $\theta.e^{-\theta x}$

57. $\frac{1}{\theta}, \frac{1}{\theta^2}$

58. Exponential distribution

59. $\frac{1}{\pi(1+x^2)}$

60. Standard cauchy

61. e-| t |

62. $\left(\sum_{i=1}^{2}\lambda i, \sum_{i}^{2}\mu i\right)$

63. Cauchy

Section-C : Multiple Choice Questions

1. (d)	2. (b)	3. (b)
4. (b)	5. (a)	6. (d)
7. (d)	8. (a)	9. (b)
10. (c)	11. (c)	12. (a)
13. (c)	14. (a)	15. (b)
16. (a)	17. (a)	18. (a)
19. (b)	20. (a)	21. (a)
22. (b)	23. (a)	24. (a)
25. (b)	26. (a)	27. (d)
28. (c)	29. (b)	30. (c)
31. (b)	32. (a)	33. (d)
34. (c)	35. (d)	36. (c)
37. (a)	38. (a)	39. (b)
40. (a)	41. (b)	42. (c)
43. (b)	44. (d)	45. (a)
46. (a)	47. (a)	48. (a)
49. (a)	50. (c)	51. (a)
52. (a)	53. (a)	54. (b)
55. (a)	56. (d)	57. (d)
58. (a)	59. (a)	60. (b)
61. (a)		

❑❑❑

Chapter - 11

Theory of Testing of Hypothesis (Preliminaries)

11.1 Section-A : Write True / False

Q.1 Some assumptions or guesses made about the unknown population parameters, is know as "Statistical hypothesis".

Q.2 For the normal distribution $N(\mu, \sigma^2)$, the statement (1) $\mu = 25.0$ and (ii) $\sigma^2 = 37.0$, are known as statistical hypotheses.

Q.3 Let $p = 0.5$ is the probability of getting a head in single throw of a coin. Then, this statement is not a statistical hypothesis.

Q.4 The statistical hypothesis under consideration (under test), is known as "Null hypothesis".

Q.5 The null hypothesis is denoted by "Ho".

Q.6 Let a statistical hypothesis is formulated as :
Ho : There is no difference between the two procedures adopted under study. Then, it is an example of null hypothesis.

Q.7 In Q3, $p = 0.5$ is an example of null hypothesis.

Q.8 Any hypothesis which differs from a given null hypothesis, is called "An alternative hypothesis or complementary hypothesis".

Q.9 Let H0 : p = 0.5, then H1 : p ≠ 0.5 is an example of alternative hypothesis.

Q.10 An alternative hypothesis is denoted by H1.

Q.11 If the statistical hypothesis completely specifies the distribution or about an unknown parameter, then it is known as "Simple hypothesis".

Q.12 A statistical hypothesis which does not specifies the distribution completely or about an unknown parameter of a distribution, is known as "Composite hypothesis".

Q.13 Let us consider a normal distribution $N(\mu, \sigma^2)$

Then $H0 : \mu = 30.0$ and $H0 : \sigma^2 = 37.0$

are the examples of simple hypotheses.

Q.14 For binomial distribution (X, n, p), let us formulate the null hypothes as ;

Ho1 : p ≠ 0.5

Ho2 : p > 0.5

Ho3 : p < 0.5

All these are composite hypotheses.

Q.15 If an event is very rare but it happens then such event is known as "Significant event".

Q.16. If an event is very common and it happens, then such event is known as "Non-significant" event.

Q.17 If it rains continuously and hevily for 3 or 4 days and skey remains cloudy in the month of April at Jabalpur, then it is a "significant event".

Q.18 An event described in Q(17) but for the month of July and August is known as "Non-Significant" event.

Q.19 The procedures which enable us to decide whether to accept or reject the hypothesis, are called "Test of hypotheses or Test of significance or rules of decision".

Q.20 If any body wants to test the hypothesis, then the number of types of decisions taken is equal to 4.

Q.21 If the null hypothesis Ho is correct but it is rejected then the error committed is known as "Type one error" or "First kind of error".

Q.22 Type one error is denoted by "α", the level of significance.

Q.23 First kind of error is also known as "Producer's risk".

Q.24 If the null hypothesis Ho is false but it is accepted then the error committed is known as "Type Two error" or "Second kind of error".

Q.25 The type two error is denoted by "β".

Q.26 Both type of errors committed are measured in term of "probability".

Q.27 The "type two error, is considered more serious than "Type one error".

Q.28 The type "two error" is always tried to be avoided.

Q.29 The type "two error" is also known as "Consumer's risk".

Q.30 Both type of errors can not be decreased simultanously.

Q.31 Generally, type one error is kept constant (fixed) and type second error is tried to be minimised.

Q.32 The probability of taking correct decision is known as the power of the test.

Q.33 The power of a test is measured by a quantity (1 - β).

Q.34 The level of significance is the probability level at which the statistical hypothesis (Ho) is declared either "significant nonsignificant".

Q.35 The level of significance is denoted by "α".

Q.36 In practice, the level of significance is generally taken as 1% or 5%.

Q.37 Type one error is also known as "α" error.

Q.38 Type two error is also known as "β" error.

Q.39 Let the sample space Rn is divided into two dis joint sets C and A such that (C + A) = Rn. If the sample (x_1, x_2, ..., x_n) falls in the region C and rule is to reject Ho hypothesis, then the region "C" is known as "Critical region or Rejection region".

Q.40 If the sample values belongs to the region A = (Rn - C) and rule is to accept the Ho hypothesis then "A" is known as "Acceptance region".

Q.41 In test of hypothesis problem, the terms "Acceptance and Rejection" must not be taken in their literal senses.

Q.42 The acceptance of H0 does not mean that Ho has been aproved to be true.

Q.43 The acceptance of Ho means only that so far as the given sample values are considered, Ho is a plausible hypothesis.

Q.44 The rejection of Ho hypothesis, does not mean that Ho does not look plausible.

Q.45 For testing each Ho, there corresponds a unique critical region (C.R.) and conversely for each C.R., there corresponds a unique test.

Q.46 The value of test statistic which seperates the critical (or rejection) region and acceptance region, is called the critical (or significant value)

Q.47 The critical value of a test statistic depends on "Level of significance" used.

Q.48 The critical value of a test statistic depends upon "Alternative hypothesis" whether it is single tailed or two tailed test :

Q.49 To test a hypothesis Ho if alternative hypothesis H1 is either right tailed or left tailed, is known as "one tailed test".

Q.50 Consider the hypotheses - Ho : $\mu = \mu_0$ V.S. H1 : $\mu > \mu_0$
Then the hypothesis H1 is known as single tailed (right tailed) test.

Q.51 To test the hypothesis H0 : $\mu = \mu_0$ V.S. H1 $\mu < \mu_0$
Then hypothesis H1 is known as one tailed (left tailed) test.

Q.52 In right tailed test, the C.R. lies entirely in the right tail of sampling distribution of test statistic.

Q.53 For the left tailed test, the C.R. lies entirely in the left side of sampling distribution, of the test statistic.

Q.54 Any statistics which is used to test the hypothesis Ho, is known as "Test statistic".

Q.55 In case of normal distribution $N(\mu, \sigma^2)$, if we have to test $H_0 : \mu = \mu_0$ V.S. $H1 : \mu \neq \mu_0$ and α is the level of significance then the area of C.R. either in left side or in right side will be equal to "$\alpha/ 2$".

Q.56 Let $Z\alpha$ be the critical value of Z statistic then area to the right of $Z\alpha$ is equal to "$\alpha/ 2$" and to the left of $- Z\alpha$ is also equal to "$\alpha/ 2$".

Q.57 For right tailed test : $P(Z > Z\alpha) = \alpha/2$.

Q.58 The critical value, Za for a single tailed test at the level of significance α is the same as the critical value of z for two tailed test but at the level "2α".

Q.59 The critical values of z statistic are given in table generally for "two tailed test" at the given level of significance "2α%".

Q.60 It is notable that if one has to apply one tailed test at the level of significance "α" then he must see (consider) the critical value Za at the level "2α" in case of z test.

Q.61 The statistical tests which deal with the parameters of the distribution are known as the "Non parametric tests".

Q.62 The statistical tests which do not make any assumption about the particular form of distribution function of population from which the sample is drawn, are known as "Parametric tests".

Q.63 Signed rank test is known as "Parametric test".

11.2 Section - B : Fill in the Blanks

Q.1 Some assumptions or guesses made about the unknown parameter of a population is known as ....

Q.2 For the normal distribution $N(\mu, \sigma^2)$, the statements $H_0 : \mu = 25.0$ and $H_0 : \sigma^2 = 37.0$ are known as ....

Q.3 The statistical hypothesis under consideration (under test), is known as ....

Q.4 The null hypothesis is denoted by " .... ".

Q.5 The hypothesis "There is no difference between two procedures adopted under comperative study" is known as .... hypothesis.

Q.6 The hypothesis "p = 0.5 is an example of ....

Q.7 Any hypothesis which differs from the hypothesis Ho, is known as ....

Q.8 In the hypotheses H0 : p = 0.5, and H1 : p = 0.5, the hypothesis H1 is an example of ....

Q.9 An alternative hypothesis is denoted by ....

Q.10 A hypothesis completely specifying about a distribution or any unknown parameter, is known as ....

Q.11 Any statistical hypothesis not specifying a distribution completely, is known as .... hypothesis.

Q.12 For normal distribution the hypotheses -
$H_0 : \mu = 30.0$ and $H_0 : \sigma^2 = 37.0$
are known as ....

Q.13 For binomial distribution b(x, n, p), the hypotheses,
$H_0 : p > 0.5$ and $H_0 : p < 0.5$ are known as ....

Q.14 If an event is very rare but it happens then such event is known as ....

Q.15 If an event is very common and it happens, then that event is known as ....

Q.16 If the weakest student of a class gets first rank in an examination of 100 examinees, then this event will be treated as ....

Q.17 If the weakest student of a class of 100 students, gets the minimum marks in an examination in the class then this result will be treated as ....

Q.18 A procedure which enables us to decide either to accept or reject the Ho hypothesis, is known as ....

Q.19 In testing a null hypothesis Ho, the number of types of decision taken is equal to ....

Q.20 If the hypothesis Ho is true but it is rejected then error committed is known as type ....

Q.21 The type one error is denoted by ....

Q.22 First kind of error is also known as .... risk.

Q.23 If the null hypothesis Ho is false but it is accepted then error committed is known as type .... error.

Q.24 The type two error is denoted by ....

Q.25 Both types of errors committed in testing of Ho, are measured in term of ....

Q.26 Type two error is considered .... than type one eror.

Q.27 The error which is always tried to be avoided, is the ....

Q.28 Second kind of error is also known as .... risk.

Q.29 Generally type .... error is kept fixed and type .... error is tried to be minimised.

Q.30 The probability of taking correct decision, is known as the .... of the test.

Q.31 The power of a test is measured by a quantity equal to ....

Q.32 The level of significance is the probability level at which H_o is declared either " .... " or " .... ".

Q.33 The level of significance in hypothesis testing is denoted by ....

Q.34 "α" error is also known as type .... error.

Q.35 "β" error is also known as type .... error.

Q.36 Critical region, acceptance region and sample space (C, A, Rn) respectively are related as Rn = ...

Q.37 Critical region "C" and acceptance region "A" are such that they make " .... " sets.

Q.38 If the sample points fall in the critical region, then the hypothesis H_0 is declared as " .... ".

Q.39 If the sample points fall in the acceptance region, then Ho is declared as " .... ".

Q.40 If the sample points fall in the region "M" and the hypothesis H0 is not acceptable then the region "M" is known as " .... ".

Q.41 If the sample points fall in the complementary region of critical region, then H0 is declared " .... ".

Q.42 The sample space Rn, critical region C and acceptance region A are defined such that

A = ...

Q.43 Acceptance of Ho does not mean that Ho has been proved to be ....

Q.44 The acceptance of H_0 means only that so for as the sample values are considered, H_0 is a .... hypothesis.

Q.45 Rejection of H_0 does not mean that Ho is not ....

Q.46 For testing each null hypothesis, there corresponds a unique

....

Q.47 For each critical region, there corresponds a ....

Q.48 The value of test statistic which separates the critical and acceptance regions, is known as " .... ".

Q.49 The level of significance affects the .... of a test statistic.

Q.50 The alternative hypothesis affects the critical value of a " .... ".

Q.51 If the alternative hypothesis H_1 is either right tailed or left tailed, is known as " .... " hypothesis.

Q.52 If the rejection region lies in both side of the critical value of a test statistic, then the test is known as " .... " tailed test.

Q.53 If the hypothesis is like H1 : $\mu > \mu_0$ then the test is known as " .... " test.

Q.54 If the hypothesis is $H_1 : \mu < \mu_0$, then the test is known as " .... " test.

Q.55 In right tailed test, the C.R. lies entirely in the .... of critical value of test statistic.

Q.56 In left tailed test, the C.R. lies entirely in the .... of sampling distribution of test statistic.

Q.57 Any statistic which is used to test the Ho, is known as " .... ".

Q.58 For $N(\mu, \sigma^2)$, if one has to test Ho : $\mu = \mu_0$, V.S. $H_1 : \mu = \mu_0$, and α is the level of significance the C.R. will be equal to " .... ".

Q.59 For right tailed test : $P(Z > Za) =$ ....

Q.60 The critical value $Z\alpha$ for a single tailed test and for two tailed test will be ....

Q.61 If for a single tailed test, the level of significance is "α" then for two tailed test, the level of significance considered to see the critical value will be equal to " .... ".

Q.62 The critical value of z statistics given in table are generally for two tailed test and are at the level of significance equal to " .... ".

Q.63 It is noticeable that if one has to apply one tailed test at the level "α" then he must see the criticla value (Za) at the level " .... " in the table.

Q.64 The statistical tests which deal with the parameters of the distribution, are known as " .... ".

Q.65 The statistical tests which do not make any assumption about the particular form of distribution function from which the sample is taken, are known as " .... ".

Q.66 The sign test belongs to " .... ".

11 .3 Section -C : Multiple Choice Questions

Q.1 An assumption or guess made about the unknown parameter of a distribution is known as -
(a) Statistical hypothesis
(b) Null hypothesis
(c) Alternative hypothesis
(d) Both (a) & (b)

Q.2 For normal distribution $N(\mu, \sigma^2)$, the statement that $\mu = 23.0$ and $\sigma^2 = 27.0$, are known as -
(a) Statistical hypotheses (b) Null hypotheses
(c) Simple hypotheses (d) All the three

Q.3 The statistical hypothesis under test, is known as
(a) Null hypothesis (b) Alternative hypotheses
(c) Composite hypothesis (d) Simple hypothesis

Q.4 The null hypothesis is denoted by
(a) H_1 (b) α
(c) β (d) H_0

Q.5 The hypothes is that -
"There is no difference between two procedures under comparison" for some study is known as
(a) Alternative hypothesis (b) Null hypothesis
(c) Composite hypothesis (d) None

Q.6 Any statistical hypothesis which differs from null hypothesis, is known as -
(a) Alternative hypothesis (b) Null hypothesis
(c) Comlementary hypothesis (d) Both (a) & (c)

Q.7 Let Ho : $p = 0.5$ and H1 : $p \neq 0.5$ then H1 is -
(a) Composite hypothesis (b) Alternative hypothesis
(c) Complementary hypothesis (d) All the three

Q.8 An alternative hypothesis is denoted by -
(a) H_1 (b) α
(c) β (d) H_0

Q. 9 If the statistical hypothesis completely specifies the distribution, then it is known as -

(a) Composite hypothesis (b) Simple hypothesis
(c) Alternative hypothesis (d) None

Q.10 A statistical hypothesis which does not specifies the distribution completely, is known as -

(a) Composite hypothesis (b) Null hypothesis
(c) Simple hypothesis (d) None

Q.11 For the binomial distribution b(x, n, p), consider the following hypotheses

$H_{01} : p \neq 0.5$, $H_{02} : p > 0.5$, $H_{03} : p < 0.5$

Then all these hypotheses are -

(a) Simple hypotheses (b) Composite hypotheses
(c) Alternative hypotheses (d) None

Q.12 If an event is very rare but it happens, then this event is known as -

(a) Non-significant (b) Significant
(c) Composite event (d) None

Q.13 If an event is very common and it happens, then it is known as

(a) Significant event (b) Composite event
(c) Non-significant (d) None

Q.14 A procedure which enables us to decide either to accept or reject the statistical hypothesis, is known as -

(a) Test of hypothesis (b) Test of significance
(c) Rule of decision (d) All the above

Q.15 If the hypothesis H_0 is correct but it is rejected then error committed is known as -

(a) α error (b) Type one error
(c) Producer's risk (d) All the above

Q.16 Type one error is denated by -

(a) β (b) α
(c) H_0 (d) H_1

Q.17 Producer's risk is denoted by
(a) H_0 (b) H_1
(c) α (d) β

Q.18 Type one error is also known as -
(a) α error (b) Producer's risk
(c) Consumer's risk (d) Both (a) & (b)

Q.19 If the hypothesis Ho is false but it is accepted then error committed is known as -
(a) Type two error (b) Second kind of error
(c) β error (d) All the above

Q.20 Type two error is also known as
(a) Consumer's risk (b) Producer;s risk
(c) β error (d) Both (a) & (c)

Q.21 Consumer's risk is denoted by
(a) β (b) α
(c) H_0 (d) H_1

Q.22 Type two error is denoted by
(a) α (b) H_0
(c) β (d) H_1

Q.23 Both type of errors committed are measured in term of
(a) Probability (b) Standard error
(c) Consumers risk (d) Relative error

Q.24 Type two error is considered
(a) More serious than type one error
(b) Less serious than type one error
(c) Equally serious
(d) No comparison

Q.25 Both type of errors can
(a) Not be decreased simultaneously
(b) Be decreased simultaneously
(c) Not be considered to decrease
(d) Be considered equally serious

Q.26 Probability of taking currect decision is known as
(a) Power of the test (b) Precison of the test
(c) Efficiency of the test (d) Merit of the test

Q.27 Power of a test is measured by the quantity
(a) α (b) β
(c) $(1 - \alpha)$ (d) $(1 - \beta)$

Q.28 The leveal of significance 'α' is the probability at which the hypothesis Ho is declared
(a) Significant
(b) Non-Significant
(c) Eigher significant or non-significant
(d) None

Q.29 The level of significance is denoted by
(a) β (b) α
(c) μ (d) λ

Q.30 In practice, the level of significance is generally taken as
(a) 5% (b) 1%
(c) Either 1% or 5% (d) (c)

Q.31 The sample space R, acceptance region A and critical region C, are related as
(a) Rn = (C + A) (b) C = (Rn - A)
(c) A = (Rn - C) (d) All the three

Q.32 The acceptance of Ho does not mean that
(a) H_0 has been proved to be true
(b) H_0 has not been proved to be true
(c) H_1 has been proved to be true
(d) None

Q.33 The acceptance of H_0 means that so for as the given sample values are considered, the hypothesis
(a) H_0 is plausible (b) H_0 is not plausible
(c) H_1 is not plausible (d) Both (a) & (c)

Q.34 The rejection of H_0 does not mean that
(a) H_0 has been disproved
(b) H_0 has not been approved
(c) H_1 has been approved (d) All the three

Q.35 The acceptance of H_1 does not mean that
(a) H1 has been approved (b) Ho has been disproved
(c) Ho has been approved (d) Both (a) & (b)

Q.36 Rejection of H_1 does not mean that
(a) Ho has been approved (b) Ho has been disproved
(c) H1 has been disproved (d) Both (a) & (c)

Q.37 The value of test statistic which separates the critical region and acceptance region, is known as
(a) Critical value (b) Acceptance value
(c) Significant value (d) Both (a) & (c)

Q.38 The critical value of a test statistic depends upon
(a) Level of significance
(b) Probability of type one error
(c) Producer's risk
(d) All the three

Q.39 The critical value of a test statistic depends upon
(a) Level of significance (b) Alternative hypothesis
(c) α error (d) All the three

Q.40 Alternative hypothis may be
(a) Left tailed (b) Right tailed
(c) Either left tailed or right tailed (d) None

Q.41 Let the hypothesis be $H_0 : \mu =$ to VS. H1 : $\mu > \mu o$, then H1 is one known as
(a) One taited test (b) Right tailed test
(c) Composite test (d) All the three

Q.42 In the right tailed test, the C.R. lies in the
(a) Right tail of distribution (b) Left tail of distribution
(c) Eigher right or left tail (d) In both tails

Q.43 The statistic which is used to test the hypothesis Ho, is known as
(a) Sufficient statistic (b) Test statistic
(c) Efficient statistic (d) Biased statistic

Q.44 In case of $N(\mu, \alpha^2)$ distribution, if one has to test Ho : $\mu = \mu_0$ V.S. H1 $\mu \neq \mu_0$ and 'α' is the level of significance. Then the area of C.R. either in left or right side of distribution will be equal to

(a) $\alpha/2$ (b) α
(c) 2α (d) None

Q.45 Let Za be the critical value of Z statistic then the area either to left or right of Za or -Za will be equal to

(a) α (b) 2α
(c) $\alpha/2$ (d) None

Q.46 For right tail test, $P(Z > Za)$ is equal to

(a) α (b) $\alpha/2$
(c) 2α (d) $\geq \alpha$

Q.47 The critical value Za are given in table generally for "two tailed test" at the level of significance equal to

(a) α (b) 2α
(c) $\alpha/2$ (d) 5%

Q.48 It is noticeable that if one has to apply one tailed Z test at the level "α", he must see the critical value Za at the level equal to

(a) $\alpha/2$ (b) 2α
(c) α (d) 5%

Q.49 The statistical tests which deal with the parameter of a distribution is, known as

(a) Student's t test (b) Parametric test
(c) Non parametric test (d) Distribution free test

Q.50 The test which does not make any assumption about the particular form of frequency function of a distribution, is known as

(a) Non parametric test (b) Parametric test
(c) Distribution free test (d) Both (a) & (c)

Key Answers

Section-A : True False

1. True	2. True	3. True
4. True	5. True	6. True

7. True	8. True	9. True
10. True	11. True	12. True
13. True	14. True	15. True
16. True	17. True	18. True
19. True	20. False	21. True
22. True	23. True	24. True
25. True	26. True	27. True
28. True	29. True	30. True
31. True	32. True	33. True
34. True	35. True	36. True
37. True	38. True	39. True
40. True	41. True	42. True
43. True	44. True	45. True
46. True	47. True	48. True
49. True	50. True	51. True
52. True	53. True	54. True
55. True	56. True	57. True
58. True	59. True	60. True
61. False	62. False	63. False

Section-B : Fill in the Blanks

1. Statistical hypothesis
2. Statistical hypothesis
3. Null hypothesis
4. Ho
5. Null
6. Null hypothesis
7. Alternative or complementary hypotheses
8. Alternative ro comlementary hypothesis
9. H1
10. Simple hypothesis
11. Alternative or complementary hypotheses
12. Simple hypotheses
13. Composite hypothesis
14. Significant event
15. Non significant event
16. Significant event
17. Non-significant
18. Test of hypothesis or test of significance
19. Two
20. One error
21. "α"
22. Producer
23. Two
24. "β"
25. Probability
26. More serious

27. Type two error
28. Consumer
29. One, Two
30. Power
31. $(1 - \beta)$
32. Significant,Non-significant
33. "α"
34. One
35. Two
36. (C + A)
37. Disjoint
38. Significant event
39. Non-significant event
40. Acceptance region
41. Non-significant
42. (Rn - C)
43. True or plausible
44. Plausible
45. Plausible
46. Critical region
47. Null hypothesis or Test of hypothesis
48. Critical or Significant value
49. Critical value
50. Test statistic
51. One tailed test
52. Two
53. Right tailed test or Single tailed test
54. Left tailed test Single tailed test
55. Right side
56. Left side
57. Test statistic
58. "α"
59. $\alpha/2$
60. Same
61. 2α
62. 2α
63. 2α
64. Parametric tests
65. Non parametric or Distribution free
66. Non-parametric test

Section-C : Multiple Choice Questions

1. (d)
2. (d)
3. (a)
4. (d)
5. (b)
6. (d)
7. (d)
8. (a)
9. (b)
10. (a)
11. (b)
12. (b)
13. (c)
14. (d)
15. (d)
16. (b)
17. (c)
18. (d)
19. (d)
20. (d)
21. (a)
22. (c)
23. (a)
24. (a)
25. (a)
26. (a)
27. (d)
28. (c)
29. (b)
30. (d)
31. (d)
32. (a)
33. (d)
34. (d)
35. (d)
36. (d)

37. (d)	38. (d)	39. (d)
40. (c)	41. (d)	42. (a)
43. (b)	44. (a)	45. (c)
46. (a)	47. (b)	48. (b)
49. (b)	50. (d)	

❐❐❐

Chapter - 12

Normal Distribution and Tests Based on It

12.1 Section-A : Write True / False

Q.1 Almost all biological data e.g. crop yield, plant height, seed weight, seed size, protein content in pulse, oil content in oil seed etc. are assumed to be normaly distributed.

Q.2 Let X be a normal variable with $N(\mu, \sigma^2)$. The random variable Z defined as $Z = (x - \mu)/\sigma$, is known as standard normal variable.

Q.3 The standard normal variable Z is denoted as $Z \sim N(0, 1)$.

Q.4 The p.d.f. of random variable Z is given as

$$f(z) = \frac{1}{\sqrt{2\pi}} e^{-\frac{1}{2}z^2} \qquad -\infty < z < \infty$$

Q.5 The proof of all the sampling distributions are based on the fundamental assumption that the population from which the samples are drawn, is a "Normal population".

Q.6 From normal probability table we have
$P(-3 \le Z \le 3) = 0.9973$

Q.7 From normal probability table, we have $P(|Z| \ge 3) = 0.0027$

Q.8 If $|Z| > 3$, then Ho will be always declared significant.

Q.9 We have the result that $P(|Z| < 1.64) = 0.90$

Q.10 $P(|Z| < 1.96) = 0.95$

Q.11 $P(|Z| < 2.58) = 0.99$

Q.12 The significant value (critical value) of Z at 5% level of significance for two tailed test is equal to 1.96.

Q.13 It is assumed that the sample mean x_n is asymptotically normally distributed even if the parent population is not normal.

Q.14 Let $X \sim N(\mu, \sigma^2)$. Then standard normal variate corresponding to sample mean x_n is given as

$$Z = \frac{|\bar{x}_n - \mu|}{\sigma / \sqrt{n}}$$

Q.15 For a single mean Z test of a $N(\mu, \sigma^2)$, the null hypothesis is written as
H0 : There is no significant difference between sample mean x_n & population mean μ.

Q.16. For a single mean Z test, the population variance σ^2 is assumed to be known.

Q.17 If σ^2 is unknown then it is estimated as

$$\hat{\sigma}^2 = \frac{1}{n-1}\sum_{i}^{n}\left(x_i - \bar{x}_n\right)^2$$

Q.18 The sampling variance of difference of two independent sample means $(x_{n1} - x_{n2})$ is given as

$$V\left(\bar{x}_{n_1} - \bar{x}_{n_2}\right) = \left(\frac{\sigma_1^2}{n_1} + \frac{\sigma_2^2}{n_2}\right)$$

Q.19 The null hypothesis for two means Z test is written as
H0 : The mean of both populations are same.

Q.20 To test $H_0 : \mu_1 = \mu_2$, the test statistic Z is defined as

$$Z = \frac{|\bar{x}_1 - \bar{x}_2|}{\sqrt{\frac{\sigma_1^2}{n_1} + \frac{\sigma_2^2}{n_2}}} \sim N(0,1)$$

Q.21 The difference $(x_1 - x_2)$ of two independent sample means also follows normal distribution.

Q.22 The mathematical expectation of $(x_1 - x_2)$ is given as

$E(x_1 - x_2) = (\mu_1 - \mu_2)$ for $Xi \sim N(\mu_i, \sigma_i^2, i = 1,2)$

Q.23 In Q (20), if $\sigma_i^2 = \sigma_2^2 = \sigma^2$, then Z is defined as

$$Z = \frac{|\bar{x}_1 - \bar{x}_2|}{\sigma\sqrt{\frac{1}{n_1} + \frac{1}{n_2}}} \sim N(0,1)$$

Q.24 For sufficiently large sample sizes (n_1, n_2), σ^2 is estimated as

$$\hat{\sigma}^2 = \frac{(n_1 - 1)s_1^2 + (n_2 - 1)s_2^2}{(n_1 + n_2 - 2)}$$

Q.25 For sufficiently large (n_1, n_2), σ^2 can be estimated in practice without any serious error as

$$\hat{\sigma}^2 = (n_1 s_1^2 + n_2 s_2^2)/(n_1 + n_2)$$

Q.26 If σ_1^2 and σ_2^2 are unknown then for sufficiently large sample sizes

sizes ((rip n2) the statistic Z is defined as

$$Z = \frac{|\bar{x}_1 - \bar{x}_2|}{\sigma\sqrt{\frac{s_1^2}{n_1} + \frac{s_2^2}{n_2}}} \sim N(0,1)$$

Q.27 Let (P, p) be the population proportion and sample proportion respectively for a certain characteristic say A, then Z statistic is defined as

$$Z = \frac{|x - np|}{\sqrt{nPQ}} \sim N(0,1)$$

Q.28 The Z statistic defined in Q(27) is applicable to test the average number of x success in n independent binomial trials.

Q.29 If one has to test the hypothesis about population proportion P i.e. Ho : Sample proportion of success say p = Population proportion P then z statistic is defined as

$$Z = \frac{|p - P|}{\sqrt{\frac{PQ}{n}}} \sim N(0, 1)$$

Q.30 If we have the sampling from finite population of size N then

$$S.E.\ (p) = \sqrt{\frac{N-n}{N-1}\frac{PQ}{n}}$$

Q.31 Probable limits for proportion of success P is given as

$$\text{Limit } (p) = P \pm 3\sqrt{\frac{PQ}{n}}$$

Q.32 If population proportion P is unknown then estimated probable limits for P are given as estimate

$$\text{Limit } (P) = p \pm 3\sqrt{\frac{pq}{n}}$$

Q.33 To test the significant difference between two population propositions say (P_1, P_2), the z statistic is defined as

$$Z = \frac{|p_1 - p_2|}{\sqrt{\frac{P_1Q_1}{n_1} + \frac{P_2Q_2}{n_2}}}$$

Q.34 If under Ho : $P_1 = P_2 = P$ then Z statistic is defined as

$$Z = \frac{|p_1 - p_2|}{\sqrt{PQ\left(\frac{1}{n_1} + \frac{1}{n_2}\right)}}$$

Q.35 If in Q(34), P is unknown then its unbiased estimate based on combined sample is obtained as

$$\hat{P} = p = \frac{(x_1 + x_2)}{(n_1 + n_2)}$$

Q.36 For situation described in Q(35), the Z statistic is defined as

$$Z=\frac{|p_1-p_2|}{\sqrt{pq\left(\frac{1}{n_1}+\frac{1}{n_2}\right)}}$$

Q.37 For the two population standard deviations say (σ_1, σ_2), to test the hypothesis - $H_0 : \sigma_1 = \sigma_2$ the z statistic is defined as

$$Z=\frac{|s_1-s_2|}{\sqrt{\frac{\sigma_1^2}{2n_1}+\frac{\sigma_2^2}{2n_2}}}$$

Q.38 In Q(37), if (σ_1^2, σ_2^2) are unknown, then Z statistic is defined as

$$Z=\frac{|s_1-s_2|}{\sqrt{\frac{s_1^2}{2n_1}+\frac{s_2^2}{2n_2}}} \sim N(0, 1)$$

12.2 Section-B : Fill in the Blanks

Q.1 Almost all kind of biological variables eg plant height, seed weight, seed size, protein content in pulse, etc. are assumed to be .... distributed.

Q.2 Biological variables like plant height, seed weight, seed size etc., are assumed to be .... variables.

Q.3 Let $X \sim N(\mu, \sigma^2)$then standard normal variate Z is defined as

$$Z=\frac{(\text{.............})}{\sigma}$$

Q.4 The standard normal variate Z has mean .... and variance ....

Q.5 The p.d.f. of standard normal variate Z is defined as

$$f(z) = \frac{1}{\sqrt{\dots}} e^{-\frac{1}{2}(\dots)^2}, \quad -\infty < z < \infty$$

Q.6 The proof of all the sampling distributions are based on the fundamental assumption that the parent population from which the samples are drawn, is a " .... " population.

Q.7 The proof of all .... distributions are based on fundamental assumption that parent population follows .... distribution.

Q.8 From normal probability table, we have
$P(-3 \le Z \le 3)$ = ...

Q.9 From normal probability table, we have
$P(|Z| \ge 3)$ = ...

Q.10 For $|Z| > 3.0$, Ho will always be declared ....

Q.11 We have the result that
$P(|Z| \le 1.64$ = ... at α = ... parent

Q.12 $P(|Z| \le 1.96$ = ... at α = 5% level of significance

Q.13 $P(|Z| <$ ....) = 0.99 at α = 1% level of significance.

Q.14 The critical value of z at 5% level of significance for two tailed test, is equal to ....

Q.15 The sample mean x_n is asymptotically .... distributed even if the parent population does not follow .... distribution.

Q.16 Let $X \sim N(\mu, \sigma^2)$, then Z statistic is defined as
Z = ...

Q.17 In Q(16), the samplwe mean $x_n \sim$ N(..............., ...).

Q.18 Let $X \sim N(\mu, \sigma^2)$, then for single mean test, we write Ho as
H0 : There is no significant difference between ....

Q.19 For single mean Z test, we assume that the σ^2 is ....

Q.20 Let $X \sim N(\mu, \sigma^2)$ and μ is unknown then

$$\hat{\sigma}^2 = \text{........................}$$

Q.21 For two independent sample means $(\bar{x}n_1 - \bar{x}n_2)$ from N $(\mu i, \sigma_i^2)$, i= 1, 2, we have

$v(\bar{x}n_1 - \bar{x}n_2)$ = (.......)

Q.22 For two mean z test, we write H_0 as
H_0 : Means of both populations are ....

Q.23 To test Ho : $\mu_1 = \mu_2$, for two normal populations
N μi, σ_i^2 = i = 1, 2), Z is defined as

$$Z = \frac{|\bar{x}_1 - \bar{x}_2|}{\sqrt{\ldots\ldots\ldots\ldots}}$$

Q.24 The difference $(x_1 - x_2)$ fo two independent sample means also follows ....

Q.25 The sum (x_1+x_2) of two independent sample means also follows ....

Q.26 In Q(23), we have E $(x_1 - x_2)$ = (.................)

Q.27 In Q(23), $\sigma_1^2 = \sigma_2^2 = \sigma^2$ then

$$Z = \frac{|\bar{x}_1 - \bar{x}_2|}{\sqrt{\ldots\ldots\ldots\ldots}} \sim N(0,1)$$

Q.28 In Q(27), for sufficiently large sample sizes (n_1, n_2), we have

$$\hat{\sigma}^2 = \frac{(n_1 - 1)s_1^2 + (n_2 - 1)s_2^2}{(\ldots\ldots\ldots\ldots)}$$

Q.29 In Q(28), σ^2 can be estimated in practice without any serious error as

$$\hat{\sigma}^2 = (\ldots\ldots\ldots)/(n_1 + n_2)$$

Q.30 If $(\sigma^2{}_1, \sigma^2)$ are unknown then for sufficiently for large sample sizes (n_1, n_2), we have

$v(\bar{x}_1 - \bar{x}_2)$ = (..........)

Q.31 In Q(30), the statistic Z is defined as

$$Z = \frac{|\bar{x}_1 - \bar{x}_2|}{\sqrt{\ldots\ldots\ldots\ldots}} \sim N(0,1)$$

Q.32 Let (p, P) be the sample proportion and population proportion respectively for a certain characteristic A, then z statistic is defined as

$$Z = \frac{|X - nP|}{\sqrt{\ldots\ldots\ldots\ldots}} \sim N(0, 1)$$

Q.32* Let z statistic defined in Q(32) is applicable to test the average number of x success in n independent .... trial.

Q.33 To test the hypothesis
$H_0 : p = P$, $H_1 : p \neq P$, the z statistic is defined as

$$Z = \frac{|p - P|}{\sqrt{\ldots\ldots\ldots\ldots\ldots}} \sim N(0, 1)$$

Q.34 If the sample of size n is drawn from a finite population of size N then

$$\text{S.E. (p)} = \sqrt{\ldots\ldots\ldots\ldots\frac{PQ}{n}}$$

Q.35 In Q(34), we have

$$V(p) = \left(\frac{N-n}{N-1}\right)(\ldots\ldots\ldots)$$

Q.36 The probable limts for proportion of success P is given as
limit (P) = P ± $\sqrt{\ldots\ldots}$

Q.37 If in Q(36), P is unknown then estimated probable limits for P are given as
Estimate limt (P) = p ± 3$\sqrt{\ldots\ldots\ldots\ldots}$

Q.38 For two population proportions (P_1, P_2), to test the hypothesis $H_0 : P_1 = P_2$, the z statistic is defined as

$$Z = \frac{|p_1 - p_2|}{\sqrt{(\ldots\ldots\ldots) + (\ldots\ldots\ldots)}} \sim N(0, 1)$$

Q.39 In Q(38), under H_0; $P_1 = P_2 = P$, the Z statistic is defined as

$$Z = \frac{|p_1 - p_2|}{\sqrt{PQ(\ldots\ldots\ldots)}} \sim N(0, 1)$$

Q.40 In Q(39), if P is unknown then its umbiased estimate based on a combined samples of sizes (n_1, n_2) is,obtained as

$$\hat{P} = \frac{(x_1 + x_2)}{(\dots\dots\dots\dots)} = p$$

Q.41 In Q(40), the Z statistic is defined as

$$Z = \frac{|p_1 - p_2|}{\sqrt{pq(\dots\dots)}} \sim N(0, 1)$$

Q.42 To test the hypothesis $H_0 : \sigma_1 = \sigma_2$, the z statistic is defined as

$$Z = \frac{|s_1 - s_2|}{\sqrt{\frac{\sigma_1^2}{\dots\dots} + \frac{\sigma_2^2}{\dots\dots}}} \sim N(0,1)$$

Q.43 In Q(42), if $\sigma_1 = \sigma_2$ and are unknown then z statistic is defined as

$$Z = \frac{|s_1 - s_2|}{\sqrt{(\dots\dots) + (\dots\dots)}} \sim N(0, 1)$$

12.3 Section-C : Multiple Choice Questions

Q.1 The biological data on plant height, seed weight, seed size, pratain content in pulse, oil content in oil seed etc. are assumed to follow -

(a) Poisson distribution (b) Normal distribution
(c) Binomial distribution
(d) Negative binomial distribution

Q.2 Biological variables like plant height, seed weight, seed size, etc. are assumed to be -

(a) Normal variables (b) Poisson variables
(c) Binomial variables (d) Rectangular variables

Q.3 For X ~ N (μ , σ^2), the standard normal variable Z is defined as

(a) $Z = \frac{(x - \mu)}{\sigma}$ (b) $Z = \frac{(x - \mu)}{\sqrt{\sigma^2}}$

(c) $Z = \frac{(x + \mu)}{\sigma}$ (d) Both (a) & (b)

Q.4 The standard normal variable has mean and variance respectively equal to -

(a) (1, 0) (b) (0, 1)

(c) (0, 2) (d) (2, 0)

Q.5 The p.d.f. of standard normal distribution is given as

(a) $f(z)=\frac{1}{\sqrt{2\pi}}e^{-\frac{1}{2}z^2}$ (b) $f(z)=\frac{1}{\sqrt{2\pi}}e^{-z^2}$

(c) $f(z)=\frac{1}{\sqrt{2\pi}}e^{\frac{1}{2}z^2}$ (d) $f(z)=\frac{1}{2\pi}e^{-\frac{1}{2}z^2}$

Q.6 The parent population from which the samples are drawn for hypothesis testing, is assumed to follow

(a) Normal distribution (b) Binomial distribution

(c) Gaussion distribution (d) Both (a) & (c)

Q.7 From normal probability table, $p(-3 \le z \le 3)$ is equal to

(a) 0.9973 (b) 0.90

(c) 0.95 (d) 0.0027

Q.8 The normal probability table states that $P(|Z| \ge 3.0)$ is equal to

(a) 0.49 (b) 0.0027

(c) 0.05 (d) 0.01

Q.9 From normal probability table, $P(|Z| \le 1.64)$ is equal to

(a) 0.95 (b) 0.90

(c) 0.99 (d) 0.49

Q.10 From normal probability curve, the $P(|Z| \le 1.96)$ is equal to

(a) 0.9973 (b) 0.95

(c) 0.90 (d) 0.49

Q.12 If $P(|Z| \le Z\alpha) = 0.9973$ then Za is equal to

(a) 1.96 (b) 2.58

(c) 3.0 (d) 1.64

Q.13 The critical value of Z at 5% level of significance for two tailed test is equal to

(a) 1.64 (b) 2.58

(c) 1.96 (d) 3.0

Q.14 If parent population does not follow normal distribution even then the sample mean x_n follows the distribution named as

(a) Asymptotically normal (b) Normal

(c) Continuous (d) Discreate

Q.15 Let x ~N(μ, σ²). The sample mean x_n follows -

(a) $N\left(\mu, \frac{\sigma^2}{n}\right)$ (b) $N\left(\mu, \frac{\sigma}{\sqrt{n}}\right)$

(c) $N\left(\frac{\sigma^2}{n}, \mu\right)$ (d) $N\left(\frac{\sigma}{\sqrt{n}}, \mu\right)$

Q.16 The sampling variance of difference of two sample means $(xn_1 - xn_2)$ is given as

(a) $v\left(\bar{x}_{n_1} - \bar{x}_{n_2}\right) = \left(\frac{\sigma_1^2}{n_1} + \frac{\sigma_2^2}{n_2}\right)$ (b) $v\left(\bar{x}_{n_1} - \bar{x}_{n_2}\right) = \left(\frac{\sigma_2^2}{n_1} + \frac{\sigma_1^2}{n_2}\right)$

(c) $v\left(\bar{x}_{n_1} - \bar{x}_{n_2}\right) = \left(\frac{\sigma_1^2}{n_2} + \frac{\sigma_2^2}{n_1}\right)$ (d) $v\left(\bar{x}_{n_1} - \bar{x}_{n_2}\right) = \left(\frac{\sigma_1^2}{2n_1} + \frac{\sigma_2^2}{2n_2}\right)$

Q.17 For the equality of two means of two normal distributions N $(\mu i, \sigma_i^2$, i= 1, 2). We write null hypothesis as -

(a) $H_0 : \mu_1 = \mu_2$ (b) $H_0 : \mu_1 \neq \mu_2$

(c) $H_0 : \mu_1 > \mu_2$ (d) $H_0 : \mu_1 < \mu_2$

Q.18 In Q(17), the Z statistic is defined as

(a) $Z = \frac{|\bar{x}_1 - \bar{x}_2|}{\sqrt{\frac{\sigma_1^2}{n_1} + \frac{\sigma_2^2}{n_2}}}$ (b) $Z = \frac{|\bar{x}_1 - \bar{x}_2|}{\left(\frac{\sigma_1^2}{n_1} - \frac{\sigma_2^2}{n_2}\right)}$

(c) $Z = \frac{|\bar{x}_1 - \bar{x}_2|}{\sqrt{\frac{\sigma_1}{n_1} + \frac{\sigma_2}{n_2}}}$ (d) $Z = \frac{|\bar{x}_1 - \bar{x}_2|}{\sqrt{\frac{\sigma_1^2}{n_2} + \frac{\sigma_2^2}{n_1}}}$

Q.19 The difference of two independent sample means (xn_1, xn_2) from two normal populations N $(\mu i, \sigma_i^2$, i= 1, 2) follows the distribution as

(a) $N\left(\mu_1-\mu_2, \frac{\sigma_1^2}{n_2}+\frac{\sigma_2^2}{n_1}\right)$ (b) $N\left(\mu_1-\mu_2, \frac{\sigma_1^2}{n_1}+\frac{\sigma_2^2}{n_2}\right)$

(c) $N\left(\mu_2-\mu_1, \frac{\sigma_1^2}{n_1}+\frac{\sigma_2^2}{n_2}\right)$ (d) $N\left(\mu_1-\mu_2, \frac{\sigma_1^2}{n_1}+\frac{\sigma_2^2}{n_2}\right)$

Q.20 The sum $(\bar{x}_{n_1} - \bar{x}_{n_2})$ in Q(19) follows the distribution as

(a) $N\left(\mu_1+\mu_2, \frac{\sigma_1}{n_1}+\frac{\sigma_2}{n_2}\right)$ (b) $N\left(\mu_1+\mu_2, \frac{\sigma_1^2}{n_1}+\frac{\sigma_2^2}{n_2}\right)$

(c) $N\left(\mu_1+\mu_2, \frac{\sigma_2^2}{n_2}+\frac{\sigma_1^2}{n_1}\right)$ (d) Both (b) & (c)

Q.21 In Q(17), if $\sigma_1^2 = \sigma_1^2 = \sigma^2$ then z is defined as

(a) $Z=\frac{|\bar{x}_1-\bar{x}_2|}{\sqrt{\sigma^2\left(\frac{1}{n_1}+\frac{1}{n_2}\right)}}$ (b) $Z=\frac{|\bar{x}_1-\bar{x}_2|}{\sqrt{\sigma\left(\frac{1}{n_1}+\frac{1}{n_2}\right)}}$

(c) $Z=\frac{|\bar{x}_1-\bar{x}_2|}{\sqrt{\left(\frac{\sigma^2}{n_1}+\frac{\sigma^2}{n_2}\right)}}$ (d) Both (a) & (c)

Q.22 In Q(21), for sufficiently large sample sizes (n_1, n_2), the pooled estimate $\hat{\sigma}^2$ of σ^2 is given by

(a) $\hat{\sigma}^2=\frac{(n_1-1)s_1^2+(n_2-1)s_2^2}{(n_1+n_2-2)}$ (b) $\hat{\sigma}^2=\frac{\sum_i^2 (ni-1)s_i^2}{\left(\sum_{i=1}^2 ni-2\right)}$

(c) $\hat{\sigma}^2=\frac{(n_1-1)s_2+(n_2-1)s_2}{(n_1+n_2-2)}$ (d) Both (a) & (b)

Q.23 In Q(22), σ^2 can be estimated in practice without any serious error by the expression

(a) $\dfrac{\sum_{i}^{2} nis_i^2}{\sum_{i=1}^{2} ni}$

(b) $\dfrac{n_1 s_1 + n_2 s_2}{n_1 + n_2}$

(c) $\dfrac{n_1 s_1^2 + n_2 s_2^2}{n_1 + n_2}$

(d) Both (a) & (c)

Q.24 If (σ_1^2 and σ_2^2) are unknown then for sufficiently large (n_1, n_2), the V (x_1 - x_2) is equal to

(a) $\left(\dfrac{s_1^2}{n_1} + \dfrac{s_2^2}{n_2}\right)$

(b) $\sum_{i=1}^{2} \dfrac{s_i^2}{ni}$

(c) $\left(\dfrac{s_1}{n_1} + \dfrac{s_2}{n_2}\right)$

(d) Both (a) & (b)

Q.25 In Q(24), the Z statistic is defined as

(a) $Z = \dfrac{|\bar{x}_1 - \bar{x}_2|}{\sqrt{\dfrac{s_1^2}{n_1} + \dfrac{s_2^2}{n_2}}}$

(b) $Z = \dfrac{|\bar{x}_1 - \bar{x}_2|}{\sqrt{\sum_{i=1}^{2} \dfrac{s_i^2}{ni}}}$

(c) $Z = \dfrac{|\bar{x}_1 - \bar{x}_2|}{\sqrt{\dfrac{s_1}{n_1} + \dfrac{s_2}{n_2}}}$

(d) Both (a) & (b)

Q.26 Let (p, P) be the sample and population proportions respectively for a certain characteristic A, then Z to test H0 : p = P, the Z statistic is defined as

(a) $Z = \dfrac{|X - nP|}{\sqrt{nPQ}}$

(b) $Z = \dfrac{|X + np|}{\sqrt{nPQ}}$

(c) $Z' = \dfrac{|X - npq|}{\sqrt{nPQ}}$

(d) None

Q.27 The z statistic defined in Q (26), is applicable to test the average number of x success in n independent

(a) Binomial trials

(b) Bernouille trials

(c) Poisson trials

(d) Both (a) & (b)

Q.28 In Q(26), to test H0 : p = P, the z statistic under Ho to is equal to

(a) $\frac{|p-P|}{\sqrt{\frac{PQ}{n}}}$ (b) $\frac{|p-P|}{\sqrt{\frac{pq}{n}}}$

(c) $\frac{|p-P|}{\sqrt{nPQ}}$ (d) $\frac{|p-P|}{\sqrt{npq}}$

Q.29 If a sample of size n is drawn from a finite population of size N then V(p) is equal to

(a) $\left(\frac{N-n}{N-1}\right)\frac{PQ}{n}$ (b) $\left(\frac{N-n}{nN}\right)\frac{PQ}{n}$

(c) $\left(\frac{N-1}{nN}\right)\frac{PQ}{n}$ (d) $\left(\frac{N-n}{nN}\right)\frac{pq}{n}$

Q.30 The probable limits for proportion of success P is given as

(a) $P\pm3\sqrt{\frac{PQ}{n}}$ (b) $P\pm\sqrt{\frac{PQ}{n}}$

(c) $P\pm1.96\sqrt{\frac{PQ}{n}}$ (d) $P\pm3\frac{PQ}{n}$

Q.31 If population proportion P is unknown the estimated probable limits for P is given by

(a) $p\pm3\frac{pq}{n}$ (b) $p\pm3\sqrt{\frac{pq}{n}}$

(c) $p\pm1.96\sqrt{\frac{pq}{n}}$ (d) $p\pm1.96\frac{pq}{n}$

Q.32 To test the hypothesis Ho : $P_1 = P_2$, the z statistic is defined as

(a) $Z=\frac{|p_1-p_2|}{\left(\frac{P_1Q_1}{n_1}+\frac{P_2Q_2}{n_2}\right)}$ (b) $Z=\frac{|p_1-p_2|}{\sqrt{\frac{P_1Q_1}{n_1}+\frac{P_2Q_2}{n_2}}}$

(c) $Z=\frac{|p_1-p_2|}{\sqrt{\frac{P_1Q_1}{n_2}+\frac{P_2Q_2}{n_1}}}$ (d) $Z=\frac{|p_1-p_2|}{\sqrt{\frac{P_1Q_1}{n_2}+\frac{P_2Q_2}{n_1}}}$

Q.33 Under H0 : $P_1 = P_2 = P$, the Z statistic is defined as

(a) $Z = \dfrac{|p_1 - p_2|}{\sqrt{PQ\left(\dfrac{1}{n_1} + \dfrac{1}{n_2}\right)}}$

(b) $Z = \dfrac{|p_1 - p_2|}{\sqrt{PQ\left(\dfrac{1}{n_2} + \dfrac{1}{n_1}\right)}}$

(c) $Z = \dfrac{|p_1 - p_2|}{PQ\left(\dfrac{1}{n_2} + \dfrac{1}{n_1}\right)}$

(d) Both (a) & (b)

Q.34 If in Q(33), P is unknown and $\hat{P} = p$ then z statistic is defined as

(a) $Z = \dfrac{|p_1 - p_2|}{\sqrt{pq\left(\dfrac{1}{n_1} + \dfrac{1}{n_2}\right)}}$

(b) $Z = \dfrac{|p_1 - p_2|}{\sqrt{\left(\dfrac{1}{n_2} + \dfrac{1}{n_1}\right)pq}}$

(c) $Z = \dfrac{|p_1 - p_2|}{\sqrt{1.96\left(\dfrac{1}{n_1} + \dfrac{1}{n_2}\right)}}$

(d) Both (a) & (b)

Q.35 If in Q(33), $P_1 = P_2 =$ P(Unknown) and $n_1 = n_2 =$ n then Z statistic is defined as

(a) $Z = \dfrac{|p_1 - p_2|}{\sqrt{pq\left(\dfrac{1}{n} + \dfrac{1}{n}\right)}}$

(b) $Z = \dfrac{|p_1 - p_2|}{\sqrt{\dfrac{2pq}{n}}}$

(c) $Z = \dfrac{|p_1 - p_2|}{\sqrt{1.96\left(\dfrac{1}{n} + \dfrac{1}{n}\right)}}$

(d) Both (a) & (b)

Q.36 For two populations N (μi, σ^2_i, i= 1, 2), to test the hypothesis $H_0: \sigma_1 = \sigma_2$, the Z statistic is defined as

(a) $Z = \frac{|s_1 - s_2|}{\sqrt{\frac{\sigma_1^2}{2n_1} + \frac{\sigma_2^2}{2n_2}}}$

(b) $Z = \frac{|s_1 - s_2|}{\sqrt{\frac{1}{2}\left(\frac{\sigma_1^2}{n_1} + \frac{\sigma_2^2}{n_2}\right)}}$

(c) $Z = \frac{|s_1 - s_2|}{\sqrt{\frac{1}{2}\left(\frac{\sigma_2^2}{n_2} + \frac{\sigma_1^2}{n_1}\right)}}$

(d) All the three

Q.37 In Q(36), if (σ_1^2, σ_2^2) are unknown and $\hat{\sigma}_i^2 = s_i^2$, then z statistic is defined as

(a) $Z = \frac{|s_1 - s_2|}{\sqrt{\frac{s_1^2}{2n_1} + \frac{s_2^2}{2n_2}}}$

(b) $Z = \frac{|s_1 - s_2|}{\sqrt{\frac{1}{2}\left(\frac{s_1^2}{n_1} + \frac{s_2^2}{n_2}\right)}}$

(c) $Z = \frac{|s_1 - s_2|}{\sqrt{\frac{1}{2}\left(\frac{s_2^2}{n_2} + \frac{s_1^2}{n_1}\right)}}$

(d) All the three

Key Answers

Section-A : True False

1. True	2. True	3. True
4. True	5. True	6. True
7. True	8. True	9. True
10. True	11. True	12. True
13. True	14. True	15. True
16. True	17. True	18. True
19. True	20. False	21. True
22. True	23. True	24. True
25. True	26. True	27. True
28. True	29. True	30. True
31. True	32. True	33. True
34. True	35. True	36. True
37. True	38. True	

Section-B : Fill in the Blanks

1. Normally
2. Normal
3. $(X - \mu)$
4. (0, 1)
5. 2π, Z
6. Normal
7. Normal
8. 0.9973
9. 0.0027
10. Significant
11. 0.99, 10%
12. 0.95
13. 3.0
14. 1.96
15. Normally, Normal
16. $(x - \mu) / \sigma$
17. $(\mu, \frac{\sigma^2}{n})$
18. Population mean and sample mean
19. Known
20. $\frac{1}{n-1} \sum^{n} (xi - \bar{x}n)^2$
21. $\left(\frac{\sigma_1^2}{n_1} + \frac{\sigma_2^2}{n_2}\right)$
22. Equal or same
23. $\left(\frac{\sigma_1^2}{n_1} + \frac{\sigma_2^2}{n_2}\right)$
24. Normal distribution
25. Normal distribution
26. $(\mu_1 - \mu_2)$
27. $\sigma^2\left(\frac{1}{n_1} + \frac{1}{n_2}\right)$
28. $(n_1 + n_2 - 2)$
29. $(n_1 s_1^2 + n_2 s_2^2)$
30. $\left(\frac{s_1^2}{n_1} + \frac{s_2^2}{n_2}\right)$
31. $\left(\frac{s_1^2}{n_1} + \frac{s_2^2}{n_2}\right)$
32. n PQ

32* Binomial

33. $\frac{PQ}{n}$
34. $\frac{(N - n)}{(N - 1)}$
35. $\frac{PQ}{n}$
36. $\frac{PQ}{n}$
37. $\frac{pq}{n}$
38. $\left(\frac{P_1Q_1}{n_1} + \frac{P_2Q_2}{n_2}\right)$
39. $\left(\frac{1}{n_1} + \frac{1}{n_2}\right)$

40. $(n_1 + n_2)$

41. $\left(\frac{1}{n_1} + \frac{1}{n_2}\right)$

42. $\left(\frac{1}{2n_1} + \frac{1}{2n_2}\right)$

43. $\left(\frac{s_1^2}{2n_1} + \frac{s_2^2}{2n_2}\right)$

Section-C : Multiple Choice Questions

1. (b)	2. (a)	3. (d)
4. (b)	5. (a)	6. (d)
7. (a)	8. (b)	9. (b)
10. (b)	11. (...)	12. (c)
13. (c)	14. (a)	15. (a)
16. (a)	17. (a)	18. (a)
19. (b)	20. (d)	21. (d)
22. (d)	23. (d)	24. (d)
25. (d)	26. (a)	27. (d)
28. (a)	29. (a)	30. (a)
31. (b)	32. (b)	33. (d)
34. (d)	35. (d)	36. (d)
37. (d)		

❑❑❑

Chapter - 13

Chi-Square Distribution and its Applications

13.1 Section-A : Write True / False

Q.1 The square of a standard normal variate is known as Chi-square variate with one d.f.

Q.2 If X ~ N(μ, σ^2) and Z = (x - μ)/σ then Z^2 is known as Chi-square variate with 1.d.f.

Q.3 If Xi ~ N(μi, σi), i = 1,2, ... , n are n independent normal variates then

$$\chi_n^2 = \sum_{i=1}^{n} \frac{(Xi - \mu i)^2}{\sigma_i^2}$$ is a chi-square variate with n d.f.

Q.4 The p.d.f. of a chisquare variate with n d.f. is given as

$$f(X^2) = \frac{1}{2^{\frac{n}{2}}\left\lceil\frac{n}{2}\right|}(\chi^2)^{\frac{n}{2}-1} e^{-\frac{\chi^2}{2}}, \quad 0 \le \chi^2 < \infty$$

Q.5 Chi-square distribution is a continuous distribution.

Q.6 The sampling distribution of a chi-square statistic x^2 is free from (μ, σ^2) and hence x^2 statistic is known as "distribution free" statistic.

Q.7 Moment generating function of x^2 variate with n d.f. is written as $M_{X2}(t) = (1 - 2t)^{-\frac{n}{2}}$, $|2t| < 1$

Q.8 The mean and variance of x^2 distribution with n d.f. are given respectively as Mean = n, $\sigma^2 = 2n$.

0.9 Skewness and kurtosis of x^2 distribution with n d.f. are given as : $\beta 1 = \frac{8}{n}$, $\beta 2 = 3 + \frac{12}{n}$

Q.10 The standard x^2 variate tends to standard normal variate as $n \rightarrow \infty$

Q.11 The x^2 distribution tends to normal distribution for large d.f.

Q.12 The characteristic function of x^2 distribution is written as

$\phi_{X2}(t) = (1 - 2it)^{-\frac{n}{2}}$

Q.13 The mode of a x^2 distribution with n d.f. is equal to (n - 2).

Q.14 The x^2 distribution is positively skewed distribution.

Q.15 The sum of independent x^2 variates is also a chisquare variate.

Q.16. *Cochran Theoram* : Let X_1, X_2, ..., Xn be independently distributed as N(0, 1). Let $\sum_{i}^{n} x_i^2 = Q_1 + Q_2 + \ldots\ldots\ldots\ldots + Qk$, where each Qi is a sum of squares of linear combinations of X_1, X_2, ... Xn with n d.f. Then if $\sum_{i}^{k} ni = n$ the quantities Q_1, Q_2, ... Q_k are independent x^2 variates with n_1, n_2, ..., nk d.f. respectively.

Q.17 If x_i^2 and x_2^2 are two independent X^2 variates with (n_1, n_2) d.f. respectively then

$$\frac{\chi_1^2}{\chi_2^2} \sim \beta_2\left(\frac{n_1}{2}, \frac{n_2}{2}\right) \text{ variate}$$

Q.18 If (x^2_i, x_2^2) are independent chi-square variates with (n_1, n_2) d.f. respectively then

$$U = \frac{\chi_1^2}{\chi_1^2 + \chi_2^2} \sim \beta_1\left(\frac{n_1}{2}, \frac{n_2}{2}\right)$$

Q.19 The degrees of freedom of chisquare distribution follow the additive property.

Q.20 For the valid use of x^2 statistic, the constraints on cell frequencies should be linear.

Q.21 The total frequency N - should be reasonably large i.e. N > 50 to maintain the continuuity property of x^2 distribution.

Q.22 Any cell frequency of contingency table to apply x^2 test of goodness of fit should not be less than 5.

Q.23 If any cell frequency is less than 5, then for valid application of x^2 test, it is pooled with the preceeding or succeeding cell frequency so that the pooled cell frequency must be > 5.

Q.24 After pooling the cell frequencies; a suitable adjustment is made for d.f. lost in pooling.

Q.25 The critical value of x^2 increases as the n, the d.f. increases.

Q.26 The critical value (table value) of x^2 statistic increases as the level of significance decreases.

Q.27 To test whether the hypothetical value of population variance σ^2 is equal to σ^2_0, the suitable test statistic, is the x^2 statistic.

Q.28 To test the goodness of fitting between observed and theoretical (expected) frequencies of the events in an experimental trial, the suitable test statistic is x^2.

Q.29 To test the independence of attributes, the plausible test statistic is x^2.

Q.30 To test the homogenity of several independent estimates of the same population variance σ^2, the test statistic is the x^2 statistic.

Q.31 To combine the various probabilities obtained from independent experiments, to give a single test of significance, the test statistic applicable is the x^2 statistic.

Q.32 To test the homogeneity of several independent estimates of the same population correlation coefficient p, the test statistic is the x^2 statistics.

Q.33 The number of independent observations which make up the statistic (x^2) is known as the "Degree of freedom" and is denoted by v.

Q.34 The d.f. is generally equal to the total number of observations minus the number of independent constraints imposed on them.

Q.35 If k is the number of independent constraints, imposed in a set of n observations used for x^2 statistic then d.f. = (n - k).

Q.36 The excess of the number of observations over the number of parameters estimated in case of applying x^2 statistic, is known as its d.f.

Q.37 In a set of n observation, usually the d.f. is equal to (n - 1)

Q.38 In a set of n observations for use of x^2 test, one degree of freedom is lost due to linear constraint $\sum_{i}^{k} Oi = \sum_{i}^{k} ei$ imposed on frequencies.

Q.39 To test the hypothesis Ho : $\sigma^2 = \sigma^2_0$, the x^2 test statistic is written as

$$\chi^2 = \sum_{i}^{n} \frac{(xi - \bar{x})}{\sigma_0^2} \sim \chi^2_{n-1}$$

Q.40 If the sample size n is large ie n > 30, then we can apply Fisher's approximation written as

$$\sqrt{2x^2} \sim N\left(\sqrt{2n - 1}, 1\right)$$

Q.41 If we use Fisher's approximation for large sample test in place of x^2 test then z statistic is defined as

$$Z = \left(\sqrt{2x^2} - \sqrt{2n - 1}\right) \sim N(0,1)$$

Q.42 For applying x^2 test of goodness of fit, the x2 statistic is written as

$$\chi^2 = \sum_{i}^{k} \frac{(Oi - ei)^2}{ei} \sim \chi^2_{k-1}$$

Q.43 If to test the independence of two attributes, the number of rows and columns are as (r, c) in a contingency tablre, then the d.f. is written as d.f. = (r - 1) (c - 1)

Q.44 In case of testing the independence of two attributes, the Ho is written as

Ho : Two types of classifications are independent of each other.

Q.45 Consider the 2 x 2 contingency table written as where (a, b, c, d) are cell frequencies. Then d.f. = (2 - 1) (2 - 1) = 1.

a	b	$a + b = R_1$
c	d	$c + d = R_2$
$a + c = C_1$	$b + d = C_2$	N

Q.46 In Q(45) x^2 statistic can be applied on observed frequencies in the expression written as

$$x_1^2 = \frac{(ad - bc)^2 N}{R_1 R_2 C_1 C_2}$$

Q.47 Consider the 2 x 2 contingency table in the form of expected frequencies written as. Then expected frequencies (e_{ij}, i = 1, 2, j = 1, 2) can be worked out from the obsered frequency table as

$$e_{ij} = E\,(.....) = \frac{RiCj}{N},\ i=1,2\quad j=1,2$$

e_1	e_2
e_3	e_4

Q.48 In case of 2 x 2 contingency table, "Yates correction for continuity" of the x^2 statistic is applied.

Q.49 To apply Yates correction in 2 x 2 contingency table, the correction consists of adding (½) to the diagonal cell frequencies whose product is smaller and (½) is substrated from the remains cell frequencies.

Q.50 Yates correction for 2 x 2 contingency table is applied in the following form -

$$\chi_1^2 = \frac{\left[|ad - bc| - \frac{N}{2}\right]^2 \times N}{R_1 R_2 C_1 C_2}$$

Q.51 The coeficient of contingency "C" measures the degree of relationship or dependence of the classification in a contingency table is defined as

$$C = \sqrt{\frac{\chi^2}{\chi^2 + N}}$$

Q.52 The larger value of coefficient of contingency indicates the high degree of association.

Q.53 In case of k x k contingency table, the maximum value of "C" is given as

$$C_{max ....} = \sqrt{\frac{k-1}{k}}$$

Q.54 For k x k contingency table, the correlation coefficient between two attributes or classification is measured by

$$r = \sqrt{\frac{\chi^2}{N(k-1)}}$$

Q.55 The coefficient r defined in Q54, lies in interval (0, 1).

13.2 Section -B : Fill in the Blanks

Q.1 The square of standard normal variate is known as a .... with 1 d.f.

Q.2 The square of a .... is known as x^2 variate one d.f.

Q.3 Let X ~ N (μ, σ^2) and Z = (X − μ)/σ then Z^2 is known as a .... with 1 d.f.

Q.4 Let Xi = N(μi, σ_1^2, i = 1,2, ... , n) be n independent normal variates. Then

$$Z^2 = \sum_{i}^{k} \frac{X_i - \mu i)^2}{\sigma_1^2}$$ is known as .... with .... d.f.

Q.5 Chisquare distribution is a .... distribution.

Q.6 The chisquare statistic is known as .... free statistic.

Q.7 The sampling distribution of .... statistic is free from (μ, σ^2).

Q.8 The M.G.F. of x^2 variate with n d.f. is written as

M_{x2} (t) = (.......)$^{-\frac{n}{2}}$, |2t| < 1

Q.9 For x^2 distribution with n d.f. we have

Mean = ..., σ^2 = ...

Q.10 For x^2 distribution with n d.f., we have

$\beta_1 =$ ....

Q.11 For x^2 distribution with n d.f., we have

$\beta_2 =$ ....

Q.12 The standard chi-square variate tends to .... variate as $n \to \infty$

Q.13 The x^2 distribution tends to normal distribution for ....

Q.14 The characteristic function of x^2 distribution is written as

$\phi x^2 (t) = (..........)^{-\frac{n}{2}}$

Q.15 The mode of chisquare distribution with n d.f. is equal to ....

Q.16 The chisquare distribution is .... skewed distribution.

Q.17 The sum of independent chisquare variates is also a .... variate.

Q.18 Let $(x^2_i,\ x^2_2)$ be two independent x^2 statistic with (n_1, n_2) d.f. respectively then $\dfrac{x^2_i}{x^2_2} \sim \beta_2$ (...............)

Q.19 In Q(18), the variate

$$U = \frac{x^2_1}{(x^2_i,\ x^2_2)} \sim \beta_1 \text{ (........)}$$

Q.20 The degress of freedom of x^2 distribution follow .... laws.

Q.21 For valid use of x^2 statistic, the constraint imposed on cell frequencies should be ....

Q.22 The total number of observations N > 50, to maintain the .... property of x^2 distribution.

Q.23 Any cell frequency of contingency table to apply .... test should not be less than 5.

Q.24 The critical value of x^2 statistic increases with the .... in its d.f.

Q.25 The critical value (table value) of x^2 statistic increases as the level of significance α ....

Q.26 To test the hypothesis H o : $\sigma^2 = \sigma^2_o$, the suitable test statistic is ....

Q.27 To test the goodness of fitting between observed frequencies and expected frequencies of events in an experimental trial, the suitable test statistic is ....

Q.28 Independence of two attributes is tested with the help of .... statistic.

Q.29 To test the homogeneity of several independent estimates of the same population variance σ^2, the suitable test statistic is ....

Q.30 To test the .... of several independent estimates of a population variance σ^2, the x^2 statistic is applicable.

Q.31 To combine the various probabilities obtained from independent experiments for a single test of significance, the suitable test statistic is ....

Q.32 To test the homogeneity of several independent estimates of the same population correlation coefficient ρ, the suitable test statistic is the ....

Q.33 The number of independent observations which make up the statistic, is known as ....

Q.34 The number of .... observations which makes up the statistic, is known as its d f.

Q.35 The d.f. is generally equal to the total number of observations minus the number of independent .... imposed on them.

Q.36 Let k be the number of independent constraints, imposed on a set of n observations to define a test statistic then d.f. = ...

Q.37 The excess of number of observations over number of parameters estimated in case of applying x^2 test, is known as its ....

Q.38 In a set of n observations for use of x^2 test, one d.f. is lost due to linear constraint imposed on it as $\sum_{i}^{n} O_i =$ ....

Q.39 To test the hypothesis Ho : $\sigma^2 = \sigma^2_0$, the test statistic x^2 is written as

$$\chi^2 = \sum_i^n \frac{(\text{........})^2}{\sigma_0^2} \sim \chi^2_{n-1}$$

Q.40 Let n > 30, then we can apply Fisher's approximation written as

$$\sqrt{2\chi^2} \sim N(\text{........}, \text{..........})$$

Q.41 If we apply Fisher's approximation in large sample test in place of x^2 test then Z statistic is defined as

$Z = (\text{............}) \sim N(0, 1)$

Q.42 For applying x^2 test of goodness of fit, the x^2 statistic is written as

$$\chi^2 = \sum_{i=1}^{k} \frac{(\text{....})^2}{ei} \sim \chi^2_{k-1}$$

Q.43 For a r x c contingency table, then d.f. to apply a x^2 statistic is equal to ....

Q.44 To test the independence of two attributes, the H_0 is written as H_0 : Two types of classifications are .... of each other.

Q.45 For 2 x 2 contingency table, the d.f. for applying x^2 statistic is equal to ....

Q.46 Consider 2 x 2 contingency table. Then without using expected frequency, we can work out x^2 statistic as

$$x_i^2 = \frac{(\text{..........})^2 N}{R_1 R_2 C_1 C_2} \quad \text{where}$$

a	b	R_1
c	d	R_2
C_1	C_2	N

Q.47 In Q(46), the expected cell frequency (eij, i = 1, 2, j = 1, 2) can be obtained as

$$eij = E(Oij) = \frac{\text{.........}}{N}, i = 1, 2, \quad j = 1, 2$$

Q.48 In 2 x 2 contingency table, Yate's correction is applied to maintain the .... property of x^2 statistic.

Q.49 In 2 x 2 consistency table, Yate's correction consists of .... ½ to the diogonal cell frequencies whose product is .... and ½ is substracted to the remaining cell frequencies.

Q.50 In 2 x 2 contingency table, Yate's correction is as

$$x_1^2 = \frac{[|ad - bc| - \text{............}]^2 N}{R_1 R_2 C_1 C_2}$$

Q.51 The coefficient of contingency "C" in a contingency table is defined as

$$C = \sqrt{\frac{\chi^2}{\text{..........}}}$$

10.3 Section-C : Multiple Choice Questions

Q.1 The square of standard normal variate is known as
(a) Beta variate
(b) Chisquare variate
(c) Binomial variate
(d) Compound binomial variate

Q.2 Let x ~ N (μ, σ^2) and Z = (x − μ)/σ, then Z^2 is known as
(a) Chisquare statistic
(b) Fisher's F statistic
(c) Students t statistic
(d) Poisson variate

Q.3 Let Xi ~ N (μ_i σ_i^2, i = 1, 2, ..., n) be n independent normal variates then

$$Z^2 = \sum_{i=1}^{n} \left(\frac{Xi - \mu i}{\sigma i} \right)^2 \text{ is a}$$

(a) Chisquare statistic with n d.f.
(b) Student's t statistic with (n - 1) d.f.
(c) F statistic with (1, n) d.f.
(d) Gamma variate with parameter n

Q.4 Chi-square distribution is a
(a) Continuous distribution
(b) Positively skewed distribution
(c) Discreate distribution
(d) Both (a) & (b)

Q.5 The sampling distribution of x^2 statistic is
(a) Independent of parameters (μ, σ^2)
(b) Dependent on parameters (μ, σ^2)
(c) Independent of g but not of σ^2
(d) Independent of σ^2 but not of μ

Q.6 The x^2 statistic is
(a) Distribution free statistic
(b) Not distribution free statistic
(c) Dependent of population parameters
(d) Both (b) & (c)

Q.7 Moment generating function of x^2 distribution with n d.f. is written as

(a) $M_{x2}(t) = (1-2t)^{\frac{n}{2}}$ (b) $M_{x2}(t) = (1-2t)^{\frac{n}{2}}$
(c) $M_{x2}(t) = (1-2t)^{n}$ (d) $M_{x2}(t) = (1-2t)^{-n}$

Q.8 For x^2 distribution with n d.f., the mean and variance are respectively equal to
(a) (2n, n) (b) (n , 2n)
(c) (1, n) (d) (n ,1)

Q.9 Skewness of chisquare distribution with n d.f. is equal to

(a) $\frac{4}{n}$ (b) $\frac{8}{n}$

(c) $\frac{n}{8}$ (d) $\frac{5}{2n}$

Q.10 The kurtosis of x^2 distribution with n d.f. is equal to

(a) $3+\frac{12}{n}$ (b) $3+\frac{8}{n}$

(c) $\frac{12}{n}+3$ (d) Both (a) & (c)

Q.11 As $n \to \infty$ the standard x^2 variate reduces to
(a) Standard cauchy variate (b) Standard normal variate
(c) Standard Poisson Variate (d) F statistic with (1, n) d.f.

Q.12 The characteristic function $\phi_{X_2}(t)$ of x^2 distribution is given as

(a) $(1-2t)^{-\frac{n}{2}}$ (b) $(1-2it)^{-\frac{n}{2}}$

(c) $(1-2t)^{\frac{n}{2}}$ (d) $(1-2it)^{\frac{n}{2}}$

Q.13 The mode of x^2 distribution with n d.f. is equal to

(a) (n - 2) (b) (n - 3)

(c) (n + 2) (d) (n + 3)

Q.14 x^2 distribution is

(a) Positively skewed distribution

(b) Negatively skewed distribution

(c) Positively asymmetrical distribution

(d) Both (a) & (c)

Q.15 The sum of independent x^2 variates is a

(a) Standard normal variate (b) x^2 variate variate

(c) Standard cauchy variate (d) Beta variate of first kind

Q.16 Let $(x^2_{n_1}, x^2_{n_2})$be two independent chi square variates then (x^2_1 / x^2_2) follows

(a) $\beta_2(n_2, n_1)$distribution (b) $\beta_2(n_1, n_2)$distribution

(c) $\beta_2\left(\frac{n_1}{2}, \frac{n_2}{2}\right)$distribution (d) $\beta_2\left(\frac{n_2}{2}, \frac{n_1}{2}\right)$distribution

Q.17 In Q (16), (x^2_1 / x^2_2) follows the distribution as

(a) $\beta_2\left(\frac{n_2}{2}, \frac{n_1}{2}\right)$ (b) $\beta_2\left(\frac{n_1}{2}, \frac{n_2}{2}\right)$

(c) $\beta_2(n_1, n_2)$ (d) $\beta_2(n_2, n_1)$

Q.18 In Q(16), the variate $U = x_1^2 / (x^2_1 / x^2_2)$ follows the distribution give as

(a) $\beta_1\left(\frac{n_1}{2}, \frac{n_2}{2}\right)$ (b) $\beta_2\left(\frac{n_2}{2}, \frac{n_1}{2}\right)$

(c) $\beta_2(n_1, n_2)$ (d) $\beta_2(n_2, n_1)$

Q.19 The d.f. of independent chi square distributions follows the
(a) Multiplicative law (b) Additive law
(c) Probabilistic law (d) Asymptotis law

Q.20 For valid use of x^2 statistic, the constraints on cell frequency must be of
(a) Linear nature (b) Quadratic nature
(c) Additive nature (d) Multiplicative nature

Q.21 Total number of observations must be > 50 to maintain that property of x^2 distribution which is known as
(a) Continuity property (b) Symmetricity
(c) Discreteness (d) Normality

Q.22 Any cell frequency of contingency table to apply x^2 test of goodness of fit should
(a) Not be < 5 (b) Not be > 5
(c) Be ≥ 5 (d) Both (a) & (c)

Q.23 In contingency table, the cell frequency < 5, must be pooled with
(a) Succeeding cell frequency (b) Preceeding cell frequency
(c) Any cell frequency (d) Either (a) or (b)

Q.24 The critical value of x^2 increases with the
(a) Increase of d.f. (b) Decrease of d.f.
(c) Decrease of level of significance
(d) Both (a) & (c)

Q.25 Critical value of x^2 increases with the
(a) Decrease in level of significance
(b) Increase in degree of freedom
(c) Increase in level of significance
(d) Both (a) & (b)

Q.26 To test the goodness of fit between observed and theortical frequency of an experimental trial, the suitable test statistic is the
(a) Chi-square statistic (b) Z statistic
(c) Student's t statistic (d) Behren-Fisher d statistic

Q.27 For a normal distribution N (μ, σ^2) to test the hypothesis Ho : $\sigma^2 = \sigma^2_0$, the appropriate test statistic is the

(a) Chi-square statistic (b) Z statistic
(c) F statistic (d) None of these

Q.28 Independence of two attributes is tested with the help of

(a) x^2 statistic (b) Z statistic
(c) Variance ratio statistic (d) Paired t statistic

Q.29 Homogeneity of several independent estimates of the same population variance σ^2 is tested with the use of

(a) Z statistic (b) Student's t statistic
(c) Chi-square statistic (d) None of these

Q.30 For the pooled probabilities obtained from several independent experiments, a single test of significance is applicable with the help of

(a) x^2 statistic (b) Z statistic
(c) Additive law of probabilities (d) Student's t statistic

Q.31 Homogeneity of several independent estimates of the same population correlation coefficient p is tested with the use of

(a) Chisquare statistic (b) Student's t statistic
(c) Variance ratio statistic (d) Z statistic

Q.32 The number of independent observations which makes up a test statistic is known as its -

(a) Critical value (b) Degree of freedom
(c) Level of significance (d) None of these

Q.33 If k is the number of independent constainsts imposed on a set of n observation to apply the x^2 statistic then its d.f. is equal to

(a) (n - k + 1) (b) (n - k)
(c) (n - k - 1) (d) (n - 2k)

Q.34 If k is the number of parameters estimated from a set of n observations then d.f. to apply the x^2 statistic is equal to

(a) (n - k (b) (n - k - 1)
(c) (n - k + 1) (d) (n - 2k)

Q.35 To test the hypothesis H 0 : $\sigma^2 = \sigma^2_0$, test statistic is written as

(a) $\chi^2 = \sum_{i}^{n} \frac{(xi - \bar{x})^2}{\sigma_0^2}$ (b) $\chi^2 = \sum_{i}^{n} \frac{(xi - \bar{x})^2}{\sigma^2}$

(c) $\chi^2 = \sum_{i}^{n} \frac{(xi - \mu)^2}{\sigma_0^2}$ (d) $\chi^2 = \sum_{i=1}^{n} \frac{(xi - \mu_0)^2}{\sigma_0^2}$

Q.36 For large sample ie n > 30, Fisher's approximation is written as

(a) $\sqrt{2\chi^2} \sim N\left(\sqrt{2n-1}, 1\right)$ (b) $\sqrt{2\chi^2} \sim N\left(\sqrt{2n}, 1\right)$

(c) $\sqrt{2\chi^2} \sim N\left(1, \sqrt{2n-1}\right)$ (d) $\sqrt{2\chi^2} \sim N\left(1, \sqrt{2n}\right)$

Q.37 In case of applying Fisher's approximation, the standard normal variate z can be defined as

(a) $Z = \left(\sqrt{2\chi^2} - \sqrt{2n}\right)$ (b) $Z = \left(\sqrt{2\chi^2} - \sqrt{2n-1}\right)$

(c) $Z = \left(\sqrt{2\chi^2} - \sqrt{2n+1}\right)$ (d) $Z = \left(\sqrt{2\chi^2} - \sqrt{n-1}\right)$

Q.38 To test the goodness of fit between observed and expected freuencies, the x^2 statistic is defined as

(a) $\chi^2_{k-1} = \sum_{i}^{k} \left(\frac{oi - ei}{ei}\right)^2$ (b) $\chi^2_{k-1} = \sum_{i=1}^{k} \frac{(oi - ei)^2}{ei}$

(c) $\chi^2_{k-1} = \sum_{i=1}^{k} \left(\frac{oi - ei}{\sqrt{ei}}\right)^2$ (d) Both (b) & (c)

Q.39 For (r x c) contingency table to apply the x^2 test, the d.f. will be equal to

(a) d.f. = (r - 1) (c - 1) (b) d.f. = (c - 1) (r - 1)

(c) d.f. = (r - 1) (c + 1) (d) Both (a) & (b)

Q.40 For testing the independence of two attributes, H_0 is written as

(a) H_0 : Classification of two attributes are independent

(b) H_0 : Classification of two attributes are dependent

(c) H_0 : Classification of two attributes are not dependent

(d) Both (a) & (c)

Q.41 Consider the given 2 x 2 contingency table. Then x^2 statistic can be defined as

a	b	R_1
c	d	R_2
C_1	C_2	N

(a) $\chi_1^2 = \frac{(ad-bc)^2 N}{R_1 R_2 C_1 C_2}$ (b) $\chi_1^2 = \frac{(ad+bc)^2 N}{R_1 R_2 C_1 C_2}$

(c) $\chi_1^2 = \frac{(bc-ad)^2 N}{C_1 C_2 R_1 R_2}$ (d) Both (a) & (c)

Q.42 In Q(41), the expected cell frequencies (eij, i = 1, 2, j = 1,2) can be worked out as

(a) $e_{ij} = \frac{RiCj}{N}$ (b) $e_{ij} = \frac{CjRi}{N}$

(b) $e_{ij} = \frac{RiCj}{N}$ (d) Both (a) & (b)

Q.43 For 2 x 2 contingency table, Yate's correction is applied to maintain the

(a) Continuity of x^2 (b) Symitry of x^2

(c) Skewness of x^2 (d) None of these

Q.44 For 2 x 2 contingency tablwe, Yate's correction is written as

(a) $\chi_1^2 = \frac{\left(|ad-bc|+\frac{N}{2}\right)^2 N}{R_1 R_2 C_1 C_2}$ (b) $\chi_1^2 = \frac{\left(|ad-bc|-\frac{N}{2}\right)^2 N}{R_1 R_2 C_1 C_2}$

(c) $\chi_1^2 = \frac{(|ad-bc|+N)^2}{R_1 R_2 C_1 C_2}$ (d) $\chi_1^2 = \frac{\left(|ad-bc|-\frac{N}{3}\right)^2}{R_1 R_2 C_1 C_2}$

Q.45 In case of x^2 statistic, the coefficient of contingency say "C" is defined as

(a) $\sqrt{\frac{\chi^2}{\chi^2+N}}$ (b) $\sqrt{\frac{\chi^2+N}{\chi^2}}$

(c) $\sqrt{\frac{2\chi^2}{\chi^2+N}}$ (d) $\sqrt{\frac{\chi^2+N}{2\chi^2}}$

Q.46 Larger value of coefficient of contingency indicates the
(a) Lower degree of association
(b) Higher degree of association
(c) Moderate degree of association
(d) None of these

Q.47 In case of (k x k) contingency table, the maximum value of "C" is given as

(a) $C_{Max} = \sqrt{\frac{k-1}{k}}$ (b) $C_{Max} = \sqrt{\frac{k}{k-1}}$

(c) $C_{Max} = \sqrt{\frac{k+1}{k}}$ (d) $C_{Max} = \sqrt{\frac{k}{k+1}}$

Q.48 In case of (k x k) contingency table, the correlation coefficient "r" can be measured as

(a) $r = \sqrt{\frac{\chi^2}{N(k-1)}}$ (b) $r = \sqrt{\frac{2\chi^2}{N(k-1)}}$

(c) $r = \sqrt{\frac{N(k-1)}{\chi^2}}$ (d) $r = \sqrt{\frac{N(k-1)}{2\chi^2}}$

Key Answers

Section-A : True False

1. True	2. True	3. True
4. True	5. True	6. True
7. True	8. True	9. True
10. True	11. True	12. True
13. True	14. True	15. True
16. True	17. True	18. True
19. True	20. False	21. True
22. True	23. True	24. True
25. True	26. True	27. True
28. True	29. True	30. True
31. True	32. True	33. True
34. True	35. True	36. True
37. True	38. True	39. True
40. True	41. True	42. True
43. True	44. True	45. True

46. True
47. True
48. True
49. True
50. True
51. True
52. True
53. True
54. True
55. True

Section-B : Fill in the Blanks

1. Chisquare variate
2. Standard normal variate
3. Chisquare variate
4. x^2 square variate, n
5. Continuous
6. Distribution
7. Chisquare
8. (1 - 2t)
9. n, 2n
10. $\beta_1 = \frac{8}{n}$
11. $\beta_2 = 3 \pm \frac{12}{n}$
12. Standard normal
13. Large d.f.
14. (1 - 2it)
15. (n - 2)
16. Positively
17. Chi-square
18. $\left(\frac{n_1}{2}, \frac{n_2}{2}\right)$
19. $\left(\frac{n_1}{2}, \frac{n_2}{2}\right)$
20. Additive
21. Linear
22. Continuity
23. x^2 test
24. Increase
25. Decreases
26. x^2 statistic
27. x^2 statistic
28. x^2
29. x^2 statistic
30. Homogeneity
31. x^2 statistic
32. x^2 statistic
33. Degree of freedom
34. Independent
35. Constraints
36. (n - k)
37. Degree of freedom
38. $\sum_{i=1}^{k} ei$
39. $(x_i - \bar{x})$
40. $\left(\sqrt{2n - 1}, 1\right)$
41. $\left(\sqrt{2x^2}, \sqrt{2n - 1}\right)$
42. (oi - ei)
43. (r - 1) (c - 1)
44. Independent

45. One
46. (ad - bc)
47. Ri Cj
48. Continuity
49. Adding, Less
50. $\frac{-N}{2}$
51. $(x^2 + N)$

Section-C Multiple Choice Questions

1. (b)	2. (a)	3. (a)
4. (d)	5. (a)	6. (a)
7. (b)	8. (b)	9. (b)
10. (d)	11. (b)	12. (b)
13. (a)	14. (d)	15. (b)
16. (c)	17. (a)	18. (a)
19. (b)	20. (a)	21. (a)
22. (d)	23. (d)	24. (d)
25. (d)	26. (a)	27. (a)
28. (a)	29. (c)	30. (a)
31. (a)	32. (b)	33. (b)
34. (a)	35. (a)	36. (a)
37. (b)	38. (d)	39. (d)
40. (d)	41. (d)	42. (d)
43. (a)	44. (b)	45. (a)
46. (b)	47. (a)	48. (a)

❑❑❑

Chapter - 14

Exact Sampling Distributions and Related Small Sample Tests (F, t)

14.1 Section-A : Write True / False

Q.1 If sample size n is small ie n < 30, then the small sample test is applicable.

Q.2 The exact sample test (small sample test) can be applied to large sample test but the converse is not plaussible.

Q.3 If $(x_1^2,\ x_2^2)$ are two independent chi-square variates with (V_1, V_2) degree of freedom then F statistic is defined as

$$F = \frac{x_1^2 / v_1}{x_1^2 / v_1} \sim F(v_1, v_2)$$

Q.4 The probability density function of F statistic is defined as

$$f(F) = \frac{\left(\frac{v_1}{v_2}\right)^{\frac{v_1}{2}} F^{\frac{v_1}{2}-1}}{\beta\left(\frac{v_1}{2}, \frac{v_2}{2}\right)\left(1+\frac{v_1}{v_2}F\right)^{\frac{v_1+v_2}{2}}}, \quad 0 \le F < \infty$$

Q.5 To define F statistic, the two population variances $(\sigma_1^2, \sigma_2)$are assumed to be unknown.

Q.6 Let the population variances (σ_1^2, σ_2^2) are unknown. Then to test Ho : $\sigma_1^2 = \sigma_2^2$, the F statistic is defined as $F = \frac{s_1^2}{s_2^2}$, $s_1^2 > s_2^2$

Q.7 If $(s_1^2 > s_2^2)$ are two independent sample variances of sizes (n_1, n_2) then d.f. of F statistic will be. $(n_1 - 1, n_2 - 1)$.

Q.8 The student's t statistic is defined with the assumption that the population standard delliation a is unknown.

Q.9 It is noticeable that the statistic 't' itself and its sampling distribution are functionaly independent of σ.

Q.10 The p.d.f. of t statistic is written as

$$f(t) = \frac{1}{\sqrt{v}\beta\left(\frac{1}{2},\frac{v}{2}\right)\left(1+\frac{t^2}{v}\right)^{\frac{v+1}{2}}}, \quad -\infty < t < \infty$$

Q.11 If we take v = 1, then we get

$$f(t) = \frac{1}{\beta\left(\frac{1}{2},\frac{1}{2}\right)(1+t^2)} = \frac{1}{\pi}\cdot\frac{1}{(1+t^2)}, \quad -\infty < t < \infty$$

which is the p.d.f. of standard cauchy distribution.

Q.12 If degree of freedom of 't' statistic become 1, then student's t distribution reduces to cauchy distribution.

Q.13 The 95% confidence limits for μ are given as

$$\bar{x} \pm t_{0.05}\left(\frac{s}{\sqrt{n}}\right) \text{ with t at (n - 1) d.f.}$$

Q.14 If $\xi \sim N(0, 1)$ and x_n^2 is an independent chi-square variate then Fisher's t is defined as

$$t = \xi / \sqrt{x^2 / n}$$

Q.15 The f(t) is symmetrical about the line t = 0 and all the odd order moments about origin vanish.

Q.16. The mean of student's t distribution is equal to zero.

Q.17 The variance of t_n distribution $\sigma^2 = \mu_2 = \frac{n}{n-2}$

Q.18 The skewnes of t_n distribution $\beta_1 = 0 = \frac{\mu_3^2}{\mu_2^3}$

Q.19 Kurtosis β_2 of t_n distribution is given as

$$\beta_2 = 3\left(\frac{n-2}{n-4}\right)$$

Q.20 For large d.f., t distribution tends to standard normal distribution.

Q.21 For critical value of t statistic, we have

$P(|t| > t_{\alpha(v)}) = \alpha$ and $P(|t| < t_{\alpha(v)}) = 1 - \alpha$

Q.22 $P(t > t_{\alpha(v)}) + P(t < -t_{\alpha(v)}) = a = P(t > t_{\alpha(v)}) = \alpha/2$

Q.23 For Q(22), we have $P(t > t_{2\alpha(v)}) = \alpha$

Q.24 It is noticeable that the significant values of t at level of significance 'α' for a single tailed test can be obtained from those of two tailed by looking the values at the level of significance "2α".

Q.25 t_8 (0.05) for single tail test = t_8 (0.10) for two tailed test = 1.86.

Q.26 Let (Xi, i = 1, 2, ... , n) be a random sample of small size n < 30 drawn from N(μ, σ²). Then student's t statistic is defined as

$$t = \frac{|\bar{x} - \mu|}{s/\sqrt{n}} \sim t_{(n-1)}.$$

Q.27 For a N(μ, σ²), to test the hupothesis

$H_0 : \mu = \mu_0$, the t statistis is defined as

$$t = \frac{|\bar{x} - \mu_0|}{s/\sqrt{n}} \sim t_{(n-1)}$$

Q.28 We conclude that if the calculated value of t statistic is greater than the tabulated value oft with corresponding d.f. at α% level of significant then H_0 is declared significant.

Q.29 Consider two norm al populations N (μ_1, σ^2_1) and $N(\mu_2, \sigma^2_2)$, Suppose $\sigma^2_1 = \sigma^2_2 = \sigma^2$ but unknown. To test the hypothesis Ho : $\mu_1 = \mu_2$, the following t statistic is defined

$$t = \frac{|\bar{x}_1 - \bar{x}_2|}{s\sqrt{\frac{1}{n_1} + \frac{1}{n_2}}} \sim t_{(n_1+n_2-2)}$$

Q.30 In Q(29), if $s^2_1, \neq s^2_2$, and are unknown then student's two mean t test fails and we apply Fisher and Behren's 'd' statistic defined as

$$d = \frac{|\bar{x}_1 - \bar{x}_2|}{\left(\sqrt{\frac{s_1^2}{n_1} + \frac{s_2^2}{n_2}}\right)}$$

Q.31 For the test statistic "d" defined in Q(30), the critical value of d statistic say d* is workedout

$$d^* = \frac{t_1 v(\bar{x}_1) + t_2 v(\bar{x}_2)}{v(\bar{x}_1) + v(\bar{x}_2)}$$

Q.32 In Q(31) we have $t_1 = t_{\alpha\%}(n_1 - 1)$, $t_2 = t_{\alpha\%}(n_2 - 1)$

Q.33 If the two samples are "not independent" and observations are taken in pairs then "paired t statistic" is applicable.

Q.34 If the observations are taken on twins and sample size of both samples are same, then paired t test is applied.

Q.35 If the two samples are suitable to apply the paired t test then the paired t test will be more efficient than the ordinary two mean t test.

Q.36 Let {(xi, yi), i = 1, 2, ... , n} be the observations taken on n pairs and di = (xi - yi), i = 1, 2, ... , n be the n differences.

Then $\bar{d} = \frac{1}{n}\sum_{i}^{n} di$

Q.37 In Q(36), $\sum_{i=1}^{n}$ di the is taken as the algebraic sum of differences Which may be positive or negative or equal to zero.

Q.38 For paired t test the null hypothesis is given as
H_0 : The population mean of differences (xi - yi) = 0

Q.39 To test the null hypothesis written in Q (38), the paired t statistic is defined as

$$t = \frac{|\bar{d}|}{sd / \sqrt{n}} \sim t_{(n-1)}.$$

Q.40 In paired t statistic, s_d^2 isd worked out as

$$s_d^2 = \frac{1}{n-1}\left[\sum^{n} di^2 - \frac{\left(\sum di\right)^2}{n}\right]$$

Q.41 Ten students were given a test in statistic. They were given one month for further practice and a second test was held at the end of it. Ho is that students have improved or not due to this practice? Then paired t test is applicable.

Q.42 Mean of $F_{(V1, V2)}$ distribution is $\mu_1' = \frac{V_2}{V_2 - 1} > 1$

Q.43 Mode of $F_{(V1, V2)} = \frac{v_2(v_1-2)}{v_1(v_2+2)} = \frac{v_2}{(v_2+2)}\left(\frac{v_2-2}{v_1}\right)$

Q.44 F distribution is highly positively skewed.

Q.45 Under Ho : The population correlation ratio η_{yx} is equal to zero, the suitable test statistic is F and is defined as

$$F = \frac{\eta^2}{1-\eta^2}\left(\frac{n-h}{h-1}\right) \sim F(h-2, n-h)$$

Where h is the number of arrays.

Q.46 For a sample of size n arranged in h arrays, from a bivariate normal population the test statistic to test the linearity of regression is

$$F = \frac{(\eta^2 - r^2)}{(1-\eta^2)}\left(\frac{n-h}{h-2}\right) \sim F(h-2, n-h)$$

Q.47 To test the hamogeneity of 3 or more than 3 means, the suitable test statistic is F.

Q.48 If a statistic t follows student's t distribution with n d.f. then t^2 folowsnedecor's F distribution with (1, n) d.f.

Q.49 In F (n_1,n_2) if we take $n_2 \to \infty$, then $n_1 F \sim x^2_{n_1}$.

14.2 Section-B : Fill in the Blanks

Q.1 If sample size n is small ie n < 30, then .... test is applicable.

Q.2 The small sample test can be applied to the .... but converse is not plaussible.

Q.3 If (x_1^2, x_2^2) are two independent chi-square variates with (v_1, v_2) df respectively then the

$$\text{ratio} = \frac{x_1^2 / v_1}{x_1^2 / v_1} \sim \ldots\ldots\ldots\ldots$$

Q.4 To define F statistic, the two population variances (σ_1^2, σ_2^2) are assumed to be ....

Q.5 Let $\hat{\sigma}_1 = s_1^2$ and $\sigma_2^2 = s_2^2$ ande $s_2^2 < s_1^2$ then F statistic is defined as

F = ...

Q.6 Let $(s_1^2 > s_2^2)$ be two independent sample estimates of (σ_1^2, σ_2^2) based on sample sizes (n_1, n_2) respectively. Then d.f. of F statistic will be equal to (............, ...)

Q.7 The student's t statistic is defined with the assumption that the population standard deviation a is assumed to be ....

Q.8 It is noticeable that the statistic 't' itself and its sampling variance is .... of σ.

Q.9 If in the sampling distribution of 't' statistic, the d.f. say v is put equal to 1, then f(t) reduces to standard ....

Q.10 The 95% confidence limits for μ in case of small sample of size n < 30, is given as
$\bar{x} \pm t_{0.05}$ (..........) with t at (n - 1) d.f.

Q.11 If $\xi \sim N(0, 1)$ and x_n^2 is an independent chi-square variate then Fisher's t is defined as
$t = \xi / \sqrt{\ldots\ldots}$

Q.12 The probability curve f(t) is .... about the line t = 0.

Q.13 All odd order moments of t distribution about origin are equal to ....

Q.14 The mean of t_n distribution is equal to ....

Q.15 The variance of t distribution with (n - 1) d.f. is given as μ_2 = ...

Q.16 The skewness of t distribution say (β_1) is equal to ....

Q.17 The kurtosis β_2 of t distribution is given as
β_2 = 3 (..........)

Q.18 For large d.f the t distribution reduces to .... distribution.

Q.19 For critical value $t_{\alpha(n)}$ of t statistic, we have

$P(|t| > t_{\alpha (n)})$ = ...

Q.20 In Q(19), we can write

$P(|t| < t_{\alpha (n)})$ = ...

Q.21 Due to symmetricity of t distribution, we have

$P (t > t_{\alpha (n)})$ = ...

Q.22 In Q(21), we have

$P(t > t_{2\alpha(v)})$ =n ....

Q.23 It is noticiable that t_α for a single tail test can be obtained from those of two tailed test for looking the t values at the level of significance " ....".

Q.24 For a small sample size n < 30, drawn from $N(\mu, \sigma^2)$, the student's t statistic is defined as

$$t = \frac{|\bar{x}_1 - \mu|}{\text{..........}} \sim t_{(n-1)}$$

Q.25 To test the hypothesis Ho : $\mu = \mu_0$, the t statistic is defined as

$$t = \frac{|\text{..........}|}{s/\sqrt{n}} \sim t_{(n-1)}$$

Q.26 If calculated $t > t_{\alpha (n)}$, (nY then Ho is declared ....

Q.27 For two normal distribution N (μ_i, σ_i^2 , i =1, 2), let σ_1^2, = σ_2^2 = σ^2 be unknown. To test the hypothesis $H_0 : \mu_1 = \mu_2$, the t statistic is defined as

$$t = \frac{|\bar{x}_1 - \bar{x}_2|}{s\sqrt{\ldots\ldots\ldots}} \sim t_{(n_1+n_2-2)}$$

Q.28 If in Q(27), $\sigma_1^2 = \sigma^2$ and are unknown, then Fisher and Behren's 'd' statistic is defined as

$$d = \frac{|\bar{x}_1 - \bar{x}_2|}{\sqrt{(\ldots\ldots\ldots)}}$$

Q.29 In Q(28), the critical value d* for calculated d is worked out as

$$d^* = \frac{t_1 v(\bar{x}_1) + t_2 v(\bar{x}_2)}{\ldots\ldots\ldots\ldots}$$

Q.30 In Q(29), we have $t_1 = t_{\alpha\%}$ (..................).

Q.31 Let two samples be "not independent" and their sizes are same then to test H : $\mu_1 = \mu_2$, the test statistic applicable is known as ....

Q.32 If two samples are suitable for applying paired t test, then it will be .... efficient than the ordinary two mean t test.

Q.33 For n paired observations (xi, yi), let di = (xi - yi). Then to apply paired t test, $\sum_i^n$ di is taken as .... sum of deviations.

Q.34 In Q(33), $\sum_i^n$ di may be equal to either .... or negative or ....

Q.35 For applying paired t test, the null hypothesis is written as H_0 : The population mean of (xi - yi) = 0 ....

Q.36 In Q(33), the paired 't' statistic is defined as

$$t = \frac{\bar{d}}{\ldots\ldots} \sim t_{(n-1)}$$

Q.37 In applying paired t statistic, the sample variance of difference di is worked out as

$$s_d^2 = \frac{1}{(n-1)}\left[\sum_i^n d_i^2 - \ldots\ldots\ldots\ldots\right]$$

Q.38 Let F (v_1, v_2) denotes the F distribution. The mean μ'_1 will be given as
μ'_1 = ... > 1.

Q.39 For F (v_1, v_2), we can write that

$$\text{Mode} = \frac{V_2}{(v_2 + 2)} \ (.....)$$

Q.40 F distribution is highly .... skewed distribution.

Q.41 To test the hypothesis -
H_0 : Population correlation ratio $\eta_{yx} = 0$ the suitable test statistic is .... and is defined as

$$F = \frac{\hat{\eta}^2_{yx}}{(1-\eta^2_{yx})} \frac{(......)}{(h-1)}$$

Q.42 To test H_0 : $\eta_{yx} = 0$, the d.f. of F statistic is equal to (................, ...).

Q.43 For a. sample of size n arranged in h arrays, to test the linearity of regression, the suitable test statistic is F and is defined as

$$F = (............)\left(\frac{n-h}{h-2}\right) \sim F(h-2, n-h)$$

Q.44 In Q (43), the d.f. of F statistic is equal to (.................).

Q.45 To test the homogeneity of three or more than three population means, the- test statistic applicable is ....

Q.46 Let t_n be a student's t statistic then $t_n^2 \sim F_{(.........)}$

Q.47 For $F(n_1, n_2)$ if we put $n_2 \to \infty$, then
$n_1 F \sim$ .... with n_1 d.f.

Section-C : Multiple Choice Questions

Q.1 If sample size n is small (say < 30), then to test the hypothesis about population mean it, the suitable test statistic is

(a) Z statistic (b) F statistic

(c) t statistic (d) x^2 statistic

Q.2 The small sample test for the population mean can
(a) Be applied to large sample
(b) Not be applied to the large sample .
(c) Z test
(d) Both (a) & (c)

Q.3 For (x_1^2, x_2^2) being two independent Chi-square variates with d.f. equal to (v_1, v_2) respectively, the ratio $\frac{x_1^2 / v_1}{x_2^2 / v_1}$ follows

(a) $\beta_2 (v_1, v_2)$ distribution
(b) $F (v_1, v_2)$ distribution
(c) $F(v_2, v_1)$ distribution
(d) $\beta_2 (v_2, v_1)$ distribution

Q.4 To define F statistic the two population variances $(\sigma_1^2, \sigma_2^2)$are assumed to be
(a) Known
(b) Unknown
(c) Unknown and independent
(d) Unknown and dependent

Q.5 Let the population variances (σ_1^2, σ_2^2). unknown and dependent then to test the hypothesis Ho : $\sigma_1^2 = \sigma_2^2$, the test statistic is defined as

(a) $F = \frac{s_1^2}{s_2^2}$ for $s_2^2 > s_2$
(b) $F = \frac{s_1^2}{s_2^2}$ for $s_1^2 > s_2^2$
(c) $F = \frac{s_1^2}{s_2^2}$ for $s_2^2 > s_1^2$
(d) Both (a) & (c)

Q.6 To test the hypothesis Ho : $\sigma_1^2 = \sigma_2^2$ the F is definwed as

$F = \frac{s_1^2}{s_2^2}$ for $s_1^2 > s_2^2$. The d.f. for F statistic will be equal to
(a) (n_1, n_2)
(b) $(n_2 -1, n_1 -1)$
(c) $(n_1 -1, n_2 -1)$
(d) $(n_2, n,)$

Q.7 The p.d.f. of t statistic is written as

$$f(t) = \frac{1}{\sqrt{v}\beta\left(\frac{1}{2},\frac{v}{2}\right)\left(1+\frac{t^2}{v}\right)^{\frac{v+1}{2}}}, -\infty < t < \infty$$

Then for V = 1, f(t) reduces to
(a) Standard cauchy distribution
(b) Standard normal distribution
(c) Beta distribution as β (½, ½)
(d) None of these

Q.8 If d.f. of student's t distribution becomes equal to 1, then it reduces to
(a) Standard cauchy distribution
(b) Standard normal distribution
(c) β (½, ½) distribution
(d) None of the above

Q.9 95% confidence limits for la in case of small sample test is given by

(a) $\overline{x} - t_{0.05}\left(\frac{s}{\sqrt{n}}\right)$ (b) $\overline{x} \pm t_{0.05}\left(\frac{s^2}{n}\right)$

(c) $\overline{x} \mp t_{0.5}\sqrt{\frac{s^2}{n}}$ (d) Both (b) & (c)

Q.10 If ξ ~ N(0, 1) and x^2 is an independent chisquare variate then Fisher t is defined as

(a) $t = \frac{\varepsilon}{(\chi^2/n)}$ (b) $t = \frac{\varepsilon}{\sqrt{\chi^2/n}}$

(c) $t = \frac{\varepsilon^2}{(\chi/n)}$ (d) $t = \frac{\varepsilon^2}{\sqrt{\chi^2/n}}$

Q.11 The mean of t_n distribution is equal to
(a) Zero (b) 2n
(c) $\sqrt{2n}$ (d) $\sqrt{n}$

Q.12 The variance of t_n distribution is equal to

(a) $\mu_2 = \frac{n}{n - 1}$ (b) $\mu_2 = \frac{n}{n - 2}$

(c) $\mu_2 = \frac{2n}{n - 2}$ (d) $\mu_2 = \frac{n^2}{n - 1}$

Q.13 Skewness β_1 of t distribution is equal to

(a) $\beta_1 = \sqrt{n}$ (b) $\beta_1 = \sqrt{\frac{n}{n+1}}$

(c) $\beta_1 = 0$ (d) $0 \le \beta_1 < 1$

Q.14 Kurtosis β_2 of to distribution is given as

(a) $\beta_2 = 3\left(\frac{n-2}{n-4}\right)$ (b) $\beta_2 = 3\left(\frac{n-1}{n-2}\right)$

(c) $\beta_2 = 3\left(\frac{n}{n-1}\right)$ (d) $\beta_2 = 3\left(\frac{n}{n+4}\right)$

Q.15 For large d.f. the t distribution reduces to
(a) Normal distribution
(b) Standard normal distribution
(c) Standard cauchy distribution
(d) None of these

Q.16 For a small sample of size n drawn from $N(\mu, \sigma^2)$, the student's t statistic is defined as

(a) $t = \frac{|\bar{x} - \mu|}{s / \sqrt{n}}$ (b) $t = \frac{|\bar{x} - \mu|}{\sqrt{s^2 / n}}$

(c) $t = \frac{|x + \mu|}{s / \sqrt{n}}$ (d) Both (a) & (b)

Q.17 For $N(\mu, \sigma^2)$, the student's t statistic to test the hypothesis $H_0 : \mu$ = [to is defined as

(a) $t = \frac{|\bar{x} - \mu|}{s / \sqrt{n}}$ (b) $t = \frac{|\bar{x} - \mu_0|}{\sqrt{s^2 / n}}$

(c) $t = \frac{|\bar{x} - \mu_0|}{s^2 / \sqrt{n}}$ (d) $t = \frac{|\bar{x} - \mu_0|}{s / n}$

Q.18 If calculated t_{n-1} value > tabulated $t_{\alpha(n-1)}$, then Ho is declared

(a) Significant (b) Non Significant

(c) Non accepted (d) Both (a) & (c)

Q.19 In case of two normal distributions, if $\sigma_1^2 = \sigma_1^2 = \sigma^2$ then to test $H_0 : \mu_1 = \mu_2$, t statistic is defined as

(a) $$t_{n_1+n_2-2} = \frac{|\bar{x}_1 - \bar{x}_2|}{\sqrt{s^2\left(\frac{1}{n_1} + \frac{1}{n_2}\right)}}$$

(b) $$t_{n_1+n_2-2} = \frac{|\bar{x}_1 - \bar{x}_2|}{s\sqrt{\left(\frac{1}{n_1} + \frac{1}{n_2}\right)}}$$

(c) $$t = \frac{|\bar{x}_1 - \bar{x}_2|}{\sqrt{\left(\frac{s^2}{n_1} + \frac{s^2}{n_2}\right)}}$$

(d) All the three

Q.20 In Q(19), if $\sigma_1^2 \neq \sigma_2^2$ and are unknown then to test $H_0 : \mu_1 = \mu_2$, Behren Fisher "d" statistic is defined as

(a) $$d = \frac{|\bar{x}_1 - \bar{x}_2|}{\sqrt{\left(\frac{s_1^2}{n_2} + \frac{s_2^2}{n_1}\right)}}$$

(b) $$d = \frac{|\bar{x}_1 - \bar{x}_2|}{\sqrt{\left(\frac{s_1^2}{n_1} + \frac{s_2^2}{n_2}\right)}}$$

(c) $$d = \frac{|\bar{x}_1 - \bar{x}_2|}{\sqrt{\left(\frac{s_1}{n_1} + \frac{s_2}{n_2}\right)}}$$

(d) $$d = \frac{|\bar{x}_1 - \bar{x}_2|}{\sqrt{s^2\left(\frac{1}{n_1} + \frac{1}{n_2}\right)}}$$

Q.21 In Q(20), for calculated value d, the critical value d* is worked out as

(a) $$d^{x} = \frac{t_1\sqrt{v(\bar{x}_1)} + t_2\sqrt{v(\bar{x}_2)}}{\sqrt{v(\bar{x}_1)} + \sqrt{v(\bar{x}_2)}}$$

(b) $$d^{x} = \frac{t_1 v(\bar{x}_1) + t_2 v(\bar{x}_2)}{v(\bar{x}_1) + v(\bar{x}_2)}$$

(c) $$d^{x} = \frac{t_1 v(\bar{x}_1) + t_2 v(\bar{x}_2)}{\sqrt{v(\bar{x}_1) + v(\bar{x}_2)}}$$

(d) None of the above

Q.22 If two small samples are dependent and observations are taken on paired sampling units, then to test the hypathesis Ho : $\mu_1 = \mu_2$, the suitable test statistic is

(a) Paired t statistic (b) Two mean t statistic
(c) Behren Fisher d statistic (d) Two mean Z statistic

Q.23 If the observations are taken on twins, and sizes of both the samples are same then to test Ho : $\mu_1 = \mu_2$, the test statistic is

(a) Two mean t statistic (b) Behren Fisher d statistic
(c) Paired t statistic (d) Z statistic

Q.24 Let two small samples be suitable to apply paired t test to test Ho : $\mu_1 = \mu_2$. Then paired t test will be

(a) Less efficient than two mean t test
(b) More efficient than two mean t test
(c) More efficient than Behren Fisher d test
(d) Equivalent to two mean z test

Q.25 In case of applying paired t test, the total of differences of paired observations will be -

(a) Positive (b) Negative
(c) Equal to zero (d) Any of three may be

Q.26 For applying paired t test, the Ho will be written as

(a) Population mean of differences di > zero
(b) Population mean of differences di < zero
(c) Population mean of differences di = zero
(d) Any one of the three

Q.27 For paired t test, the test statistic 't' is defined as

(a) $t_{n-1} = \frac{|\bar{d}|}{\frac{s^2}{\sqrt{n}}}$ (b) $t_{n-1} = \frac{|\bar{d}|}{\sqrt{\frac{s^2}{n}}}$

(c) $t_{n-1} = \frac{|\bar{d}|}{\frac{s}{\sqrt{n}}}$ (d) Both (b) & (c)

Q.28 The mean μ_1 of $F(v_1, v_2)$ distribution given as

(a) $\mu_1' = \frac{v_2}{v_2 + 1}$ (b) $\mu_1' = \frac{v_2}{v_2 - 1}$

(c) $\mu_1' = \frac{v_2}{v_1 + 1}$ (d) $\mu_1' = \frac{v_2 + 1}{v_2}$

Q.29 F distribution is
(a) Highly positively skewed
(b) Highly negatively skewed
(c) Symmetrically distributed
(d) None of the three

Q.30 For population correlation ratio η_{yx} to test the hypothesis H0 : $\eta_{yx} = 0$, the F statistic is defined as

(a) $F = \frac{\eta}{1 - \eta} \cdot \frac{n - h}{h - 1} \sim F_{(h - 1, n - h)}$

(b) $F = \frac{\eta^2}{1 - \eta^2} \cdot \frac{n - h}{h - 1} \sim F_{(h - 1, n - h)}$

(c) $F = \frac{\eta^2}{1 - \eta^2} \cdot \frac{n - h}{h - 1} \sim F_{(n - h, h - 1)}$

(d) None of the above, where h is the number of arrays

Q.31 For a sample of n values arranged in h arrays, from a bivariate normal population, to test H_0 : Regression is linear, the test statistic F is defined as

(a) $F = \left(\frac{\eta^2 - r^2}{1 - \eta^2}\right)\left(\frac{n - h}{h - 2}\right) \sim F_{(h-2, n-h)}$

(b) $F = \left(\frac{\eta^2 - r^2}{1 - \eta^2}\right)\left(\frac{h - 2}{n - h}\right) \sim F_{(h-2, n-h)}$

(c) $F = \left(\frac{\eta - r}{1 - \eta^2}\right)\left(\frac{n - h}{h - 1}\right) \sim F_{(h-1, n-h)}$

(d) None of these

Q.32 To test the homogeneity of more than two population means, the suitable test statistic is
(a) Student's t statistic (b) Fisher's F statistic
(c) Bartlett test statistic (d) Mahalanob's D^2 statistic

Q.33 If t statistic follows student's t distribution with n degree of freedom then t^2 follows -

(a) Snedecor's F distribution with d.f. (1, n)
(b) Snedecor's F distribution with d.f. (n, 1)
(c) Chi square distribution with d.f. "n"
(d) Chi-square distribution with d.f. "n^2"

Key Answers

Section-A : True False

1. True	2. True	3. True
4. True	5. True	6. True
7. True	8. True	9. True
10. True	11. True	12. True
13. True	14. True	15. True
16. True	17. True	18. True
19. True	20. True	21. True
22. True	23. True	24. True
25. True	26. True	27. True
28. True	29. True	30. True
31. True	32. True	33. True
34. True	35. True	36. True
37. True	38. True	39. True
40. True	41. True	42. True
43. True	44. True	45. True
46. True	47. True	48. True
49. True		

Section-B : Fill in the Blanks

1. Small	2. Large sample
3. $F(V_1, V_2)$	4. Unknown
5. s_2^2 / s_1^2	6. $(n_1\text{-}1, n_2\text{-}1)$
7. Unknown	8. Independent
9. Normal distribution	10. $\sqrt{s^2/n}$
11. x^2 / n	12. Symmetric
13. Zero	14. Zero

15. $\frac{n}{n-2}$

16. Zero

17. $\left(\frac{n-2}{n-4}\right)$

18. Standard normal

19. α

20. $(1 - \alpha)$

21. $\alpha /2$

22. α

23. 2α

24. $\sqrt{\frac{s^2}{n}}$

25. $(\bar{x} - \mu_0)$

26. Significant

27. $\left(\frac{1}{n_1}+\frac{1}{n_2}\right)$

28. $\left(\frac{s_1^2}{n_1}+\frac{s_2^2}{n_2}\right)$

29. $v(\bar{x}_1) + v(\bar{x}_2)$

30. $(n_1, - 1)$

31. Paired t statistic

32. More

33. Algebraic

34. Positive, zero

35. Zero

36. $\sqrt{\frac{s_d^2}{n}}$

37. $\frac{\left(\sum_i^n di\right)^2}{n}$

38. $\frac{v_2}{v_2 - 1}$

39. $\left(\frac{v_2-1}{v_1}\right)$

40. Positively

41. F, (n - h)

42. (h - 1, n - h)

43. $\left(\frac{\eta^2-r^2}{1-\eta^2}\right)$

44. (h - 2, n - 2)

45. F statistic

46. (1, n)

47. x^2 statistic

Section -C : Multiple Choice Questions

1. (c)	2. (a)	3. (b)
4. (c)	5. (d)	6, (c)
7. (a)	8. (a)	9. (d)
10. (b)	11. (a)	12. (c)

13. (c) 14. (a) 15. (b)
16. (d) 17. (b) 18. (c)
19. (d) 20. (b) 21. (b)
22. (a) 23. (c) 24. (b)
25. (d) 26. (c) 27. (d)
28. (b) 29. (a) 30. (b)
31. (a) 32. (b) 33. (a)

❑❑❑

Chapter - 15

Simple Correlation and Regression Analysis

15.1 Section-A : Write True / False

Q.1 In bivariate distribution if the change in one variable affects a change in the other variable, the variables are said to be correlated.

Q.2 If the change in both the variables takes place in the same direction then the variables are said to have a positive or direct correlation.

Q.3 The pair of variables say, Income and expenditure in house hold or the height and weight of a group of children possess the direct correlation between them.

Q.4 If the change in both the variables takes place in oposite directions, then the variables are said to have "Negative or inverse" correlation.

Q.5 The pairs of variables say '(i) Price and demand of a commodities or (ii) Volume and pressure of a perfect gas" have a negative correlation.

Q.6 The two variables are said to have a perfect positive correlation if the deviation in one variable is followed by a corresponding and proportionate deviation in the other variable.

Q.7 There are some funny combinations of variables between which the correlation are said to be "spurious or nonsense".

Q.8 The pairs of variables like - (i) "Number of runs scored by a batsman and the consumption of fertilizer in the local market" or (ii) "The number of space flights and the population of tiger", are the example of spurious correlation.

Q.9 The scatter diagram is the simplest way of diagramatic representation of bivariate population.

Q.10 In scatter diagram, if the points are very dense, we should expect for existance of high correlation between the variables.

Q.11 If in the scatter diagram, the points are widely scattered then we expect the existance of very poor correlation.

Q.12 Karl pearson was a "British-Biometrician" who developed the formula to measur the correlation coefficient.

Q.13 The simple correlation between two variables indicates the existense of "linear relationship" between them.

Q.14 The simple correlation coefficient p xy give a measure of "Intensity or Degree of" linear relationship between two variables.

Q.15 The simple correlation coefficient between two variables (X, Y) is denoted by pxy and is defined as

$$\rho_{x,y} = \frac{COV(x,y)}{\sigma_x \sigma_y}$$

Q.16. The samplecovariance (X, Y) is defined as

$$COV(X, Y) = \frac{1}{n}\sum_{i}^{n}(xi - \bar{x})(yi - \bar{y})$$

Q.17 The sample COV (X, Y) can also be expressed as

$$COV(X, Y) = \left(\frac{1}{n}\sum_{i}^{n} xiyi - \bar{x}_n \bar{y}_n\right)$$

Q.18 If γxy for a standard data is equal to. zero than the scatter diagram is as:

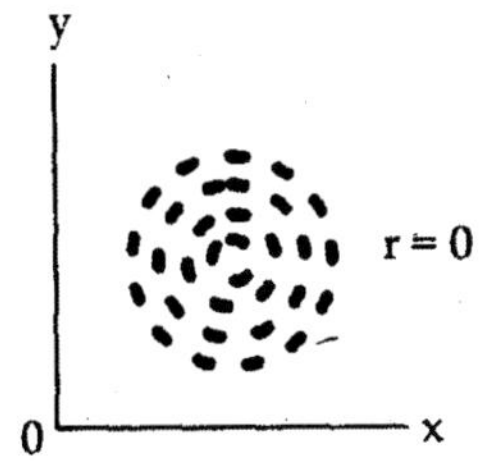

Q.19 For γxy < 0, the scatter diagram for such standard data is

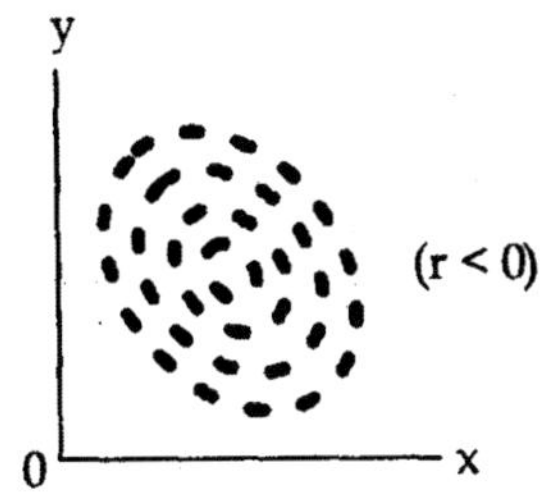

Q.20 If $r_{xy} > 0$, then the scatter diagram for such standard data is like

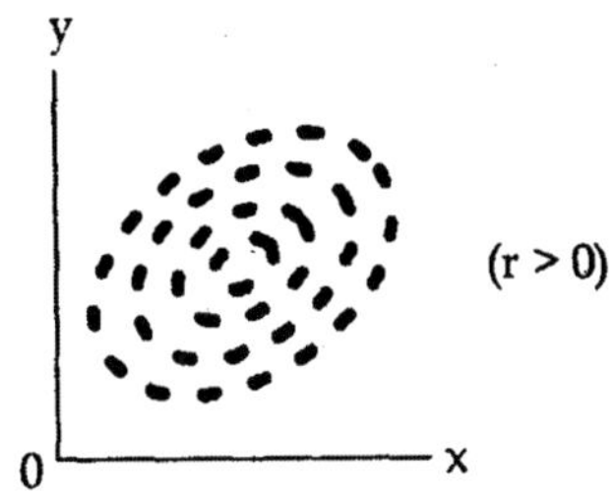

Q.21 If $r_{xy} = -1$ then the scatter diagram for such data is like (straight line with slope < 0).

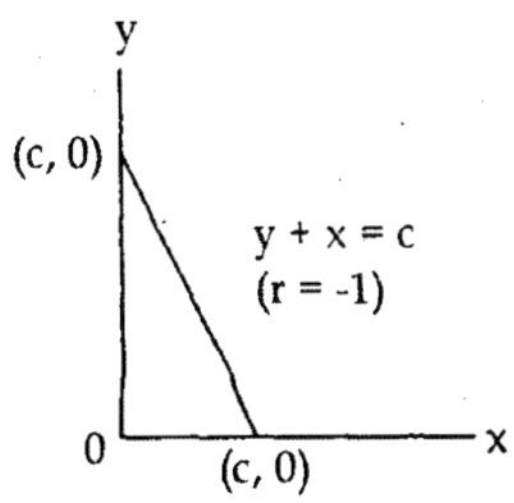

Q.22 If γxy = + 1 then the scatter diagram for such standard data is like (Straight line with slope > 0).

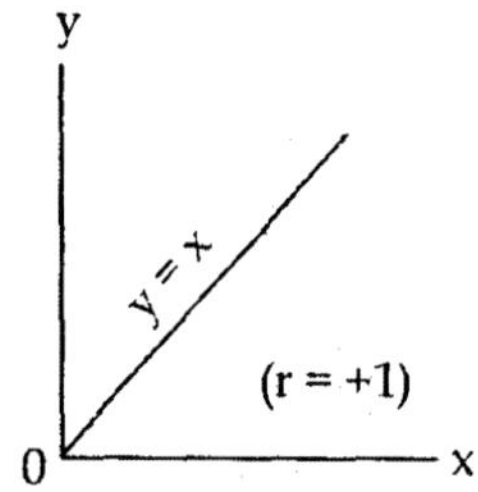

Q.23 Sometimes, we can write COV (X, Y) = $\sigma_{xy} = \sigma_{yx}$.

Q.24 Karl Pearson correlation coefficient is also known as "Product moment correlation coefficient".

Q.25 In terms of mathematical expectation, the COV (X, Y) is written as

COV (X, Y) = E [{X - E(X)} (Y - E(Y)}]

Q.26 The limit for correlation coefficient r_{xy} is as

$(-1 \leq r_{xy} \leq 1)$

Q.27 If $r_{xy} = 1$, then correlation is perfect and positive.

Q.28 If $r_{xy} = -1$, then correlation is perfect but negative.

Q.29 The correlation coefficeient γxy is independent of change of origin and scale both.

Q.30 If we have u = a x + b, and V = cy + d for a > 0, c > 0, then $r_{uv} = r_{xy}$.

Q.31 Two independent variables x and y are always uncorrelated.

Q.32 If $r_{xy} = 0$ then it is not necessary that (X, Y) will be independent.

Q.33 If $r_{xy} = 0$ then it simply indicates the complete absense of linear relationship between (X, Y). But in that situation some other kind of relationship say, quadratic, cubic, trigonometric may exist between (X, Y).

Q.34 If X ~ N (0, 1) and $y = x^2$ then γxy = 0

Q.35 Let X be a r.v. with p.d.f. f(x) = ½, -1 < x < 1 and $Y = X^2$ then γ(X, Y) = 0.

Q.36 If (X, Y) are jointly normally distributed with ρ (X, Y) = 0, then both the variables (X, Y) are independent.

Q.37 Let the bivariate data is like
X = -3, -2, -1, 0, 1, 2, 3
Y = 9, 4, 1, 0, 1, 4, 9
then $y = x^2$ but r (X, Y) = 0 i.e. Y and X have zero correlation but they are related.

Q.38 If r is the correlation coefficient between two r.v. (X, Y) based on a sample of size n then its standard error is equal to
S.E. (r) = $(1 - r^2) / \sqrt{n}$

Q.39 To test the reliability of an observed correlation coefficient γ , the probable error (P.E.(r)) is given by P.E.(r) = $0.6745 \dfrac{(1-r^2)}{\sqrt{n}}$

Q.40 If r < P.E. (r), then correlation is not at all significant .

Q.41 If r < 6 P.E. (r), the r is definitely significant.

Q.42 The population correlation coefficient ρ is expected to vary in the limit given as
r - P.E. (r) ≤ p ≤ r + P.E. (r)

Q.43 Let (Xi, Yi, i = 1, 2, ... , n) be the ranks of the ith individual in two characteristics A & B respectively. then correlation coefficient between (Xi, Yi) is known as rank correlation between A and B.

Q.44 In general the ranks Xi ≠ Yi, i = 1, 2, ..., n and let di = (xi - yi). Then spearman' s rank correlation coefficient (ρ) is given by

$$\rho = \left(1 - \frac{6\sum^{n} d_i^2}{n(n^2-1)} \right)$$

Q.45 The rank correlation. Coefficient p lies in the limit given as
(-1 ≤ ρ ≤ 1)

Q.46 The term regression" means "stepping back towards the average".

Q.47 Regression analysis is the mathematical measure of the average relationship between two or more variables.

Q.48 In regression analysis, independent variable is also known as regressor or predictor or explanatory variable.

Q.49 In regression study, the dependent variable is also known as "regressed or explained or study" variable.

Q.50 The line of regression is the line which gives the best estimate of the dependent variable for any specific value of an independent variable.

Q.51 The line of regression is the line of "best fit" and is obtained by the "principle of least squares".

Q.52 In case of two r.v's (X, Y), there are two lines of regression.

Q.53 The line of regression of Y on Xis given by the relation $Y = y_n + byx (X - x_n)$.

Q.54 The coefficient byx is known as the regression coefficient of Y on X.

Q.55 The coefficient byx gives the amount of change in the dependent variable Y for the unit change in an independent variable x.

Q.56 The coefficient byx gives the "rate of change" in the dependent variable Y.

Q.57 The coefficient byx gives the "measure of slope or gradient" of regression line of Y on X.

Q.58 The coefficient byx can be calculated from the formula -

$$byx = \frac{COV(x,y)}{v(x)} = \frac{\sum_{i}^{n} xiyi - \left(\sum xi \sum yi\right)/n}{\sum^{n} x_i^2 - \left(\sum xi\right)^2/n}$$

Q.59 The coefficient byx can also be calculated from the relation

$$byx = r\frac{\sigma_y}{\sigma_x}$$

Q.60 The sign of both the regression coefficient i.e. byx and bxy are always same.

Q. 61 If one of the regression coefficients is greater than 1 then other is less than 1 such that byx bxy $\leq$ 1.0.

Q.62 The following relation exists between regression coefficients and correlation coefficient.

(byx.bxy = r^2_{xy})

Q.63 The two lines of regression of (Y on X) and of (X on Y) pass through the point ($\overline{x}$, y).

Q.64 The point ($\overline{x}$, y) can be obtained as the point of intersection of two lines of regression.

Q.65 In case of perfect correlation ie r = ± 1, the equation of Y on X becomes

$$y = y_n \pm \frac{\sigma_y}{\sigma_x}(x - \overline{x}_n)$$

Q.66 In case of perfect correlation ie r = ± 1, we have

$\left(\frac{y-\overline{y}}{\sigma_y}\right) = \pm\left(\frac{x-\overline{x}}{\sigma_x}\right)$ and both the lines of regression conside and we get only one line.

Q.67 We have the relation that

$r = \pm\sqrt{byxbxy}$ and we consider the same sign for "r" as is of regression coefficients.

Q.68 The arithematic mean of two regression coefficients is greater than correlation coefficient ie

$$\left|\frac{byx + bxy}{2}\right| > |r|$$

Q.69 Regression coefficient is independent of change of origin but not of scale.

Q.70 Let $U = \frac{X - a}{h}$, $V = \frac{Y - b}{k}$ then we have

$byx = \frac{k}{a}$. bvu and $b_{xy} = \frac{h}{a}$. buv

Q.71 If θ is the acute angle between two lines of regression then

$$\theta = \tan^{-1}\left[\frac{1-r^2}{|r|}\left(\frac{\sigma_y\sigma_x}{\sigma_x^2+\sigma_y^2}\right)\right]$$

Q.72 If r = 0 then tan 0 = ∞ and $\theta = \frac{\pi}{2}$.

Q.73 If two variables are uncorrelated, then lines of regression become perpendicular to each other.

Q.74 If r = ± 1, tan θ = 0 and θ = 0 or π.

Q.75 In case of r = ± 1, the two lines of regression concide.

Q.76 If θ_1 is the acute angle between two lines of regression then

$$\theta_1 = \tan^{-1}\left[\frac{1-r^2}{|r|}\left(\frac{\sigma_y\sigma_x}{\sigma_x^2+\sigma_y^2}\right)\right]$$ and obtuse angle $\theta_2 = (\pi - \theta_1)$.

Q.77 When γ = 0 then lines of regression are as
$Y = \bar{y}$ and $X = \bar{x}$

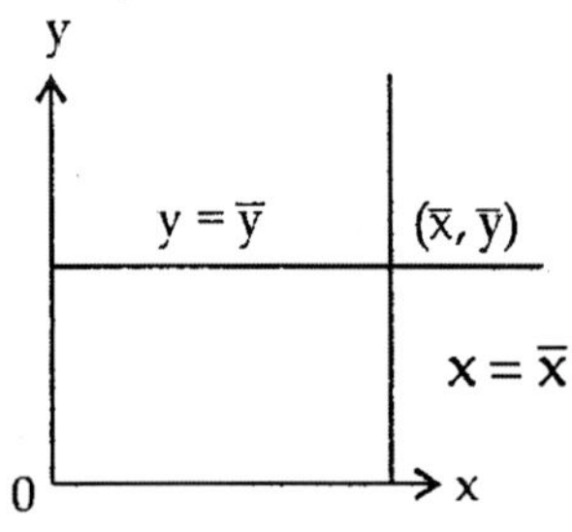

Q.78 When two variables are uncorrelated then two lines of regression of Y on X & X on Y are parallel to X axis and Y axis respectively.

Q.79 For higher degree of correlation between two variables, the angle between two lines is smaller and vice versa.

Q.80 The standard error of an estimated value say $\hat{\bar{Y}}$ is given as
S.E. ($\hat{\bar{Y}}$) = $\sigma_y \sqrt{(1 - r^2)}$

Q.81 The departure of r^2 from unity, indicates the departure of relationship from linearity.

Q.82 $r(Y, \hat{Y}) = r(X, Y)$

Q.83 The conditional mean E (Y/X = x) for a continuous distribution is called the regression function of Y an X and the graph of this function of X is known as the "Regression curve of Y on X or the regression curve for the mean of Y".

Q.84 Geometrically, the regression function represents the y co-ordinate of the centre of mass of bivariate probability mass in the infinitesimal vertical strip bounded by x and (x + dx).

Q.85 If one of the regressions is linear, it does not mean that other will also be linear.

Q.86 If regression of Y on X is linear then it is expressed as

$$E(y/x) = \bar{y} + r\frac{\sigma_y}{\sigma_x}(x - \bar{x})$$

Some More Topies on Correlation & Regression

Q.87 Correlation ratio "η_{xy}" is an appropriate measure of curvilinear relationship between two variables (X, Y).

Q.88 The correlation ratio "η" measures the concentration of points about the curve of best fit.

Q.89 If curvilinear relationship exists between two variables (X, Y) then $|\eta| > |r|$.

Q.90 The correlation ratio of Y on X usually denoted by η_{yx} is given by

$$\eta_{yx}^2 = \left(1 - \frac{\sigma_{ey}^2}{\sigma_y^2}\right)$$

where $\sigma_{ey}^2 = \frac{1}{N}\sum_i\sum_j fij\left(y_{ij} - \bar{y}_i\right)^2$ and $\sigma_y^2 = \frac{1}{N}\sum_i\sum_j f_{ij}\left(y_{ij} - \bar{y}\right)^2$

Q.91 A convenient expression for η_{yx} can be obtained as

$$\eta_{yx}^2 = \frac{\sigma_{my}^2}{\sigma_y^2} \text{ where } N\sigma_{my}^2 = \sum_i ni\left(\bar{y}_i - \bar{y}\right)^2 = \sum_i \frac{T_i^2}{ni} - \frac{T^2}{N}$$

where $T = \sum_i\sum_j f_{ij}y_{ij}$

Q.92 We have the relation between $N\sigma_{em}^2$, σ_y^2 and em η_{yx} as $\sigma_{em}^2 = \sigma_y^2(1 - \eta_{yx}^2)$

Q.93 The lira it for correlation ratio η_{yx} is as $(-1 \leq \eta_{yx} \leq 1)$

Q.94 It is noticeable that η_{yx} is not independent of the classification of the data.

Q.95 As the class intervals become narrower η_{yx} approaches unity, since in that case $\sigma^2_{my} \to \sigma^2_y$

Q.96 The correlation ratio η^2_{yx} is independent of change of origin and scale of measurements.

Q.97 In general we notice that $\eta_{yx} \neq \eta_{yx}$.

Q.98 In term of expectation, correlation ratio η_{yx} is defined as

$$\eta^2_{yx} = \frac{E_y[E(y/x)-E(y)]^2}{[y-E(y)]^2}$$

Q.99 η_{yx} can be defined as

$$\eta^2_{yx} = \frac{E_x[E(x/y)-E(x)]^2}{[x-E(x)]^2} = \frac{E_y[E(x/y)-E(x)]^2}{\sigma^2_x}$$

Q.100 For completely random scattering of dots with no trend, both $r = \eta = 0$.

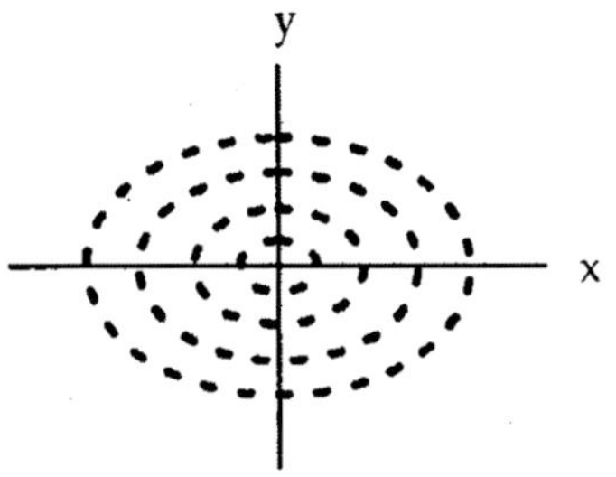

Q.101 If dots precisely lie on the line then $r = \eta = 1.0$

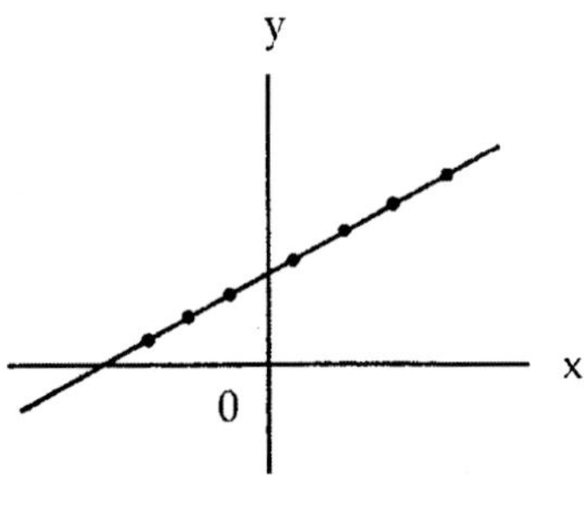

Q.102 If dots lie on a curve such that no ordinate cuts it more than once, $\eta_{yx} = 1$, and if further more dots are syjmmetrically placed about y axis, then $\eta_{yx} = r = 0$.

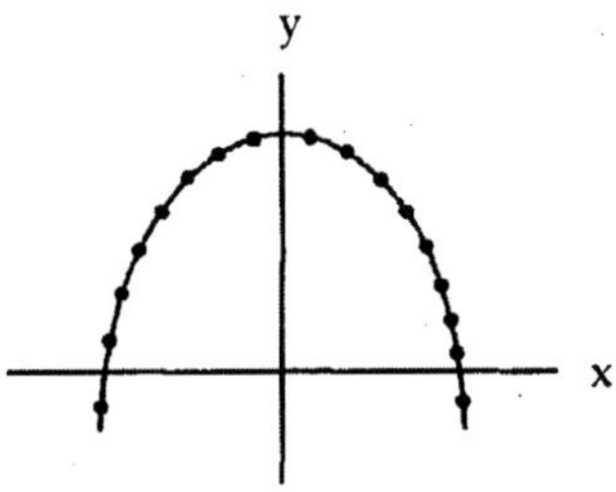

Q.103 If $\eta_{yx} > r$ then the dots are scattered definitely around a curved trend line.

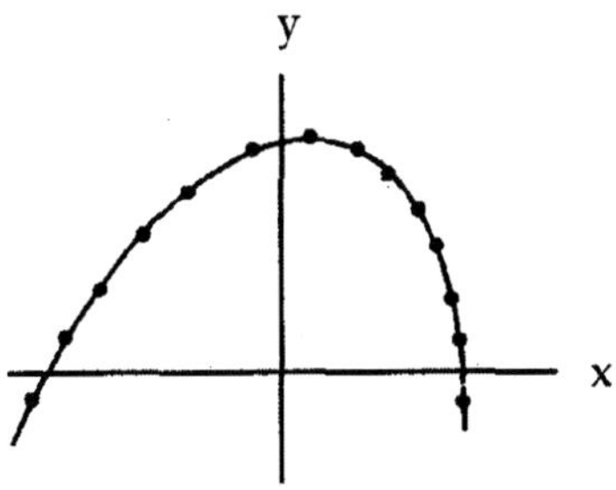

Q.104 Intraclass correlation means within class correlation.

Q.105 In case of study of intraclass correlation, both the variables measure the same characteristics.

Q.106 Let xij; (i = 1, 2, ...n, j = 1, 2, ...k) denotes the measurement on the jth member in the ith family. Then intraclass correlation coefficient f_w is defined as

$$\rho_\omega = \frac{1}{(k-1)}\left\{k.\frac{\sigma_m^2}{\sigma^2} - 1\right\}$$

where σ^2 denotes the variance of X and σ^2_m the variance of means of families.

Q.107 The limits for ρ_ω w is like $\left(-\frac{1}{(k-1)} \leq \rho_\omega \leq 1\right)$

Q.108 In case of bivariate normal distribution, the regression of y on x is linear.

Q.109 In case of bivariate normal distribution, the arrays are homoscedastic ie the common variance of estimate of y in each array is given by $\sigma_y^2\,(1 - \rho^2)$.

Q.110 The distribution of Y in different arrays of bivariate normal distn is normal.

Q.111 The p.d.f. f (x, y) of bivariate normal distribution is written as

$$f_{xy}(x,y)=\frac{1}{2\pi\sigma_1\sigma_2\sqrt{1-\rho^2}}e\left[-\frac{1}{2(1-\rho^2)}\left\{\frac{(x-\mu)^2}{\sigma_1^2}-2\rho\frac{(x-\mu_1)(y-\mu_2)}{\sigma_1\sigma_2}+\frac{(y-\mu_2)}{\sigma_2^2}\right\}\right]$$

Q.112 In bivariate normal population, $(\mu_1, \mu_2, \sigma_1^2, \sigma_2^2, \rho)$ are five parameters.

Q.113 In bivariate normal distribution, the variables (X, Y) are said to be normally correlated and the surface Z = f(x, y) is known as "normal correlation surface" which is in three dimension.

Q.114 Bivariate normal distribution is abbriviated as

$(X, Y) \sim N(\mu_1, \mu_2, \sigma_1^2, \sigma_2^2, \rho)$.

Q.115 In bivariate normal distribution, if $\mu_1 = \mu_2 = 0$,

$\sigma_1^2 = \sigma_2^2 = 1$, then $(X, Y) \sim N(0, 0, 1, 1, p)$.

Q.116 The m.g.f. of bivariate normal distribution is given as

$$M_{x,y}(t_1, t_2) = e^{\left[\frac{1}{2}(t_1^2+t_2^2+2\rho t_1 t_2)\right]}$$

Q.117 Let $(X, Y) \sim N(\mu_1, \mu_2, \sigma_1^2, \sigma_2^2, \rho)$ then (X, Y) are independent iff $p = 0$.

Q.118 The (X, Y) possesses a bivariate normal distribution fff every linear combination of x and y viz. ax + by, $a \neq 0 \neq b$, is a normal variate.

Q.119 In case of bivariate normal distribution, the marginal distribution of Y for given X = x is given as

$Y \sim N(\mu_y, \sigma_y^2)$

Q.120 We may have joint p.d.f. f(x, y) of (x, y) which is "not normal" but the marginal p.d.f. f(x) or f(y) may still be normal.

Q.121 The conditional distribution of y for fixed x is normal and is given as

$$(Y/X = x) \sim N\left[\mu_y + \rho\frac{\sigma_y}{\sigma_x}(x-\mu_x),\ \sigma_y^2(1-\rho^2)\right]$$

Multiple Correlation and Regression

Q.122 If we are interested in studying the joint effect of a group of variables on a variable which is not included in that group then our study is known as the "study of multiple correlation and multiple regression".

Q.123 The correlation and regression between only two variates after the elimination of linear effect of the other variates in them is known as the "partial correlation and partial regression".

Q.124 Let us consider a distribution of three random variables say X_1, X_2 & X_3. We assume that all the three variables have been measured from their respective means. Then a plane of regression of X_1 on X_2, X_3 is written as

$$\hat{X}_1 = b_{12.3}X_2 + b_{13.2}X_3$$

Q.125 In Q(124), the coefficients $b_{12.3}$ and $b_{13.2}$ are known as partial regression coefficients of x_1 on x_2 and x_1 on x_3 respectively.

Q.126 The quantity say

$e_{1.23} = b_{12.3}\ x_2 + b_{13.2}\ x_3$ is called "an estimate of x_1" as given by the plane of x_1 on x_2 and x_3.

Q.127 The quantity say $X_{1.23} = (X_1 - e_{1.23})$ is called the "error of estimate or residual".

Q.128 The subscripts before the dot(.) are known as primary subscripts.

Q.129 The subscripts after the dot(.) are known as secondary subscripts.

Q.130 The number of secondary subscripts in the partial regression coefficient is known as the "order of regression coefficients".

Q.131 The order of the partial regression $b_{12.34\,..........\,n}$ is equal to (n - 2).

Q.132 It is noticeable that the order in which the primary subscripts are written in partial regression coefficient has a significant meaning but the same thing in secondary subscripts is meaning less.

Q.133 In regression coefficient $b_{12.34 \ldots\ldots n}$, the variable x_1 is dependent but X_2 is an independent variable.

Q.134 In the residual $X_{1.234 \ldots\ldots n}$, the order of residual is equal to (n - 1).

Q.135 The sum of the product of any residual of order zero, with any other residual of higher order is zero, provided the subscript of the former occurs in the secondary subscripts of the latter.

Q.136 We can write that

$\Sigma X_3 X_{1.23} = 0$ and Σ xi $x_{1.23 \ldots\ldots n} = 0 \ \underline{V} \ i = 2, 3, \ldots\ldots\ldots\ldots n.$

Q.137 The multiple correlation coefficient R1.23 is the correlation coefficient between X_1 and its estimated value e1.23 as given by the plane of regression of X_1 on X_2 and X_3.

Q.138 In mathematical notations, the multiple correlation coefficient $R_{1.23}$ can be worked out from the expression -

$$R_{1.23} = \frac{\text{COV}(X_1, e_{1.23})}{\sqrt{v(X_1)v(e_{1.23})}}$$

Q.139 The multiple correlation coefficient $R_{1.23}$ in terms of product moment correlation coefficients between the pairs of variables is expressed as

$$R^2_{1.23} = \frac{r^2_{12} + r^2_{13} - 2r_{12}r_{23}r_{31}}{(1 - r^2_{23})}$$

Q.140 Multiple correlation coefficient measures the closeness of the association between observed values and expected value of a variable obtained from the multiple regression of that variable on other variables.

Q.141 Multiple correlation coe-efficient between observed values and expected values when worked out from a linear relation of the variables determined by the method of least square, is always greater than that obtained through any other linear combination of the variables.

Q.142 The limits of $R_{1.23}$ lies as -
$0 < R_{1.23} < 1.0$

Q.143 If $R_{1.23} = 1.0$ then association is perfect and all the regression residuals are equal to zero and as such $\sigma_{1.23}^2 = 0$.

Q.144 If $R_{1.23} = 0$, then all total partial correlations involuing x_1, are zero. So X_1 is completely uncorrelated with all other variables.

Q.145 $R_{1.23}$ is not less than any total correlation coefficient i.e. $R_{1.23} \geq r_{12}, r_{23}, r_{31}$.

Q.146 The correlation between X_1 and X_2 if the effect of X_3 on each of X_1 and X_2 were eliminated, is known as "partial correlation".

Q.147 The regression coefficient between X_1 and X_2 after the linear effect of X_3 on each of them has been eliminated, is called the "partial regression coefficient".

Q.148 The residual $X_{1.3} = (X_1 - b_{13} X_3)$ may be regarded as that part of the variable X_1 which remains after the linear effect of X_3 has been eliminated.

Q.149 The partial correlation coefficient $r_{1.23}$ can be calculated from the relation

$$r_{12.3} = \frac{COV(x_{1.3}, x_{2.3})}{\sqrt{v(x_{1.3})v(x_{2.3})}}$$

Q.150 $COV(x_{1.3}, x_{2.3}) = \sigma_1 \sigma_2 (r_{12} - r_{13} r_{23})$.

Q.151 $V(X_{1.3}) = \sigma_1^2 (1 - r_{13}^2)$.

Q.152 The partial correlation coefficient $r_{1.23}$ can be calculated from an expression written in terms of pairwise product moment correlation coefficients written as

$$r_{12.3} = \frac{(r_{12} - r_{13} r_{23})}{\sqrt{(1 - r_{13}^2)(1 - r_{23}^2)}}$$

Q.153 For partial regression coefficient $b_{12.3}$ we can write that

$b_{12.3} = \dfrac{\sum X_{1.3} X_{2.3}}{\sum X_{22}^2}$ which states that $b_{12.3}$ is the coefficient of regression of $X_{1.3}$ on $X_{2.3}$

Q.154 Like simple correlation and regression coefficients, we have the result that $r_{12.3}^2 = b_{12.3} b_{21.3}$

Q.155 If $r_{12.3} = 0$, then we have $r_{12} = r_{13} r_{23}$. This means that r_{12} will not be zero if x_3 is correlated with both x_1 and x_2.

Q.156 Partial correlation coefficient helps in deciding whether to include or not an additional independent variable in regression analysis.

Q.157 The values $\sigma_1^2 (1 - r_{12}^2)$ and $\sigma_1^2 (1 - r_{13}^2)$ are the residual variances if X_1 is estimated from X_2 and X_3 individually.

Q.158 The quantity $\sigma_1^2 (1 - R_{1.23}^2)$ is the residual variance if X_1 is estimated from X_2 and X_3 taken together.

Q.159 Let R be the observed multiple correlation coefficient of a variate Y with k other variates in a random sample of size n drawn from a (k + 1) variate population. Prof. R.A. Fisher proved that under Ho: R = 0, the statistic.

$$F = \frac{R^2}{1-R} \cdot \frac{n-(k+1)}{k} \sim F_{k,\, n-(k+1)}$$

Q.160 Let $r_{12.34 \ldots\ldots\ (k+2)}$ be the partial correlation coefficient of order k in a sample of size n drawn from a (k + 2) variate population. Prof. R.A. Fisher proved that under Ho : $p_{12.34 \ldots\ldots\ (k+2)} = 0$, the statistic

$$t_{n-(k+2)} = \frac{|r_{12}|}{\sqrt{1-r_{12}^2}} \cdot \sqrt{n-(k+2)} \sim t_{n-(k+2)}$$

Q.161 The coefficient of determination R^2 is defined as

$$R2 = \frac{\text{Variance explained in y}}{\text{Total variance in y}}$$

Q.162 The coefficient $(1 - R^2)$ is known as the "coefficient of non-determination".

Q.163 If $R^2 = 0.64$, it means that only 64% of variation present in the dependent variable Y has been explained due to effect of independent variates X's.

Q.164 A test which gives quite satisfactory result inspite of some departure from the basic assumptions, is called a "Robust test".

Q.165 The student's t test is a robust test.

Q.166 Let (yi, xi) be a small sample of size n < 30, drawn from a bivariate normal population with correlation coefficient p_{yx}. Then to test the null hypothesis Ho : $P_{yx} = 0$, the appropriate test statistic is student's statistic.

Q.167 To test H0 : $\rho_{yx} = 0$, the test statistic "t" is defined as

$$t = \frac{|r|\sqrt{n-2}}{\sqrt{1-r^2}} \sim t_{(n-2)}$$

where r is the sample correlation coefficient.

Q.168 In case of bivariate normal population to test the H0 : byx = k, (k is a constant) the suitable student's t statistic is defined as

$$t = \frac{|byx-k|\sqrt{(n-2)\sum^{n}(xi-\bar{x}n)^2}}{\sqrt{\sum_{i}^{n}(yi-\bar{y}n)^2 - b_{yx}^2\sum_{i}^{n}(xi-\bar{x}n)^2}}$$

Q.169 In Q(168) to test the hypothesis Ho : $b_{yx} = 0$, simply replace k by zero in 't' statistic defined in Q(168).

Q.170 To test the null hypotheses $H_{01} : r_{xy} = 0$ and $H_{02} : b_{yx} = 0$, the d.f. of student's 't' statistic is equal to (n - 2).

Q.171 In case of bivariate normal population, SS due to regression from a sample of size n is given

SS (Due to regression)

$$= b_{yx}^2 \sum_{i}^{n} (xi - \bar{x}n)^2 \text{ or } = b_{yx} \sum_{i=1}^{n} (xi - \bar{x}n(yi - \bar{y}n)) \text{ with 1 d.f.}$$

Q.172 Residual SS or SS about to regression or Error SS is given by

$$\sum_{i}^{n}(yi - yn)^2 - b^2{}_{yx}\sum (xi - \bar{x}n)^2 \text{ or } \sum_{i}^{n}(yi - \bar{y}n)^2 - b_{yx}\sum_{i=1}^{n}(yi - \bar{y}n)(xi - \bar{x}n)$$

with (n - 2) d.f.

Q.173 In bivariate regression analysis, the residual variance is given as

$$\hat{\sigma}^2_{y.x} = \frac{1}{(n-2)}\left[\sum_i^n (yi - \bar{y}n)^2 - b^2 \sum_i^n (xi - \bar{x}n)^2\right]$$

Q.174 The sampling variance of sample regression coefficient b_{yx} is given by

$$v(b_{yx}) = \frac{\sum_i^n (yi - \bar{y}n) - b^2_{yx} \sum (xi - \bar{x}n)^2}{(n-2)\sum_i^n (xi - \bar{x}n)^2}$$

Q.175 The sampling variance of sample correlation coefficient rxy is given as $V(r_{xy}) = (1 - r^2) / (n - 2)$

15.2 Section- B : Fill in the Blanks

Q.1 If the change in one variable affects the change in other variable then, both the variables are said to be ....

Q.2 If the change in two variables takes place in the same direction then the variables are said to have .... or .... correlation.

Q.3 Pairs of variables say (income and expenditure on house-hold or the heights and weight of a group of children), posses a .... corelation.

Q.4 If the change in two variates X and Y takes place in oposite direction then the variates are said to have a " .... or ...." corretation.

Q.5 The pairs of variables say "price and demand of a commodity or volume and pressure of a perfect gass" have a " .... or inverse' correlation.

Q.6 Two variablex are said to have a perfect positive correlation if they deviate in the .... direction with .... amount of deviation.

Q.7 There are some funny combination of variables between which the correlation is said to be "spurious or .... " .

Q.8 Two variables say (No. or runs scored by a batsman and consumption of a fertiliser in a local community) have" ...." correlation.

Q.9 The scatter diagram is the simplest way of diagrametic representation of " .... " data.

Q.10 In scatter diagram, if the points are very dense, we should expect for the existence of " .... " correlation between the variables.

Q.11 If in the scatter diagram, the points are widely scattered then we expect for an existance of very " .... " correlation.

Q.12 Karl pearson was a "British Biometrician" who developed the formula to measure the " .... " between two variables.

Q.13 The simple correlation coefficient between two variables indicates the existence of " .... " between them.

Q.14 The coefficient γxy between two variables (X, Y) gives a measure of "Intensity or .... " of linear relationship.

Q.15 The correlation coefficient γxy between two variables (X, Y) is defined as

$$r_{xy} = \frac{\ldots\ldots}{\sqrt{v(x)\ldots\ldots}}$$

Q.16 The covariance (X, Y) is defined with n pairs of values as COV (X, Y) = ...

Q.17 The COV (X, Y) can also be expressed as

$$\text{COV}(X, Y) = \frac{1}{n}\sum_{i}^{n}(\ldots\ldots - \bar{x}n)(\ldots\ldots)$$

Q.18 For the scatter diagram given as, the γxy for a standard data is equal to ....

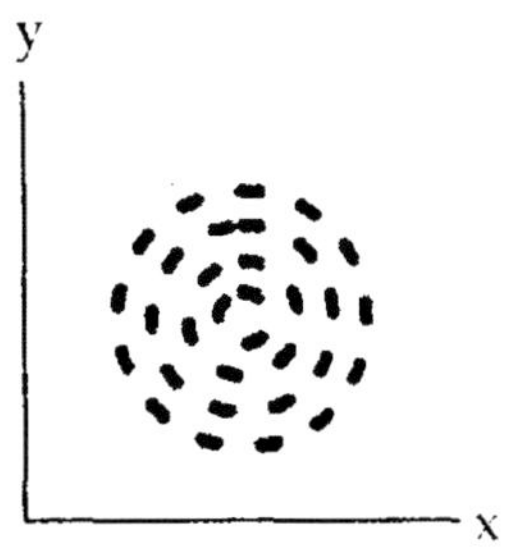

Q.19 The scatter diagram given as, the γxy for such standard data is given as r_{xy} .... zero

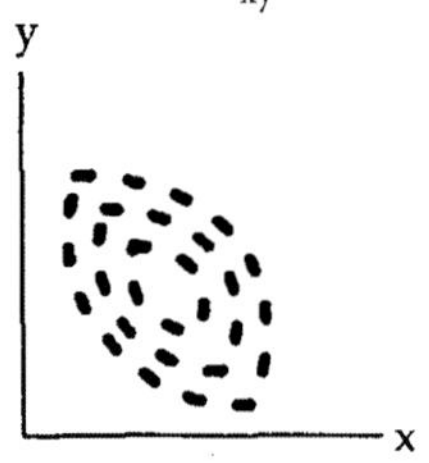

Q.20 For the scatter diagram given as, the value of r_{xy} lies as r_{Xy} .... zero.

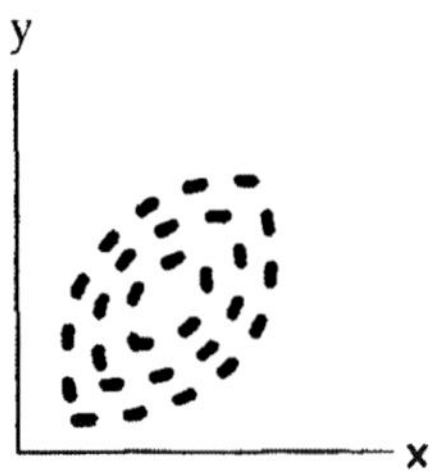

Q.21 For the straight line given in scatter diagram, the value of γxy is equal to ....

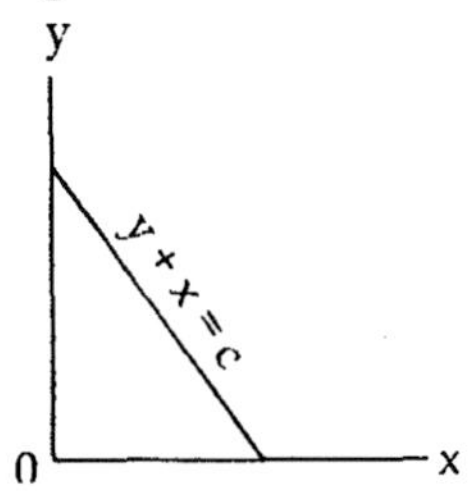

Q.22 For the straight line given in scatter diagram, the value of γxy is equal to ....

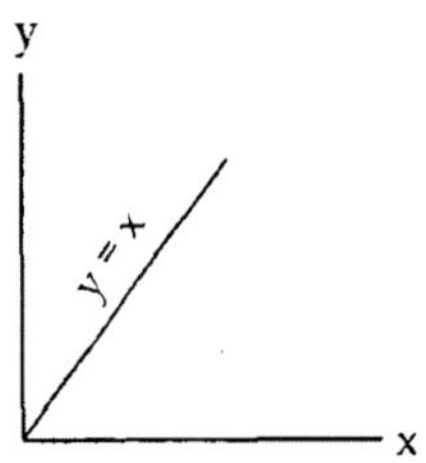

Q.23 Sometimes, we can write
COV (X, Y) = $\sigma_{xy} = \sigma_{.........}$

Q.24 Karl pearson correlation coefficient γxy is also know as " .... "moment correlation coefficient.

Q.25 In terms of mathematical expectation, the COV (X, Y) is written as

COV (X, Y) = E[{ ......} {.......}]

Q.26 The limit of correlation coefficient rxy lies as

.... $\leq r_{xy} \leq$ ....

Q.27 If r_{Xy} = 1, then correlation is .... and ....

Q.28 If r_{xy} = - 1, then correlation is negative and ....

Q.29 Correlation coefficient is " .... " of change of origin and " .... " both.

Q.30 If we have u = ax + b and v = cy + d for (a, b, c, d) > 0, r_{uv} = ...

Q.31 Two independent variables (X, Y) are always ....

Q.32 If r_{xy} = 0 then it is necessary that COV (X, Y) is equal to ....

Q.33 If r_{xy} = 0, then it simply indicates that the complete absense of " .... " relationship between (X, Y).

Q.34 If X ~ N(0, 1) and Y = X^2 then rxy = ...

Q.35 If (X, Y) are jointly normally distributed with ρ (X, Y) = 0, then both variables are ....

Q.36 If two variables (X, Y) are as

X : -3, -2, -1, 0, 1, 2, 3

Y : 9, 4, 1, 0, 1, 4, 9

and Y = X_2 then correlation (X, Y) = ...

Q.37 If r is sample correlation coefficient based on a sample size n then its -

S.E. (r) = ...

Q.38 The probable error P.E. (r) is given by

P.E. (r) = 0.6745 (.........) / $\sqrt{n}$

Q.39 If r < P.E. (r) then correlation is not at all ....

Q.40 If r > 6 P.E. (r) then correlation is definitely ....

Q.41 The population correlation coefficient p is expected to lie in the limits given as r - P.E. (r) < ρ < ....

Q.42 Let (Xi, Yi, i = 1, 2, ...n) be the rank of the ith individual in two characteristics (A, B) respectively then correlation between (A, B) is known as " .... " correlation.

Q.43 In Q(42), generally xi $\neq$ yi, and let di = (xi - yi). Then rank correlation coefficient ρ is given by

$$\rho = \left(1 - \frac{6\sum^{n} d_i^2}{n(\text{.........})}\right)$$

Q.44 The limit of rank correlation coefficient ρ is given as (.............. $\leq$ p $\leq$ ....)

Q.45 The term "regression" means "steping back towards the ...."

Q.46 Regression analysis is the mathematical measure of the average " .... " between two or more variables.

Q.47 In regression analysis, the independent variable is also known as "regressor" or ...., or ...." variable.

Q.48 In regression study, the dependent variable is also known as "..................... or explained or ...." variable.

Q.49 The line of regression is the line which gives the best estimate of the " .... " variable for any specific value of an " .... " variable.

Q.50 The line of regression of best fit, is obtained by the principle of " .... ".

Q.51 In case of two variables (X, Y), the number of lines of regression is equal to ....

Q.52 The line of regression of Y on X is given by the relation $Y = \overline{y}_n +$ ....

Q.53 The coefficient byx is known as the " .... " of .... on ....

Q.54 The coefficient byx gives the amount of change in the " .... " variable Y for the unit change in an " .... " variable X.

Q.55 The coefficient byx gives the "rate of change" in the " .... " variable Y.

Q.56 The coefficient byx gives the measure of "slope or .... " of regression line of .... on ....

Q.57 The coefficient byx can be calculated from the formula b_{yx} = COV (X, Y) / ....

Q.58 The byx can also be worked out from the relation

$$byx = r \frac{....}{....}$$

Q.59 The coefficient (byx, bxy) are always of the " .... " sign.

Q.60 If byx > 1.0 then bxy .... 1.0 such that byx bxy 1.0

Q.61 The following relation between (byx, bxy) and rxy exists as byx bxy = ...

Q.62 Two lines of regression of (Y on X) and (X on Y) passes through the point (......., ...).

Q.63 Two lines of regression of (Y on X) and of (X on Y) intersects at the point (..........., ...)

Q.64 In case of perfect correlation i.e. r = ±, the equation of Y on X becomes $y = \bar{y}_n \pm$ .... $(x - \bar{x}_n)$.

Q.65 In case of perfect correlation ie r = ± 1, we have

$$\frac{Y - \bar{y}}{\sigma_y} = \pm(........)$$

Q.66 In case of r = ± 1, both lines of regression concides and the number of line obtained becomes equal to ....

Q.67 We have the relation that $r = \pm\sqrt{...........}$

Q.68 We have the relation among b_{yx}, b_{xy} and r_{xy} as

$$\left|\frac{b_{yx}+b_{xy}}{2}\right| \ldots\ldots |r_{xy}|$$

Q.69 The change of scale of original data " .... " the regression coefficient.

Q.70 Let $U = \frac{x-a}{h}$, $V = \frac{y-b}{k}$ then we have

$$b_{yx} = \frac{k}{h}\ .b\ldots.\ \text{and}\ b_{xy} = \frac{h}{k}\ .b\ldots\ldots$$

Q.71 If θ is the acute angle between two lines of regression then

$$\theta = \tan^{-1}\left[\frac{1-r^2}{|r|}(\ldots\ldots\ldots)\right]$$

Q.72 If r = 0 then tan θ = - ...., and θ = ...

Q.73 If two variables (X, Y) are uncorrelated then lines of regression becomes " .... " to each other.

Q.74 If r = ± 1, then two lines of regression " .... ".

Q.75 If (θ_1, θ_2) are the acute and obtuse angles respectively between two lines of regression then

$$\theta_2 = (\pi - \theta_1)\ \text{and}\ \theta_1 = \tan^{-1}\left[\frac{1-r^2}{|r|}(\ldots\ldots\ldots)\right]$$

Q.76 If r = ± 1, then tan θ = 0 and θ = 0 or ....

Q.77 When r = 0 then lines of regression are as

Y = ... and X = ...

Q.78 When two variables are uncorrelated then two lines of regression of y on x and of x on y are " .... " to x axis and y axis respectively.

Q.79 For higher degree of correlation between two variables, the angle between two lines is " .... " .

Q.80 The S.E. of an estimated value say $\hat{Y}$ is given as

$$S.E.(\hat{y}) = \sigma_y \surd(\ldots\ldots\ldots)$$

Q.81 The departure of r^2 from unity, indicates the departure of relationship from " ...."

Q.82 We have the relation $r(y, \hat{y}) = r$ (........., ...)

Q.83 The conditional mean $E(Y/X = x)$ for continuous distribution is called the " .... " of y on x.

Q.84 If one of the regression line is linear, it " .... " other will also be linear.

Q.85 If regression of Y on X is linear then it is expressed as $E(Y/x) = \bar{y} + r(x - \bar{x})$.

Some More Topics on Correlation & Regression

Q.86 Correlation ratio "η" is an appropriate measure of " .... " relationship between two variables (X, Y).

Q.87 The correlation ratio "η" measures the concentration of points about the " .... " of best fit.

Q.88 If curvilinear relationship exists between two variables (X, Y) then $|\eta|$ ..... $|r|$

Q.89 The correlation ratio of Y on X can be calculated from

$$\eta_{yx}^2 = \left(1 - \frac{\sigma_{ey}^2}{\ldots\ldots}\right) \text{ where } \sigma_{ey}^2 = \frac{1}{N}\sum_i \sum_j f_{ij}\left(y_{ij} - \bar{y}_i\right)^2$$

Q.90 A convenient expression for n yx can be obtained as

$$\eta_{yx}^2 = \frac{\sigma_{my}^2}{\ldots\ldots\ldots}$$

Q.91 We have relation among σ_{em}^2, σ_y^2 and η_{yx} as $\sigma_{em}^2 = \sigma_y^2$ (..........)

Q.92 The limit for correlation ratio η_{yx} is given as

(....$\leq \eta_{yx} \leq$ ....)

Q.93 It is noticeable that η_{yx} is not " .... " of the classification of data.

Q.94 As the class intervals become narrower, η_{yx} approaches to "...............".

Q.95 If η_{yx} approaches to unity then $\sigma^2_{my} \to$ ....

Q.96 The coefficient η^2_{yx} is " .... " of change of origin and scale of measurements.

Q.97 In general, we notice that η_{yx} .... η_{xy}

Q.98 In terms of mathematical expectation, η_{xy} is defined as

$$\eta^2_{yx} = \frac{E_x\left[E(y/x) - E(y)\right]^2}{E[............]^2}$$

Q.99 We can write $\eta^2_{yx} = \dfrac{E_y\left[E(x/y) - E(x)\right]^2}{E[............]^2}$

Q.100 For completely random scattering of data with no tread, the value of γxy and η are equal to ....

Q.101 If in scatter diagram, the dots precisely lie on the line then $\gamma = n =$ ....

Q.102 If in scatter diagram, the dots lie on a curve such that no ordinate cuts it more than once then $\eta_{xy} =$ ....

Q.103 If in a scatter diagram, the dots are symmetrically placed about y axis on a curved line then $\eta_{yx} = r_{xy} =$ ....

Q.103*If $\eta_{yx} > r_{xy}$, then in a scatter diagram, the dots lie definitely about " .... " in place of a straight line.

Q.104 Intraclass correlation means " .... " class correlation.

Q.105 In case of intraclass correlation, both the variables measures the " .... " characteristic.

Q.106 Let xij, (i = 1, 2, n, j = 1, 2, ..., k) denotes the measurement on the jth member in the ith family. Then ρ_w is defined as

$$\rho_\omega = \frac{1}{(k-1)}\left\{k\frac{......}{\sigma^2} - 1\right\}$$

Q.107 The limits for ρ_w is written as $\left(-\dfrac{1}{(k-1)} \leq \rho_\omega \leq ......\right)$

Q.108 In case of bivariate normal distribution, the regression line of y on x is " .... ".

Q.109 In case of bivariate normal distribution, the common variance of estimate of y in each array is given by V ($\hat{y}$) = σ_y^2 (......)

Q.110 The distribution of y in different array of a bivariate normal distribution is also " .... ".

Q.111 For bivariate normal distribution, we have

$$f(x,y)=\frac{1}{2\pi\sigma_1\sigma_2\sqrt{1-p^2}}e\left[-\frac{1}{....}\left\{\frac{(x-\mu_1)}{\sigma_1^2}-2\rho\frac{(x-\mu_1)(y-\mu_2)}{\sigma_1\sigma_2}+\frac{(y-\mu_2)^2}{\sigma_2^2}\right\}\right]$$

Q.112 In bivariate normal distribution the five parameters are given as (μ_1 .....,........, σ_2^2 ,.....)

Q.113 In bivariate normal distribution, the surface Z = f(x, y) is known as " .... " surface.

Q.114 Bivariate normal distribution is abbreviated as (X, Y) ~ N (μ, ..., ..., $\sigma^2{}_2$, ...)

Q.115 In bivariate normal distribution, if $\mu_1 = \mu_2 = 0$, $\sigma_1^2 = \sigma^2 = 1$, then (X, Y) ~ N(0, 0 ...., ρ)

Q.116 The m.g.f. of bivariate normal distribution is given as

$$M_{x,y}(t_1,t_2)=e^{\left[\frac{1}{2}\left(t_1^2+t_2^2+........\right)\right]}$$

Q.117 Let (X, Y) ~ N(μ_1, μ_2, $\sigma^2{}_1$, $\sigma^2{}_2$, ρ) then (X, Y) are independent iff ρ = ...

Q.118 The (X, Y) possesses a bivariate normal distribution iff every linear function of x and y, is a "................ " variate.

Q.119 In case of bivariate normal distribution, the marginal distribution of Y for given X = x, is given as Y ~ N (μ_y,..........).

Q.120 The conditional distribution of Y for fixed X is normal and is and

$$(y/x=x)\sim N\left[\mu_y+\rho\frac{\sigma_y}{\sigma_x}(x-\mu_x),\ \sigma_y^2(.........)\right]$$

Q.121 In study of multiple correlation and multiple regression analysis, the joint effect of a group of " .... " variables is considered on a regressed variable Y.

Q.122 The correlation and regression between only two variables after the elimination of linear effect of other variables or them, is known as " .... " correlation and partial " .... " .

Q.123 After assuming the three variables (X_1, X_2, X_3) being measured from their respective means, the plane of regression of X_1 on X_2 and X_3 is written as ,

$\hat{X}_1 = b_{12.3} + b_{13.2}$ ....

Q.124 In Q(123), the coefficients b12.3 and b13.2 are known as partial regression coefficients of " .... " and of X_1 on .... respectively.

Q.125 The quantity say

$e_{1.23} = b_{12.3} X_2 + b_{13.2} X_3$ is known as "An .... given by the plane of X_1 on X_2 and X_3.

Q.126 The quantity say $X_{1.23} = (X_1 - e_{1.23})$

is known as " .... of estimate or .... " .

Q.127 In partial correlation and regression coefficients, the subscripts before the dot (.) are known as " .... " subscripts.

Q.128 In multiple correlation and regression analysis, the subscripts after the dot(.) is known as " .... " subscripts.

Q.129 The number of secondary subscripts in partial regression coefficient is known as the " .... " of regression coefficient.

Q.130 The order of the partial regression coefficient $b_{12.34........n}$ is equal to (..............)

Q.131 It is noticeable that the order in which the primary subscripts are written in partial " .... " has a significant meaning.

Q.132 It is noticeable that the ofder in which the secondary subscripts are written in " .... " is meaning less.

Q.133 The order in whch either primary subscripts or secondary subscripts are written in "partial correlation" is ....

Q.134 In the regression coefficient $b_{12.34 \ldots\ldots n}$, the variable X_1 is " .... " and X_2 is " .... " respectively.

Q.135 In the residual $X_{1.23 \ldots\ldots n}$, the order of residual is equal to ....

Q.136 The product sum $\sum_i Xi \; X_{1.23 \ldots\ldots n}$ = ...
for all i = 2, 3, 4, ... n.

Q.137 R1.23 is the correlation coefficient between X_1 and its " .... " value obtained from the regression of X_1 on .... and ....

Q.138 $R_{1.23}$ can be worked out from the expression

$$R_{1.23} = \frac{COV\,(X_1 , e_{1.23})}{\sqrt{\ldots\ldots\ldots\ldots\ldots\ldots}}$$

Q.139 $R_{1.23}$ in terms of pairwise correlation coefficients is expressed as

$$R^2_{1.23} = \frac{r^2_{12} + \ldots\ldots -2r_{12}\, r_{23}\, r_{31}}{(1 - \ldots\ldots\ldots)}$$

Q.140 The multiple correlation coefficient measures the closeness of the association between " .... " values and " .... " values of a variable obtained from multiple regression.

Q.141 Multiple correlation determined by the method of least square is always " .... " than that obtained through any other linear combination of variables.

Q.142 The limits of $R_{1.23}$ lies as
$(\ldots\ldots \leq R_{1.23} \leq \ldots\ldots\ldots)$

Q.143 If $R_{1.23}$ = 1.0, then all the regression residuals are equal to ....

Q.144 If $R_{1.23}$ = 0 then X_1 is completely " .... " with X_2 and X_3.

Q.145 $R_{1.23}$ is not " .... " than any total correlation r_{12}, r_{23} or r_{31} .

Q.146 The correlation between (X_1, X_2) if the effect of X_3 on them seperately were eliminated, is known as " ...."

Q.147 The regression coefficient between X_1 and X_2 after the linear effect of X_3 on each of them has been eliminated, is known as " .... " coefficient.

Q.148 The residual $X_{1.3} = (X_1 - b_{13} X_3)$ may be regardede as that part of the variable X_1 which remains left after the " .... " on x_1 has ben eliminated.

Q.149 The coefficient $r_{12.3}$ can be calculated from the relation -

$$r_{12.3} = \frac{COV\ (.........,\ ....)}{\sqrt{v\ (X_{1.3})v(X_{2.3})}}$$

Q.150 We have the relation that
$COV\ (x_{1.3}, x_{2.3}) = \sigma_1\ \sigma_2\ (r_{12}\ ....)$

Q.151 We have the relation
$V(x_{1.3}) = \sigma_1^2\ (................)$

Q.152 The coefficient r 123 can be expressed in the form of pairwise product moment correlation coefficients as

$$b_{12.3} = \frac{(r_{12}\ ....)}{\sqrt{(1 - r_{13}^2)(1 - r_{23}^2)}}$$

Q.153 The coefficient $b_{12.3}$ can be worked out from

$$b_{12.3} = \frac{\sum^{n} X_{1.3} X_{2.3}}{\sum ....}$$

Q.154 The coefficient $b_{12.3}$ indicates that it is the regression coefficient of ($x_{1.3}$ on ....)

Q.155 The relation between partial correlation and regression coefficients exists as $r_{12.3}^2 =$ ....

Q.156 If $r_{12.3} = 0$ then we have $r_{12} =$ ....

Q.157 If $r_{12} = r_{13}\ r_{23}$, this means that r_{12} will not be equal to zero if X_3 is correlated with .... and respectively.

Q.158 Partial correlation coefficient helps in deciding whether to include or not an additional " .... " variables in regression analysis.

Q.159 The values σ_1^2 $(1 - r_{12}^2)$ and σ_1^2 $(1 - r_{13}^2)$ are the residual variances it X_1 is estimated from (..........) and (.............) individually.

Q.160 The quantity σ^2_1 $(1 — R^2_{1.23})$ gives the residual variance of X_1 if it is estimated from (........, ...) taken together.

Q.162 Let R be the observed multiple correlation coefficient of variate Y with k other variates from a random sample of size n. Then to test $H_0 : R = 0$, the test statistic F is defined as

$$F = \frac{R^2}{1 — R^2} \cdot \frac{n - (......)}{k}$$

Q.163 In Q (162), to test Ho: R = 0, the F statistic was defined by Prof. ....

Q.164 In Q (162), to test Ho : R = 0, the F statistic defined by Prof. R.A.Fisher follows F distribution with degree of freedom equal to (.............., ...).

Q.165 Leg $\gamma_{12.34}$ .... (k+2) be the partial correlation coefficient of order (k + 2). Obtanied from a sample of size n. Then to test Ho : $P_{12.34 \text{............} (k+2)} = 0$, the statistic 't' is defined as

$$t = \frac{r_{12}}{\sqrt{1 - r_{12}^2}} \sqrt{.......}$$

Q.166 In Q (165), to test $H_0 : p_{12.34 \text{.......}(k+2)} = 0$, the test statistic "t" was defined by Prof. ....

Q.167 In Q(165), to test H0 : $P_{12.3 \text{.......} (k+2)}$ 0, the test statistic "t" defined by Prof. R.A. Fisher follows student's t distribution with d.f. equal to (..............).

Q.168 The result of testing the hypothesis $H_0 : b_{12.34 \text{.......} (k+2)} = 0$, will be the same as the result of testing Ho : $p_{12.34 \text{.......} (k+2)} = 0$ as obtained by using the test statistic known as " .... " with degree of freedom equal to (n-(k+2)).

Q.169 The coefficient of determination R^2 is defined as

$$R^2 = \frac{\text{Variance explained in Y}}{\text{.....................in Y}}$$

Q.170 The coefficient $(1 - R^2)$ is known as the " .... " .

Q.171 If $R^2 = 0.64$, it means that only " .... " present in the " .... "variable y has been explained due to effect of " .... " variates x's.

Q.172 A test which gives quite satisfactory result inspite of some departure from the basic assumptions, is known as " .... ".

Q.173 Let rxy be an observed product moment correlation coefficient obtained from a random sample of size n. Then to test the null hypothesis $H_0 : \rho yx = 0$, the suitable test statistic used is known as " .... " .

Q.174 For testing $H_0 : \rho_{yx} = 0$, the student's t statistic is defined as

$$t = \frac{|r|\sqrt{\dots\dots}}{\sqrt{\dots\dots\dots}}$$

Q.175 For testing $H_0 : p_{yx} = 0$, the student's t statistic has d.f. equal to (............)in case of sample size n.

Q.176 In bivariate population to test Ho : $\beta yx = k$, the student's t statistic is defined as

$$t = \frac{|b_{yx} - k|\sqrt{(n-2)\sum(xi - \bar{x}n)^2}}{\sqrt{\sum^{n}(yi - \bar{y}n)^2 - \dots\dots\dots(\dots\dots)^2}}$$

Q.177 In Q(176), to test H_o : $\beta yx = 0$, the constant k will be simply replaced by " .... " in the expression of t statistic.

Q.178 To test the null hypothesis either $H_o : \rho_{yx} = 0$ or Ho : $\beta_{yx} = 0$, the d.f. used will be equal to " .... "

Q.179 From a bivariate random sample (yi, xi) of size n, the SS due to regression can be worked out from

$$\text{SS (Due to regression)} = b_{yx} \sum_{i}^{n} (\dots\dots\dots)(\dots\dots\dots)$$

Q.180 In Q (179), SS (Due to regression) can also be worked out from

$$\text{SS (Due to regression)} = b_{yx}^2 \sum^{n} (\dots\dots.)^2$$

Q.181 In Q(179) we have - SS (About regression) = Residual SS

$$= \sum_{i}^{n} (y_i - \bar{y}_n)^2 - b_{yx} \sum_{i}^{n} (\dots\dots)(\dots\dots\dots)$$

Q.182 In Q (181), residual SS can also be worked out as

$$\text{Residual SS} = \sum_{i}^{n}(y_i - \bar{y}_n)^2 - b_{yx}\sum^{n}(.....)^2$$

Q.183 In Q(182), the residual d.f. is equal to (................) .

Q.184 In Q(182), the residual variance can be worked out from the expression -

$$\textbf{Residual variance} = \frac{1}{(n-2)}\left[........ - b_{yx}^2\sum^{n}(........)^2\right]$$

Q.185 The $V(b_{yx})$ is worked out from -

$$V(b_{yx}) = \frac{......(........) - b_{yx}^2\sum^{n}(.........)^2}{(n-2)\sum_{i}^{n}(x_i - \bar{x}_n)^2}$$

Q.186 The variance of sample correlation coefficient rxy is given as $V(r) = (1 - r^2) / (\quad)$.

15.3 Section-C : Multiple Choice Questions

Q.1 If the change in one variable affects the change in other variable then, both the variables are said to be
(a) Independent (b) Correlated
(c) Uncorrected (d) Both (a) & (b)

Q.2 If the change in two variables (X, Y) takes place in the same direction then the variables are said to possess
(a) Positive correlation (b) Reverse correlation
(c) Direct correlation (d) Both (a) & (c)

Q.3 The income and expenditure of a household has a
(a) Negative correlation (b) Positive correlation
(c) Direct correlation (d) Both (b) & (c)

Q.4 If the change in two variables takes place in two apposite directions, then they are said to have
(a) Positive correlation (b) Negative correlation
(c) Both (b) & (d) (d) Inverse correlation

Q.5 The price and demond of a commodity have
(a) Positive correlation (b) Negative correlation
(c) Both (b) & (d) (d) Inverse correlation

Q.6 If two variables (X, Y) deviate in the same direction with proportionate amount of deviation, then they are said to have
(a) Positive correlation (b) Perfect positive
(c) Direct correlation (d) All the three

Q.7 There are some funny combination of variates between which the correlation is said to be
(a) Spurious (b) Nonsense
(c) Positive (d) Both (a) & (b)

Q.8 The number of runs scored by a batsman and consumption of a fertiliser in a local community have a
(a) Spurious correlation (b) Direct correlation
(c) Inverse correlation (d) None

Q.9 Scatter diagram is the simplest way of diagrametic representation of
(a) One dimenstional data (b) Bivariate data
(c) Frequency distribution (d) Distribution function

Q.10 If points in scatter diagram are very dense, we should expect for the existence of
(a) High correlation
(b) Weak correlation
(c) Moderate correlation
(d) Non significant correlation

Q.11 If in a scatter diagram, the points are widely scattered, then we expect for existense of
(a) Non significant correlation (b) Poor correlation
(c) High correlation (d) Both (a) & (b)

Q.12 Who developed the formula to measure the correlation between two variables (X, Y).
(a) Prof. R.A. Fisher (b) Prof. Karl Pearson
(c) Prof. R.C. Bose (d) Prof. P.C. Mahalanobis

Q.13 The simple correlation coefficient between two variables (X, Y) indicates the existence of

(a) Linear relationship (b) Curvilinear relation

(c) No curvilinear relationship (d) Both (a) & (c)

Q.14 The coefficient γ_{xy} of linear relation between (X, Y) gives the measure of

(a) Intensity (b) Extent

(c) Rate of change (d) Both (a) & (b)

Q.15 The coefficient rxy between two variables (X, Y) is defined as

(a) $r_{xy} = \frac{COV(x,y)}{v(x)v(y)}$ (b) $r_{xy} = \frac{COV(x,y)}{\sqrt{v(x)v(y)}}$

(c) $r_{xy} = \sqrt{\frac{v(x)v(y)}{COV(x,y)}}$ (d) Both (b) & (c)

Q.16 With n pairs of values (xi, yi), COV (X, Y) is defined as

(a) $\frac{1}{n}\sum^{n}(x_i - \bar{x}_n)(y_i - \bar{y}_n)$ (b) $\frac{1}{n}\sum^{n}(x_i - \bar{x}_n)(y_i - \bar{y}_n)^2$

(c) $\frac{1}{n}\sum(x_i - \bar{x}_n)^2(y_i - \bar{y}_n)$ (d) $\sqrt{\frac{1}{n^2}\sum_{i}^{n}(x_i - \bar{x}_n)^2(y_i - \bar{y}_n)^2}$

Q.17 COV (X, Y) can also be expressed as

(a) $\frac{1}{n}\sum^{n}(y_i - \bar{y}_n)(x_i - \bar{x}_n)$ (b) $\frac{1}{n}\sum^{n}x_iy_i - \bar{x}_n\bar{y}_n$

(c) $\frac{1}{n}\left[\sum^{n}x_iy_i - n\bar{x}_n\bar{y}_n\right]$ (d) All the three

Q.18 For the scatter diagram given as, for a standard data, the γxy is equal to

(a) 0

(b) 1

(c) -½

(d) ½

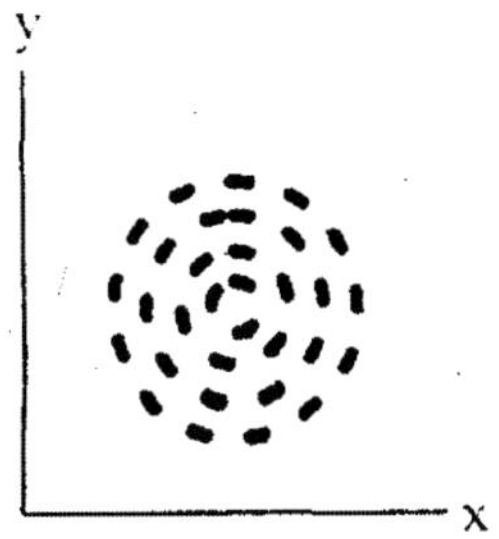

Q.19 For the scatter diagram given as, the r_{xy} for such standard diagram is equal to

(a) -1

(b) 1

(c) ½

(d) - ½

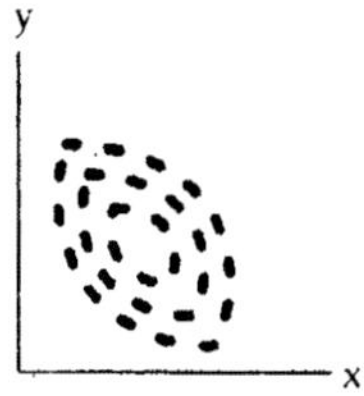

Q.20 For the scatter diagram given as, the value of r_{xy} lies as

(a) $r > 0$

(b) $r < 0$

(c) $r = 1$

(d) $r = -1$

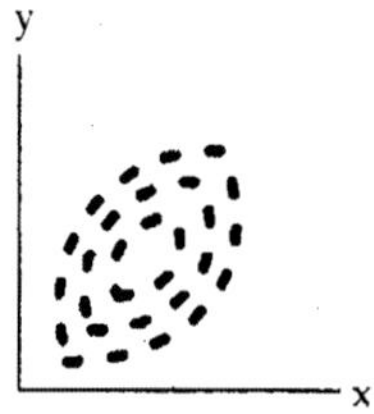

Q.21 For the straight line given in scatter diagram, the value of r_{xy} is equal to

(a) -1

(b) 0

(c) 1

(d) <0

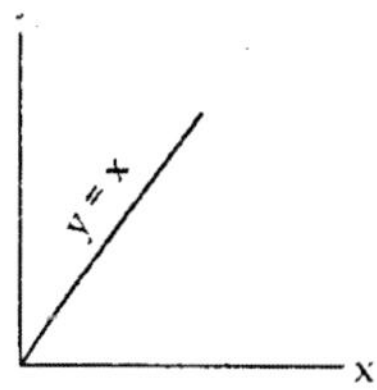

Q.22 For the straight line, given in scatter diagram, the value of r_{xy} is equal to

(d) -0

(b) 1

(c) 0

(d) >0

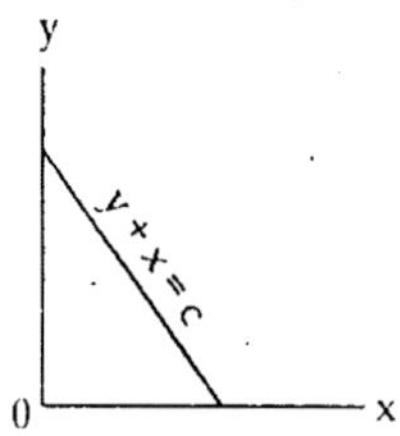

Q.23 Karl pearson correlation coefficient rxy is also known as -

(a) Product moment correlation (b) Spurious correlation

(c) Inverse correlation (d) Correlation ratio

Q.24 In terms of mathematical expectation, the COV (X, Y) can be written as

(a) E [(x - E(x))(y - E(y))] (b) E [(E (x) — x)(E (y) — y)]

(c) E [(y — E(y))(x — E(x))] (d) Both (a) & (c)

Q.25 The limit for coefficient r_{yx} exists as

(a) (-1, 1)
(b) (-1, 0)
(c) (0, 1)
(d) (- ½, ½)

Q.26 If r_{yx} = 1, the correlation is said to be

(a) Positive
(b) Perfect
(c) Perfect positive
(d) all the three

Q.27 If r_{yx} = - 1, the correlation is said to be

(a) Perfect
(b) Negative
(c) Perfect negative
(d) All the three

Q.28 Correlation coefficient r_{yx} is independent of change of

(a) Origin
(b) Scale
(c) Origin & Scale both
(d) None

Q.29 If we have U = ax + b and V = cy + d for (a, c > 0) then

(a) $r_{yx} = r_{uv}$
(b) $r_{xy} = r_{vu}$
(c) $r_{yx} = r_{uv}$
(d) All the three

Q.30 Two independent variables (X, Y) are always -

(a) Un-correlated
(b) Correlated
(c) $r_{yx} = 0$
(d) Both (a) & (c)

Q.31 if rxy are equal to zero then it is necessary that

(a) COV (X, Y) = 0
(b) XandYareindependent
(c) $b_{yx} = 0$
(d) Both (a) & (c)

Q.32 If rxy = 0 then it simply indicates the absence of

(a) Linear relationship between (X, Y)
(b) Curvilinear relationship between (X, Y)
(c) Complete independence between (X, Y)
(d) Both (a) & (c)

Q.33 If (X, Y) are jointly normally distributed with ρ (X, Y) = 0, then

(a) (X, Y) are independent
(b) (X, Y) are linearly uncorrelated
(c) COV (X, Y) = 0
(d) Both (a) & (c)

Q.34 If γ is the simple correlation coefficient calculated from a sample of size n, then its S.E. (r) is equal to

(a) $\sqrt{(1-r^2)}/\sqrt{n-2}$ (b) $\sqrt{\frac{1-r^2}{n-2}}$

(c) $\sqrt{\frac{1-r^2}{n}}$ (d) Both (a) & (b)

Q.35 The probable error P.E.(1) is equal to

(a) $0.6745\frac{(1-r^2)}{\sqrt{n}}$ (b) $0.4545\frac{(1-r^2)}{\sqrt{n}}$

(c) $0.6775\frac{(1-r^2)}{\sqrt{n}}$ (d) $0.6745\sqrt{\frac{(1-r^2)}{n}}$

Q.36 If r < P.E.(r) then correlation is
(a) Not significant at all (b) Significant at all
(c) Moderately significant (d) None

Q.37 If r > 6 P.E. (r) then correlation is definitely
(a) Significant (b) Non significant
(c) Highly positive (d) Both (a) & (c)

Q.38 Population correlation coefficient ρ_{yx} is expected to lie in the interval
(a) r - PE (r) $\leq\rho\leq$r+PE(r) (b) r +PE(r)$\leq\rho\leq$r-PE(r)
(c) r - 6PE(r)$\leq\rho\leq$ r +6PE(r) (d) None

Q.39 Let (xi, yi, i = 1, 2, ... n) be the rank of the ith individual in the two characteristics (A, B) respectivelly then correlation (A, B) is known the
(a) Rank correlation (b) Grade correlation
(c) Auto correlation (d) Serial correlation

Q.40 In Q(38), generally xi = yi and let di = (xi - yi). Then rank correlation (A, B) is given by an expression

(a) $\rho_{yx}=\frac{1-\sum^{n} d_i^2}{n(n^2-1)}$ (b) $\rho_{yx}=1-\frac{6\sum^{n} d_i^2}{n(n^2-1)}$

(c) $\rho_{yx}=1+\frac{6\sum^{n} d_i^2}{n(n^2-1)}$ (d) $\rho_{yx}=1-\frac{6\sum^{n} d_i^2}{n(n-1)}$

Q.41 The range of correlation coefficient ρ_{yx} is

(a) (0, 1)
(b) (-1, 0)
(c) (-1, 1)
(d) None

Q.42 The term "regression" means steping back toward the

(a) Median
(b) Average
(c) A. Mean
(d) Both (b) & (c)

Q.43 Regression analysis between two or more variables measures the

(a) Linear relationship
(b) Curvilinear relation
(c) Exponential relation
(d) None

Q.44 In regression analysis, the independent variables X is also known as

(a) Explanatory variable
(b) Covariates
(c) Regressor variable
(d) All the three

Q.45 In regression analysis, the dependent variable Y is also known as

(a) Study variable
(b) Regressant variable
(c) Explained variable
(d) All the three

Q.46 The line of regression gives the best estimate of dependent variable for any specific value of

(a) Independent variable
(b) Explanatory variable
(c) Regressor variable
(d) All the three

Q.47 The line of regression of best fit is obtained by the principle of

(a) Least square
(b) Optimization
(c) Randomization
(d) Replication

Q.48 In bivariate distribution, the number of lines of regression is equal to

(a) One
(b) Two
(c) Three
(d) Both (a) & (b)

Q.49 The line of regression of Y on X is given by the relation

(a) $Y = \bar{y}_n + b_{yx}(X - \bar{x}_n)$
(b) $Y = y_n + r\frac{\sigma_y}{\sigma_x}(X - \bar{x}_n)$
(b) $Y = \bar{y}_n + r\frac{\sigma_x}{\sigma_y}(X - \bar{x}_n)$
(d) Both (a) & (b)

Q.50 The coefficient byx is known as coefficient of
(a) Regression of y of x
(b) Regression of x on y
(c) Correlation between (X, Y)
(d) Correlation ratio of Y on X

Q.51 The coefficient byx give the amount of change
(a) In y for unit change in x
(b) In x for unit change in y
(c) In correlation coefficient
(d) In correlation ratio

Q.52 The corefficient byx gives the rate of change in the
(a) Varaite Y
(b) Variate X
(c) Correlation
(d) None

Q.53 The coefficient byx measures the
(a) Slope of the regression line with x axis
(b) Slope of the regression line with y axis
(c) Gradent of the regression line with x axis
(d) Both (a) & (c)

Q.54 The coefficient byx can be calculated from the formula -

(a) $b_{yx} = \dfrac{COV\,(x, y)}{\sigma_x \sigma_y}$
(b) $r_{yx} \cdot \dfrac{\sigma_y}{\sigma_x}$

(c) $b_{yx} = r_{yx} \cdot y\ \dfrac{\sigma_y}{\sigma_x}$
(d) Both (a) & (b)

Q.55 The quantity (byx . bxy) will be
(a) Positive
(b) Negative
(c) Either positive or negative
(d) Greater than one

Q.56 The coefficients byx and bxy will be such that
(a) byx . bxy > 1.0
(b) byx bxy ≤ 1.0
(c) byx . bxy < 0.0
(d) byx . bxy = 1.0

Q.57 The relation among (byx, bxy & rxy) exists as
(a) byx bxy > r_{xy}^2
(a) byx bxy < r_{xy}^2
(b) byx bxy = r_{xy}^2
(d) byx bxy = $2r_{xy}^2$

Q.58 The two lines of regression passes through the point

(a) $(\bar{x}, \bar{y})$ (b) $(\bar{y}, \bar{x})$

(c) (0, 0) (d) Both (a) & (b)

Q.59 In bivariate population, the lines of regression of Y on X can be written as

(a) $Y = \bar{y}_n + b_{yx}(X - \bar{x}_n)$ (b) $Y = \bar{y}_n - b_{yx}(X - \bar{x}_n)$

(c) $Y = \bar{y}_n \pm b_{yx}(X - \bar{x}_n)$ (d) $Y = \bar{x}_n \pm b_{xy}(Y - \bar{y}_n)$

Q.60 In case of perfect correlation ie 1= ± 1, the line of regression of y on x can be written as

(a) $Y = \bar{y}_n \pm \frac{\sigma_y}{\sigma_x}(X - \bar{x}_n)$ (b) $Y = \bar{y}_n \pm \frac{\sigma_y}{\sigma_x}(X - \bar{x}_n)$

(c) $Y = \bar{y}_n \pm (X - \bar{x}_n)$ (d) $Y = \bar{x}_n \pm (Y - \bar{y}_n)$

Q.61 In case of perfect correlation ie 1= ± 1, both lines of regression can be written as

(a) $\frac{(Y - \bar{y}_n)}{\sigma_y} = \pm\frac{(X - \bar{x}_n)}{\sigma_x}$ (b) $\frac{(Y - \bar{y}_n)}{\sigma_x} = \pm\frac{(X - \bar{x}_n)}{\sigma_y}$

(c) $\frac{(Y - \bar{y}n)}{\sigma_y} = \frac{(X - \bar{x}n)}{\sigma_x}$ (d) $\frac{(Y - \bar{y}n)}{\sigma_y} = -\frac{(X - \bar{x}n)}{\sigma_x}$

Q.62 If both the lines of regression concide then the value of rxy become equal to

(a) 1.0 (b) - 1.0

(c) ± 1.0 (d) 0.0

Q.63 We have the relation among (byx, bxy & γ_{xy}) as

(a) $r_{xy} = \sqrt{byx\ bxy}$ (b) $r_{xy} = -\sqrt{byx\ bxy}$

(c) $r_{xy} = \pm\sqrt{byx\ bxy}$ (d) $r_{xy} = \pm\sqrt{byx + bxy}$

Q.64 We have the relation among (byx, bxy & γxy) as

(a) $\left|\frac{byx + bxy}{2}\right| > |r_{xy}|$ (b) $\left|\frac{byx + bxy}{2}\right| < |r_{xy}|$

(c) $|byx + bxy| > |r_{xy}|$ (d) $|byx + bxy| < |r_{xy}|$

Q.65 The regression coefficient byx is
(a) Independent of change of origin
(b) Dependent of change of scale
(c) Independent of change of scale
(d) Both (a) & (b)

Q.66 Let $U = \frac{x - a}{h}$, $V = \frac{y - b}{k}$, then we have

(a) $b_{xy} = \frac{h}{k} buv$ (a) $b_{xy} = \frac{k}{h} buv$

(a) $b_{xy} = \frac{k}{h} buv$ (d) Both (a) & (b)

Q.67 If θ is the acute angle between two lines of regression then, we have

(a) $\tan\theta = \frac{1-r^2}{|r|}\left(\frac{\sigma_x\sigma_y}{\sigma_x^2+\sigma_y^2}\right)$ (b) $\tan\theta = \frac{1-r^2}{|r|}\left(\frac{\sigma_x^2+\sigma_y^2}{\sigma_x\sigma_y}\right)$

(c) $\tan\theta = \frac{|r|}{1-r^2}\left(\frac{\sigma_x\sigma_y}{\sigma_x+\sigma_y}\right)$ (d) $\tan\theta = \frac{1-r^2}{|r|}\left(\frac{\sigma_x\sigma_y}{\sigma_x+\sigma_y}\right)$

Q.69 If the two lines of regression concide than the value of rxy becomes equal to
(a) 0.0   (b) 1.0
(c) -1.0   (d) - ½

Q.69 If the two lines of regression concide than the value of r_{xy} becomes equal to
(a) 1.0   (b) -1.0
(c) ± 1.0   (d) 0.0

Q.70 If correlation rxy = 0 then line of regression of y on x becomes as
(a) $Y = \overline{y}$ (b) $y = \overline{x}^2$
(c) $y = 2\overline{x}$ (d) $y = \overline{x}$

Q.71 When two variables (X, Y) are uncorrelated then two lines of regression are
(a) Parallel to x and y axes respectively
(b) Perpendicular to x and y axes respectively
(c) Concides to each other   (d) None

Q.72 For higher degree of correlation between two variables, the angle between two lines is

(a) Smaller (b) Larger

(c) 90° (d) Greater than 90°

Q.73 The S E y I, an estimated value, say y is given by

(a) $\sigma_y \sqrt{1 - r^2}$ (b) $\sigma_y^2(1 - r^2)$

(c) $\sigma_1^2 \sqrt{1 - r^2}$ (d) $\sigma_y(1 - r^2)$

Q.74 Correlation coefficient between observed and estimated values say $(Y, \hat{Y})$ is given as

(a) $r(Y, \hat{Y}) = r(X, Y)$ (b) $r(Y, \hat{Y}) = r(X, \hat{Y})$

(c) $r(Y, \hat{Y}) = r(\hat{X}, Y)$ (d) $r(Y, \hat{Y}) = r(\hat{X}, \hat{Y})$

Q.75 The conditional mean $E(Y/X = x)$ for a continuous distributio is called the

(a) Regression function of Y on X (b) Regression line of Y on X

(c) Regression curve of Y on X (d) Both (a) & (b)

Q.76 If one of the regression lines is linear, then

(a) Other will also be linear

(b) Other will not be linear

(c) Other may or may not be linear

(d) None of the above

Q.77 If regression of Y on X is linear than it can be written as

(a) $E(Y / x) = \overline{y} + b_{yx}(X - \overline{x})$ (b) $E(Y / x) = \overline{y} + b_{xy}(X - \overline{x})$

(c) $E(Y / x) = \overline{y} + r\frac{\sigma_y}{\sigma_x}(X - \overline{x})$ (d) Both (a) & (c)

Some Topics on Correlation & Regression

Q.78 The correlation ratio "η_{yx}" between two variables (X, Y) measures the

(a) Linear relationship (b) Curvilinear relationship

(c) Conditional mean of Y on X (d) Partial regression

Q.79 The correlation ratio η_{yx} measures the concentration of points (x, y) about the

(a) Curve of best fit (b) Line of best fit

(c) Less than ogive (d) More than ogive

Q.80 If curvilinear relationship exists between (X, Y) then.

(a) $|\eta_{yx}| > |r_{yx}|$ (b) $|\eta_{yx}| < |r_{yx}|$

(c) $|\eta_{yx}| = |r_{yx}|$ (d) $|\eta_{yx}| = 0$

Q.81 The correlation ratio iyxcan be calculated from

(a) $\eta_{yx}^2 = \left(1 - \frac{\sigma_{ey}^2}{\sigma_y^2}\right)$ (b) $\eta_{yx}^2 = \left(1 - \frac{\sigma_y^2}{\sigma_{ey}^2}\right)$

(c) $\eta_{yx}^2 = \left(1 - \left(\frac{\sigma_{ey}^2}{\sigma_y^2}\right)^2\right)$ (d) Both (a) & (c)

Where $\sigma_{ey}^2 = \frac{1}{N}\sum_i \sum_j fij(yij - yi)^2$

Q.82 A convenient expression for η_{yx} can be obtained from

(a) $\eta_{yx}^2 = \frac{\sigma_{my}^2}{\sigma_y^2}$ (b) $\eta_{yx}^2 = \frac{\sigma_y^2}{\sigma_{my}^2}$

(c) $\eta_{yx}^2 = \left(\frac{\sigma_{my}}{\sigma_y}\right)^2$ (d) Both (a) & (c)

Where $N\sigma_{my}^2 = \sum_i ni(y_i - \bar{y}..)^2$

Q.83 We have the relation among

(a) $\sigma_{em}^2 = \sigma_y^2 (1 - \eta_{yx}^2)$ (a) $\sigma_{em}^2 = \sigma_y (1 - \eta_{yx})$

(c) $\sigma_{em}^2 = \sigma_y^2 (1 - \eta_{yx})^2$ (a) $\sigma_{em}^2 = \sigma_y (1 - \eta_{yx}^2)$

Q.84 An interval for the existence of possible value of is η_{yx} is

(a) $(-1 \le \eta_{yx} \le 1)$ (b) $(0 \le \eta_{yx} \le 1)$

(c) $\left(-1 \le \eta_{yx} \le \frac{1}{2}\right)$ (d) $\left(-\frac{1}{2} \le \eta_{yx} \le \frac{1}{2}\right)$

Q.85 It is noticeable that η_{yx} is
(a) Dependent of classification of data
(b) Not independent of classification data
(c) Independent of classification of data
(d) Both (a) & (b)

Q.86 As class interval becomes narrower then
(a) $\eta_{yx} \to 1.0$
(b) $\eta_{yx} \to 0.0$
(c) $\eta_{yx} \to r_{yx}$
(d) $\eta_{yx} > 1.0$

Q.87 In terms of expectation, correlation ratio $\eta_{...}$ is defined as
(a) $\eta^2_{yx} = \dfrac{E_x\left[E(y/x)-E(y)\right]^2}{E\left[y-E(y)\right]^2}$
(b) $\eta^2_{yx} = \dfrac{E_y\left[E(y/x)-E(y)\right]^2}{E\left[x-E(x)\right]^2}$
(c) $\eta^2_{yx} = \dfrac{E_y\left[E(x/y)-E(x)\right]^2}{E\left[y-E(y)\right]^2}$
(d) None

Q.88 η^2_{xy} can be defined as
(a) $\eta^2_{xy} = \dfrac{E_y\left[E(x/y)-E(x)\right]^2}{E\left[X-E(x)\right]^2}$
(b) $\eta^2_{xy} = \dfrac{E_y\left[E(x/y)-E(x)\right]^2}{EY\left[X-E(y)\right]^2}$
(c) $\eta^2_{xy} = \dfrac{E_x\left[E(y/x)-E(y)\right]^2}{E\left[Y-E(y)\right]^2}$
(d) None

Q.89 For completely random scattering of dots with no trend, we have
(a) $r_{xy} = \eta_{yx} = 0$
(b) $r_{xy} = \eta_{yx} = 1.0$
(c) $\eta_{yx} > r_{xy}$
(d) $r_{xy} > \eta_{yx}$

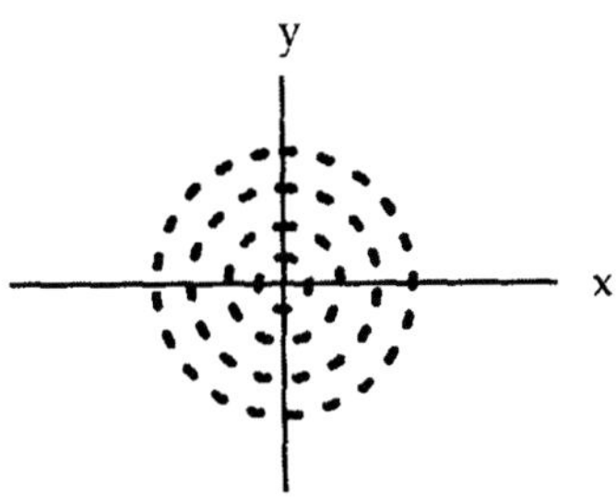

Q.90 If in a scatter diagram, the dots (x, y) precisely lie on the line shown then

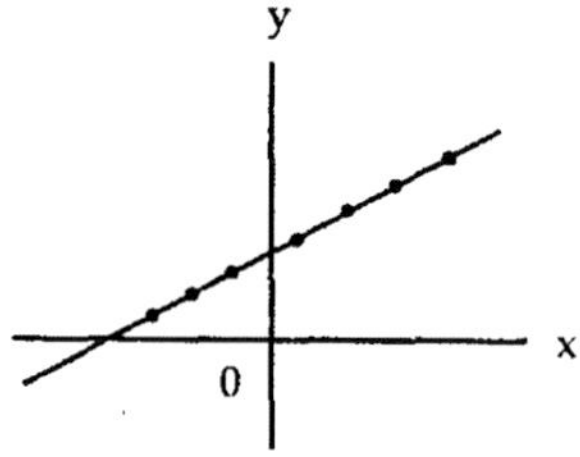

(a) $r_{yx} = r_{xy} = 1.0$
(b) $r_{xy} > n_{yx}$
(c) $r_{xy} < n_{yx}$
(d) $r_{xy} = n_{yx} = 0$

Q.91 If in a scatter diagram the dots lie on a curve that no ordinate cuts it more than once then

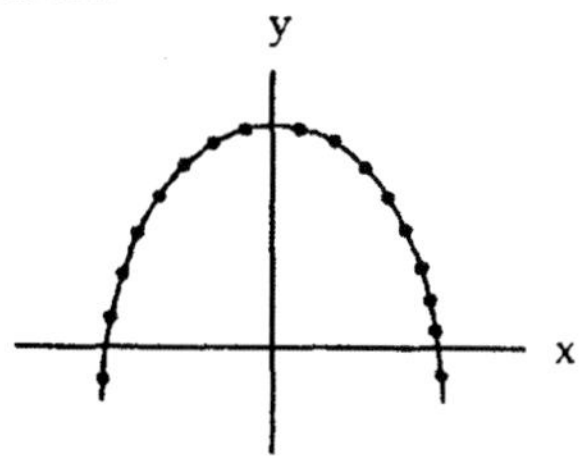

(a) $\eta_{yx} = 1.0$
(b) $\eta_{yx} = 0$
(c) $\eta_{yx} < \frac{1}{2}$
(a) $\eta_{yx} < -\frac{1}{2}$

Q.92 If in Q(91), the dots are symmetrically placed about y axis then
(a) $\eta_{yx} = 0$
(b) $\eta_{yx} = 1.0$
(c) $\eta_{yx} > 0$
(d) $\eta_{yx} < 0$

Q.93 If the dots are scattered definitely around a curved trend then

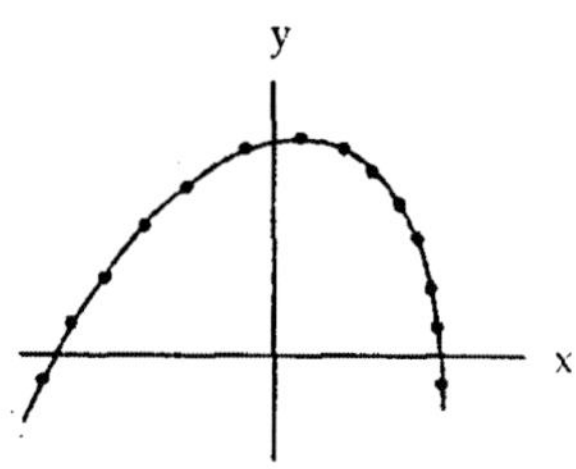

(a) $\eta_{yx} > r_{xy}$
(b) $\eta_{yx} = r_{xy}$
(c) $\eta_{yx} < r_{xy}$
(d) $\eta_{yx} \neq \frac{1}{2}$

Q.94 Intraclass correlation means
(a) Within class correlation
(b) Between class correlation
(c) Within class correlation ratio
(d) Within class variation

Q.95 In intraclass correlation study, we study about relationship
(a) Different characteristics
(b) Within same characteristic
(c) Consistency of them
(d) None

Q.96 Let x_{ij}, (i = 1, 2, ...n, j = 1, 2, ..., k) be the measurement on ith member of ith family. Then intraclass correlation coefficient ρ_w is worked out from the expression

(a) $\rho_w = \frac{1}{k-1}\left(\frac{k\sigma_m^2}{\sigma^2} - 1\right)$ (b) $\rho_w = \frac{1}{(k-1)}\left(k\left(\frac{\sigma_m}{\sigma}\right)^2 - 1\right)$

(c) $\rho_w = \frac{1}{k-1}\left(k.\frac{\sigma_m}{\sigma} - 1\right)$ (d) Both (a) & (b)

Q.97 ρ_w lies in the range

(a) $\left(-\frac{1}{k-1}, 1\right)$ (b) $\left(\frac{1}{k-1}, 1\right)$

(c) $\left(\frac{1}{k+1}, 1\right)$ (d) $\left(-\frac{1}{1+k}, 1\right)$

Q.98 In case of bivariate normal distribution, the regression of y on x is

(a) Linear (b) Curvilinear

(c) Straight line (d) Both (a) & (c)

Q.99 In case of bivariate normal distribution, the arrays have

(a) Common variance

(b) Different variance

(c) Same coefficient of variation

(d) Different coefficient of variation

Q.100 In case of bivariate normal distribution, the variance of estimate of y in each array is equal to

(a) σ_1^2 (1- ρ^2) (b) σ_y^2 (1- ρ)

(c) σ_1^2 (1- ρ)^2 (d) σ_y (1- ρ^2)

Q.101 In bivariate normal distribution, the distribution of y in each array is

(a) Skewed (b) Normal

(c) Discrete (d) Continuous

Q.102 In bivariate normal distribution, the number of parameter is equal to

(a) 2 (b) 3

(c) 4 (d) 5

Q.103 In bivariate normal distribution, the surface $z = f(x, y)$ is known as

(a) Normal regression surface

(b) Normal correlation surface

(c) Bivariate density function

(d) Bivariate frequency function

Q.104 Bivariate normal distribution is abbreviated as

(a) $(X,Y) \sim N(\mu_1, \mu_2, \sigma_1^2, \sigma_2^2, \rho)$ (b) $(X,Y) \sim N(\mu_1, \sigma^2{}_1, \mu_2, \sigma_2^2, \rho)$

(c) $(X,Y) \sim N(\sigma_1^2, \mu_1, \sigma_2^2, \mu_2, \rho)$ (d) $(X,Y) \sim N(\rho, \mu_1, \sigma^2{}_1, \mu_2, \sigma_2^2)$

Q.105 In bivariate normal distribution of $\mu_1 = \mu_2 = 0$, $\sigma_1^2 = \sigma_2^2 = 1$, then, we have

(a) $(X, Y) \sim N(0,0, \rho,1,1)$ (b) $(X,Y) \sim N(0,0,1,1,\rho)$

(c) $(X, Y) \sim N(1,1,0,0, \rho)$ (d) $(X, Y) \sim N(\rho,0,0,1,1)$

Q.106 The m.g.f. of bivariate normal distribution is given as

(a) $M_{x,y}(t_1, t_2) = e^{\frac{1}{2}(t_1^2 + t_2^2 + 2\rho t_1 t_2)}$ (b) $M_{x,y}(t_1, t_2) = e^{\frac{1}{2}(t_1^2 + t_2 + 2\rho t_1 t_2)}$

(c) $M_{x,y}(t_1, t_2) = e^{\frac{1}{2}(t_1 + t_2^2 + 2 t_1 t_2)}$ (d) $M_{x,y}(t_1, t_2) = e^{\frac{1}{2}(t_1^2 + t_2^2 + 2 t_1 t_2)}$

Q.107 The (X, Y) possesses a bivariate normal distribution iff every linear combination of (X, Y) is a

(a) Normal variate (b) Chisquare variate

(c) Binomial variate (d) Poisson variate

Q.108 In case of bivariate normal distribution, the marginal distribution of Y for given X = x, is given by

(a) $Y \sim N(\mu_y, \mu_y^2)$ (b) $Y \sim N(\mu_x, \sigma_x^2)$

(c) $Y \sim N(\mu_y^2, \mu_y^2)$ (c) $Y \sim N(\sigma_x^2, \sigma_x^2)$

Q.109 From a bivariate normal distribution, the conditional distribution of Y for fixed X = x, is given as

(a) $(Y/X=x) \sim N\left[\mu_y + \rho\frac{\sigma_y}{\sigma_x}(x-\mu_x), \sigma_y^2(1-\rho^2)\right]$

(b) $(Y/X=x) \sim N\left[\mu_y + \rho\frac{\sigma_x}{\sigma_y}(x-\mu_x), \sigma_y^2(1-\rho^2)\right]$

(c) $(Y/X=x) \sim N\left[\mu_y + \rho\frac{\sigma_y}{\sigma_x}(x-\mu_x), \sigma_y(1-\rho)\right]$

(d) $(Y/X=x) \sim N\left[\mu_y + \rho\frac{\sigma_x}{\sigma_y}(x-\mu_x), \sigma_y(1-\rho^2)\right]$

Q.110 Let (xi, yi) be a small bivariate sample of size n < 30. Then to test Ho : $p_{yx} = 0$, the appropriate test statistic is

(a) Z statistic (b) Student's t statistic
(c) X^2 statistic (d) F statistic

Q.111 To test H0 : pyx = 0, the student's t statistic is defined as

(a) $t = \frac{|r|\sqrt{n-2}}{\sqrt{(1-r)}}$ (b) $t = \frac{|r|\sqrt{n-2}}{\sqrt{(1-r^2)}}$

(c) $t = \frac{|r|\sqrt{n-1}}{\sqrt{(1-r^2)}}$ (d) $t = \frac{|r|(n-2)}{\sqrt{(1-r^2)}}$

Q.112 For a bivariate normal distribution to test the Ho : byx = k, the suitable test statistic is

(a) Student's t statistic (b) Chi-square statistic
(c) Z statistic (d) F statistic

Q.113 For a bivariate population, to test the Ho : byx = 0, the student's t statistic is defined as

(a) $$t = \frac{|byx|\sqrt{(n-2)\sum^{n}(xi-\bar{x}_n)^2}}{\sqrt{\sum^{n}(yi-\bar{y}_n)^2 - b_{yx}^2\sum^{n}(xi-\bar{x}_n)^2}}$$

(b) $$t=\frac{|byx|\sqrt{(n-2)\sum^{n}(x_i-\bar{x}_n)^2}}{\sqrt{\sum^{n}(y_i-\bar{y}_n)^2-b_{yx}\sum^{n}(y_i-\bar{y}_n)(x_i-\bar{x}_n)}}$$

(c) Both (a) & (b)

(d) $$t=\frac{|byx|\sqrt{(n-1)\sum^{n}(xi-\bar{x}n)^2}}{\sum^{n}(yi-\bar{y}n)^2-b^2_{yx}\sum^{n}(xi-\bar{x}n)^2}$$

Q.114 In Q(113), to test Ho : bxy = 0 the student's statistic will be defined as

(a) $$t=\frac{|bxy|\sqrt{(n-2)\sum^{n}(yi-\bar{y}_n)^2}}{\sqrt{\sum^{n}(xi-\bar{x}_n)^2-b^2_{xy}\sum^{n}(yi-\bar{y}_n)^2}}$$

(b) $$t=\frac{|bxy|\sqrt{(n-2)\sum^{n}(yi-\bar{y}_n)^2}}{\sqrt{\sum^{n}(xi-\bar{x}_n)^2-b_{xy}\sum^{n}(yi-\bar{y}_n)(xi-\bar{x}_n)}}$$

(c) Both (a) & (b)

(d) $$t=\frac{|byx|\sqrt{(n-1)\sum^{n}(yi-\bar{y}_n)^2}}{\sum^{n}(xi-\bar{x}_n)^2-b^2_{xy}\sum^{n}(yi-\bar{y}_n)^2}$$

Q.115 To test the null hypothesis either H_{01} : p_{yx} = 0 or H_{02} : βyx = 0, the d.f. of student's t statistic used is equal to

(a) n - 1 (b) n - 2

(c) $\sqrt{n-2}$ (d) $\sqrt{n-1}$

Q.116 In case of bivariate distribution, from a random sample of size n, the S.S. (due to regression) is equal to

(a) $b_{yx}^2 \sum_{i=1}^{n} (xi - \bar{x}_n)^2$

(b) $b_{yx}^2 \sum_{i}^{n} (yi - \bar{y}_n)^2$

(c) $b_{yx} \sum_{i}^{n} (xi - \bar{x}_n)(yi - \bar{y}_n)$

(d) Both (a) & (c)

Q.117 In a bivariate distribution, the d.f. for S.S. (due to regression) is equal to

(a) 1

(b) n-1

(c) n - 2

(d) $\sqrt{n-2}$

Q.118 From a bivariate random sample (yi, xi) of size n, the residual SS or SS (about regression) or error (SS) can be worked out from the expression

(a) $\sum_{i}^{n} (y_i - \bar{y}_n)^2 - b_{yx}^2 \sum^{n} (x_i - \bar{x}_n)^2$

(b) $\sum^{n} (y_i - \bar{y}_n)^2 - b_{yx} \sum^{n} (x_i - \bar{x}_n)^2$

(c) $\sum^{n} (y_i - \bar{y}_n)^2 - b_{yx} \sum_{i}^{n} (x_i - \bar{x}_n)(y_i - \bar{y}_n)$

(d) Both (a) & (c)

Q.119 In Q(118), the error d.f. is equal to

(a) n -1

(b) n - 2

(c) $\sqrt{n-2}$

(d) None

Q.120 In a bivariate regression analysis, the residual variance is equal to

(a) $\hat{\sigma}_{y.x}^2 = \frac{1}{(n-2)} \left[\sum_{i}^{n} (y_i - y_n)^2 - b_{yx} \sum^{n} (x_i - \bar{x}_n)^2 \right]$

(b) $\hat{\sigma}_{y.x}^2 = \frac{1}{(n-2)} \left[\sum_{i}^{n} (y_i - \bar{y}_n)^2 - b_{yx} \sum^{n} (x_i - \bar{x}_n)(y_i - \bar{y}_n) \right]$

(c) $\hat{\sigma}_{y.x}^2 = \frac{1}{(n-1)} \left[\sum_{i}^{n} (y_i - \bar{y}_n)^2 - b_{yx}^2 \sum_{i}^{n} (x_i - \bar{x}_n)^2 \right]$

(d) $\hat{\sigma}_{y.x}^2 = \frac{1}{(n-2)} \left[\sum_{i}^{n} (x_i - \bar{x}_n)^2 - b_{yx}^2 \sum^{n} (y_i - \bar{y}_n)^2 \right]$

Q.121 The sampling variance of byx is given by

(a) $$v\left(b_{yx}\right)=\frac{\sum_{i}^{n}\left(y_i-\bar{y}_n\right)^2-b_{yx}\sum^{n}\left(x_i-\bar{x}_n\right)^2}{(n-2)\sum_{i}^{n}\left(x_i-\bar{x}_n\right)^2}$$

(b) $$v\left(b_{yx}\right)=\frac{\sum_{i}^{n}\left(y_i-\bar{y}_n\right)^2-b_{yx}\sum^{n}\left(x_i-\bar{x}_n\right)\left(y_i-\bar{y}_n\right)}{(n-2)\sum_{i}^{n}\left(x_i-\bar{x}_n\right)^2}$$

(c) $$v\left(b_{yx}\right)=\frac{\sum_{i}^{n}\left(y_i-\bar{y}_n\right)^2-b_{yx}^2\sum^{n}\left(x_i-\bar{x}_n\right)^2}{(n-2)\sum_{i}^{n}\left(y_i-y_n\right)^2}$$

(d) $$v\left(b_{yx}\right)=\frac{\sum_{i}^{n}\left(x_i-\bar{x}_n\right)^2-b_{yx}^2\sum^{n}\left(y_i-\bar{y}_n\right)^2}{(n-2)\sum_{i}^{n}\left(x_i^2-\bar{x}_n\right)^2}$$

Q.122 The sam p ling variance of pxy from a random sample of size n can be calculated from

(a) $v\left(r_{xy}\right)=\frac{(1-r)}{(n-2)}$ (b) $v\left(r_{xy}\right)=\frac{\left(1-r^2\right)}{(n-2)}$

(c) $v\left(r_{xy}\right)=\left(\sqrt{\frac{1-r^2}{n-2}}\right)^2$ (d) Both (b) & (c)

Multiple Correlation & Regression

Q.123 If we are interested in studying the joint effect of a group of variables on a variable which is not included in that group, then our study is known as the study of

(a) Multiple correlation and regression analysis
(b) Partial correlation and regression analysis
(c) Simple correlation and regression analysis
(d) Partial regression analysis

Q.124 The correlation and regression between only two variables after the elimination of effect of other variables on them, is known as

(a) Multiple correlation and regression
(b) Partial correlation and regression
(c) Simple correlation and regression
(d) Simple and partial correlation & regression

Q.125 Assume that all the three variables (X_1, X_2, X_3) of a trivariate distribution have been measured from their respective means. Then the plane of regression of X_1 on X_2 & X_3 is written as

(a) $\hat{X}_1 = b_{12.3} X_2 + b_{13.2} X_3$
(b) $\hat{X}_1 = b_{21.3} X_2 + b_{23.1} X_3$
(C) $\hat{X}_1 = b_{31.2} X_2 + b_{32.1} X_3$
(d) $\hat{X}_1 = b_{31.2} X_3 + b_{32.1} X_2$

Q.126 In Q(125), the coefficient $b_{1.23}$ & $b_{13.2}$ are known respectively as partial regression coefficient of

(a) X_1 on X_2 and X_1 on X_3
(b) X_2 on X_3 and X_2 on X_3
(c) X_3 on X_1 and X_3 on X_2
(d) X_1 on X_3 and X_1 on X_2

Q.127 Let $\hat{X}_1 = e_{1.23}$ then $X_{1.23} = (X_1 - e_{1.23})$ is known as

(a) Residual of X_1
(b) Residual of X_2
(c) Residual of X_3
(d) None

Q.128 In partial regression coefficient, the subscript before the dot, is known as

(a) Secondary subscript
(b) Primary subscript
(c) Original subscript
(d) Residual subscript

Q.129 In partial correlation coefficient, the subscript after the dots, is known as

(a) Secondary subscripts
(b) Primary subscript
(c) Residual subscript
(d) Original subscript

Q.130 The number of secondary subscripts in either partial regression coefficient or correlation coefficient is known as

(a) Order of coefficient
(b) Power of coefficient
(c) Suffics of coefficient
(d) Indicator of coefficient

Q.131 The order of partial regression coefficient $b_{12.34 \ldots\ldots\ldots n}$ is equal to

(a) n (b) n - 1

(c) n - 2 (d) 2n

Q.132 It is noticeable that the order in which the primary subscripts are written in partial coefficient has

(a) Significant meaning (b) No meaning

(c) Non significant meaning (d) Ordinary meaning

Q.133 In regression coefficient $b_{12.34 \ldots\ldots, n}$, the variable x_2 will be treated as

(a) Independent variable (b) Explanatory variable

(c) Regressor variable (d) All the three

Q.134 In regression coefficient b12.34 ....n, the variable x1, will be known as

(a) Study variable (b) Dependent variable

(c) Response variable (d) All the three

Q.135 In the residual $x_{1.234 \ldots\ldots\ldots\ldots n}$, the order of residual is equal to

(a) n - 1 (b) n

(c) n - 2 (d) 2(n - 1)

Q.136 We can write that

(a) $\Sigma X_3 X_{1.23} = 0$ (b) $X_2 . X_{1.23} = 0$

(c) $\Sigma XiX_{1.23 \ldots\ldots\ldots n} = 0 \;\forall i \; = 2,3,\ldots.$ (d) All the three

Q.137 The coefficient $R_{1.23}$ is the correlation coefficient between observed values and its estimated values given by the plane of regression of

(a) X_1 on X_2 & X_3 (b) X_2 on X_1 & X_3

(c) X_3 on X_1 & X_2 (d) None

Q.138 The coefficient $R_{1.23}$ can be worked out from the expression

(a) $R_{1.23} = \frac{COV(X_1, e_{1.23})}{v(X_1)v(e_{1.23})}$ (b) $R_{1.23} = \frac{COV(X_1, e_{1.23})}{\sqrt{v(X_1)v(e_{1.23})}}$

(c) $R_{1.23} = \frac{COV(X_1, e_{1.23})}{v(X_1)v(e_{1.23})}$ (d) $R_{1.23} = \frac{COV(X_1, e_{1.23})}{\sqrt{v(X_1)v(e_{2.13})}}$

Q.139 The coefficient $R_{1.23}$ can be expressed in terms of pairwise correlation coefficient as

(a) $R_{1.23} = \frac{COV(X_1, e_{1.23})}{v(X_1)v(e_{1.23})}$

(b) $R_{1.23} = \frac{COV(X_1, e_{1.23})}{\sqrt{v(X_1)v(e_{1.23})}}$

(c) $R_{1.23} = \frac{COV(X_1, e_{1.23})}{v(X_1)v(e_{1.23})}$

(d) $R_{1.23} = \frac{COV(X_1, e_{1.23})}{\sqrt{v(X_1)v(e_{2.13})}}$

Q.140 The value of $R_{1.23}$ lies in an interval

(a) (0, 1.0) (b) (-1, 0)

(c) (-1, 1) (d) (-½, ½)

Q.141 If R1.23 = 1.0 then all the regression residuals are equal to

(a) 1.0 (b) 0.0

(c) 100.0 (d) None

Q.142 If $R_{1.23} = 0$ then the value of $\sigma_{1.23}^2$ will be equal to

(a) 0.0 (b) 100.0

(c) 1.0 (d) None

Q.143 If $R_{1.23} = 0$ then following holds

(a) $r_{12} = 0$ (b) $r_{13} = 0$

(c) $r_{23} = 0$ (d) Both (a) & (b)

Q.144 We have the relation between $R_{1.23}$, r_{12}, r_{23} & r_{31} as

(a) $R_{12.3} > r_{12}$ (b) $R_{1.23} > r_{23}$

(b) $R_{1.23} < r_{12}$ (d) Both (a) & (b)

Q.145 Correlation (X_1, X_2), if the linear effect of X_3 on them were eliminated is known as

(a) Multiple correlation (b) Partial correlation

(c) Simple correlation (d) Correlation ratio

Q.146 The coefficient $b_{12.3}$, if the linear effect of X_3 on X_1 and X_2 were elimnated, is called

(a) Partial regression (b) Multiple regression

(c) Simple regression (d) Correlation ratio

Q.147 The coefficient $r_{12.3}$ can be calculated from the relation

(a) $r_{1.23} = \dfrac{COV(X_{1.3}, X_{2.3})}{v(X_{1.3})v(X_{2.3})}$ (b) $r_{12.3} = \dfrac{COV(X_{1.3}, X_{2.3})}{\sqrt{v(X_{1.3})v(X_{2.3})}}$

(c) $r_{12.3} = \dfrac{COV(X_{3.1}, X_{3.2})}{v(X_{3.1})v(X_{3.2})}$ (d) $r_{12.3} = \dfrac{COV(X_{3.1}, X_{3.2})}{\sqrt{v(X_{3.1})v(X_{3.2})}}$

Q.148 The $COV(X_{1.3}, X_{2.3})$ is equal to

(a) $\sigma_1\sigma_2 (r_{12} - r_{13}r_{23})$ (b) $\sigma_1\sigma_2 (r_{21} - r_{31}r_{32})$

(c) Both (a) & (b) (d) None

Q.149 The $V(X_{1.3})$ is equal to

(a) $\sigma_1^2(1 - r_{13}^2)$ (b) $\sigma_1^2(1 - r_{31}^2)$

(c) Both (a) & (b) (d) None

Q.150 The coefficient $r_{12.3}$ can be expressed in terms of pairwise simple coefficient rij, as follows

(a) $r_{12.3} = \dfrac{(r_{12} - r_{13}r_{23})}{\sqrt{(1-r_{13}^2)(1-r_{23}^2)}}$ (b) $r_{12.3} = \dfrac{(r_{12} - r_{13}r_{23})}{(1-r_{13}^2)(1-r_{23}^2)}$

(c) $r_{12.3} = \dfrac{(r_{23} - r_{12}r_{13})}{(1-r_{13}^2)(1-r_{23}^2)}$ (d) None

Q.151 The regression coefficient $b_{12.3}$ can be defined as

(a) $b_{12.3} = \dfrac{\sum X_{1.3}.X_{2.3}}{\sum X_{2.3}^2}$ (b) $b_{12.3} = \dfrac{COV(X_{1.3}, X_{2.3})}{V(X_{2.3})}$

(c) $b_{12.3} = \dfrac{COV(X_1, X_2)}{V(X_2)}$ (d) Both (a) & (b)

Q.152 We have in trivariates, the relation amongst partial correlation and regression as

(a) $r_{12.3}^2 = b_{12.3}b_{21.3}$ (b) $r_{12.3} = \pm\sqrt{b_{12.3}b_{21.3}}$

(c) Both (a) & (b) (d) $r_{12.3}^2 = b_{23.1}b_{32.1}$

Q.153 If $r_{12.3} = 0$ then it implies that

(a) $r_{12} = r_{13}\, r_{23}$
(b) $r_{21} = r_{31}\, r_{32}$
(c) Both (a) & (b)
(d) $r_{12}^2 = r_{13}\, r_{23}$

Q.154 Partial correlation coefficient helps in deciding whether to include or not an additional

(a) Independent variable
(b) Regres sor variable
(c) Covariate
(d) All the three

Q.155 The values $\sigma_1^2\,(1 - r_{12}^2)$ and $\sigma_1^2\,(1 - r_{13}^2)$ are the residual variances of X_1 if it is estimated from

(a) X_2 and X_3 individualy
(b) X_3 and X_2 respectively
(c) Both (a) & (b)
(d) None

Q.156 The quantity $\sigma_1^2\,(1 - R_{1.23}^2)$ is residual variance of X_1 if it is estimated from

(a) X_2 & X_3 taken together
(b) X_2 & X_3 taken jointly
(c) Both (a) & (b)
(d) None

Q.157 Let R be the observed multiple correlation coefficient of a vartiate Y with k other variates from a random sample of size n drawn from (k + 1) variates population, then to test the $H_0 : R = 0$, the suitable statistic is defined as

(a) $F = \dfrac{R^2}{1 - R^2} \cdot \dfrac{n - (k+1)}{k} \sim F_{k,n - (k+1)}$

(b) $F = \dfrac{R^2}{1 - R^2} \cdot \dfrac{k}{n - (k+1)} \sim F_{k,n - (k+1)}$

(c) $F = \dfrac{1 - R^2}{R^2} \cdot \dfrac{n - (k+1)}{k} \sim F_{n - (k+1), k}$

(d) None

Q.158 Let $r_{12.34 \ldots\ldots (k+2)}$ be the partial correlation coefficient of order k from a random sample of size n. Then to test the H0 : $\rho_{12.34 \ldots\ldots (k+2)} = 0$, the test statistic "t" is defined as

(a) $t = \dfrac{|r_{12}|\sqrt{n-(k+2)}}{\sqrt{1-r_{12}^2}}$
(b) $t = \dfrac{|r_{12}|\sqrt{n-k-2}}{\sqrt{1-r_{12}^2}}$

(c) $t = \dfrac{|r_{12}|\sqrt{n-k-2}}{\left(1-r_{12}^2\right)}$
(d) Both (a) & (b)

Q.159 The coefficient of determination R^2 is defined as

(a) $R^2 = \dfrac{\text{Variance explained in Y by regression}}{\text{Total var fiance incurred in Y}}$

(b) $R^2 = \dfrac{\text{Total Variance incurred in Y}}{\text{Variance explained in Y by regression}}$

(c) $R^2 = \dfrac{\text{COV(Y,Y)}}{\text{v(Y)}}$

(d) None

Q.160 The coeffcient $(1 - R^2)$ is known as coefficient of

(a) Non-determination (b) Determination

(c) Regression (d) Non-association

Key Answers

Section-A : True False

1. True	2. True	3. True
4. True	5. True	6. True
7. True	8. True	9. True
10. True	11. True	12. True
13. True	14. True	15. True
16. True	17. True	18. True
19. True	20. False	21. True
22. True	23. True	24. True
25. True	26. True	27. True
28. True	29. True	30. True
31. True	32. True	33. True
34. True	35. True	36. True
37. True	38. False	39. True
40. True	41. True	42. True
43. True	44. True	45. True
46. True	47. True	48. True
49. True	50. True	51. True
52. True	53. True	54. True
55. True	56. True	57. True
58. True	59. True	60. True
61. True	62. True	63. True
64. True	65. True	66. True

67. True
68. True
69. True
70. True
71. True
72. True
73. True
74. True
75. True
76. True
77. True
78. True
79. True
80. True
81. True
82. True
83. True
84. True
85. True
86. True
87. True
88. True
89. True
90. True
91. True
92. True
93. True
94. True
95. True
96. True
97. True
98. True
99. True
100. True
101. True
102. True
103. True
104. True
105. True
106. True
107. True
108. True
109. True
110. True
111. True
112. True
113. True
114. True
115. True
116. True
117. True
118. True
119. True
120. True
121. True
122. True
123. True
124. True
125. True
126. True
127. True
128. True
129. True
130. True
131. True
132. True
133. True
134. True
135. True
136. True
137. True
138. True
139. True
140. True
141. True
142. True
143. True
144. True
145. True
146. True
147. True
148. True
149. True
150. True
151. True
152. True
153. True
154. True
155. True
156. True
157. True
158. True
159. True
160. True
161. True
162. True
163. True
164. True
165. True
166. True
167. True
168. True
169. True
170. True
171. True
172. True
173. True
174. True
175. True

Section-B : Fill in the Blanks

1. Correlated
2. Positive or Direct
3. Positive
4. Negative or Inverse
5. Negative
6. Same, proportionate
7. Non sense
8. Spurious
9. Bivariate
10. High
11. Poor or Low
12. Correlation
13. Linear-relationship
14. Magnitude
15. COV(X, Y), V(Y)
16. $\frac{1}{n}\sum^{n}(xi - \bar{x}_n)(yi - \bar{y}_n)$
17. $x_i\ y_i, \bar{y}_n$
18. Zero
19. Less than
20. Greater than
21. - 1.0
22. 1.0
23. Y, X
24. Product
25. {Y-E(Y)} {X-E(X)}
26. (-1, 1)
27. Perfect, Positive
28. Perfect
29. Independent, Scale
30. rxy
31. Uncorrelated
32. Zero
33. Linear
34. Zero
35. Independent
36. Zero
37. $\sqrt{\frac{1-r^2}{n-2}}$
38. $(1 - r^2)$
39. Significant
40. Significant
41. 1 + PE (r)
42. Rank
43. $(n^2 - 1)$
44. (-1, 1)
45. Average
46. Relationship
47. Independent, covariate
48. Study, Response
49. Dependent, Independent
50. Least square
51. Two
52. $byx\ (X - \bar{x}_n)$
53. Regression coefficient, Y, X
54. Dependent, Independent
55. Dependent
56. Gradient, Y, X
57. V(X)
58. $\sigma y/\sigma x$
59. Same
60. $<, \leq$
61. r^2_{xy}
62. $(\bar{X}, \bar{Y})$

63. $(\bar{x}, y)$
64. $\frac{\sigma_y}{\sigma_x}$
65. $(xi - \bar{x}) / \sigma x$
66. One
67. byx . bxy
68. $\geq$
69. Affects
70. yx, xy
71. $\sigma_x \sigma_y / (\sigma_1^2 + \sigma_2^2)$
72. $\infty, \frac{\pi}{2}$
73. Perpendicular
74. Concides
75. $\sigma_x \sigma_y / (\sigma_x^2 + \sigma_2^2)$
76. π
77. $\overline{Y}, \overline{X}$
78. Parallel
79. Smaller
80. $(1 - r^2)$
81. Linearity
82. (X, Y)
83. Regression
84. Does not mean
85. (σ_y / σ_x)
86. Curvilinear
87. Curve
88. $>$
89. σ_y^2
90. σ_y^2
91. $(1 - \eta_{yx}^2)$
92. (-1, 1)
93. Independent
94. Unity
95. σ_y^2
96. Independent
97. $\neq$
98. [Y - E(Y)]
99. [X - E(X)1
100. Zero
101. 1.0
102. 1.0
103. 0.0
103* Curve
104. Within
105. Same
106. σ_m^2
107. 1.0
108. Linear
109. $(1 - \rho^2)$
110. Normal
111. $2(1 - \rho^2)$
112. $(\mu_2, \sigma_1^2, \rho)$
113. Normal, correlation
114. $(\sigma_1^2, \mu_2, \rho)$
115. (1, 1)
116. $2\rho\, t_1\, t_2$
117. 0

118. Normal
119. (σ^2_y)
120. $(1 - \rho^2)$
121. Independent
122. Partial, Regression
123. (X_2, X_3)
124. X_1 on X_2, X_3
125. Estimate of X_1
126. Error, Residual
127. Primary
128. Secondary
129. Order
130. n - 2
131. Regression coefficient
132. Regression coefficient
133. meaningless
134. Dependent, Independent
135. n -1
136. Zero
137. Estimated, X_2, X_3
138. $V(X_1). V(e_{1.23})$
139. r^2_{13}, r^2_{23}
140. Observed, Estimated
141. More
142. (0, 1)
143. Zero
144. Uncorrelated
145. Less
146. Partial correlation
147. Partial regression
148. Linear effect of X_3
149. $(X_{1.3}, X_{2.3})$
150. $r_{13}\, r_{23}$
151. $(1 - r^2_{23})$
152. $r_{13}\, r_{23}$
153. $X^2_{2.3}$
154. $X_{2.3}$
155. $b_{12.3}\, b_{21.3}$
156. $r_{13}\, r_{23}$
157. X_1, X_2
158. Explanatory
159. X_2, X_3
160. X_2, X_3
162. (k + 1)
163. R.A. Fisher
164. (k, n - (k + 1))
165. n - (k + 2)
166. R.A. Fisher
167. n - (k + 2)
168. Student's t
169. Total variance present in Y
170. Coefficient of non-determination
171. 64%, Dependent Independent
172. Robut test
173. Student's t statistic
174. $(n - 2), (1 - r^2)$
175. (n - 2)
176. $b^2_{yx} \sum^{n} (x_i - \bar{x}_n)$
177. Zero
178. (n - 2)
179. $(x_i - \bar{x}_n), (y_i - \bar{y}_n)$
180. $(x_i - \bar{x}_n)$
181. $(x_i - \bar{x}_n)(y_i - \bar{y}_n)$
182. $(x_i - \bar{x}_n)$

183. (n - 2)

184. $\sum^{n}(y_i - \bar{y}_n)^2, (x_i - \bar{x}_n)$

185. $\sum^{n}(y_i - \bar{y}_n)^2, (x_i - \bar{x}_n)$

186. (n - 2)

Section-C : Multiple Choice Questions

1. (b)	2. (d)	3. (d)
4. (c)	5. (d)	6. (d)
7. (d)	8. (a)	9. (b)
10. (a)	11. (d)	12. (b)
13. (a)	14. (d)	15. (d)
16. (a)	17. (d)	18. (a)
19. (a)	20. (a)	21. (c)
22. (a)	23. (a)	24. (d)
25. (a)	26. (d)	27. (c)
28. (c)	29. (d)	30. (d)
31. (d)	32. (d)	33. (d)
34. (d)	35. (a)	36. (a)
37. (d)	38. (a)	39. (a)
40. (b)	41. (c)	42. (d)
43. (a)	44. (d)	45. (d)
46. (d)	47. (a)	48. (b)
49. (d)	50. (a)	51. (a)
52. (a)	53. (d)	54. (d)
55. (a)	56. (b)	57. (c)
58. (a)	59. (a)	60. (a)
61. (a)	62. (c)	63. (c)
64. (a)	65. (d)	66. (d)
67. (a)	68. (a)	69. (c)
70. (a)	71. (a)	72. (a)
73. (a)	74. (a)	75. (d)
76. (c)	77. (d)	78. (b)
79. (a)	80. (a)	81. (d)

82. (d)	83. (a)	84. (a)
85. (d)	86. (a)	87. (a)
88. (a)	89. (a)	90. (a)
91. (a)	92. (a)	93. (a)
94. (a)	95. (b)	96. (d)
97. (a)	98. (d)	99. (a)
100. (a)	101. (b)	102. (d)
103. (b)	104. (a)	105. (b)
106. (a)	107. (a)	108. (a)
109. (a)	110. (b)	111. (b)
112. (a)	113. (c)	114. (c)
115. (b)	116. (d)	117. (a)
118. (d)	119. (b)	120. (b)
121. (b)	122. (d)	123. (a)
124. (b)	125. (a)	126. (a)
127. (a)	128. (b)	129. (a)
130. (a)	131. (c)	132. (a)
133. (d)	134. (d)	135. (a)
136. (d)	137. (a)	138. (b)
139. (c)	140. (a)	141. (b)
142. (a)	143. (d)	144. (d)
145. (b)	146. (a)	147. (b)
148. (c)	149. (c)	150. (a)
151. (d)	152. (c)	153. (c)
154. (d)	155. (a)	156. (c)
157. (a)	158. (d)	159. (a)
160. (a)		

❑❑❑

Bibliography

Gupta, H.C. and Kapoor, V.K. (1980), Fundamental of Mathematical Statistics, 11th edition, S. Chand & Sons, New Delhi.

Kushwaha, K.S. and Kumar, Rajesh (2009) : The Theory of Sample Surveys and Statistical Decisions, New India Publishing Agency, New Delhi.

Kushwaha, K.S. and Kumar, Rajesh (2009) : Basic Concepts in Statistics, New India Publishing Agency, New Delhi.

Kushwaha, K.S. (2011) : Objective Agriculture Statistics, New India Publishing Agency, New Delhi

Mood, A.M., Graybill, F.A. and Bose, D.D. (1974) : Introduction to the Theory of Statistics, McGraw Hill Koga Kusha Ltd., London.

Murry, R. Spiegel (1961) : Schaum Outline of Theory and Problems of Statistics, Schaum Publishing Company, New York.

Panse, V.G. and Sukhatme, P.V. (1978) : Statistical Methods for Agricultural Workers, I.C.A.R., New Delhi.

Vijay, K. Rohatgi, A.K. Md Ehsanes Saleh (2009) : An Introduction to Probability and Statistics, 2nd edition, Wiley Student Edition, Wiley India Pvt. Ltd. N. Delhi.

About the Author

Dr. K.S. Kushwaha : He was an intelligent student and secured first rank in the "District Board Exam" (8t 1, Class) in 1969 and achieved 5th rank on the basis of state level merit of U.P. State prepared during 1969. He was awarded a fellowship Rs. 100/- month during High School and Rs. 150/- month during Intermediate classes of his study. He was asked by the Honourable Governor, Dr. M. Chennareddy to take admission in Govt.'s Queen's Intermediate College, Varanasi. He passed High School Exams with honour and obtained B.Sc. degree (1975) And M.Sc. (Statistics) in 1977 from Science College, B.H.U., Varanasi in first division.

Dr. Kushwaha completed one year "Professional Statistician Certificate Cource" (P.S.C.C.) from Indian. Agricultural Statistics Research Institute, New Delhi in 1979. He worked as "Research Fellow" for six months with fellowship Rs. 1000/- month in "Ford Foundation" New Delhi and completed his Ph. D. degree in 1991 from "Indian School of Mines, Dhanbad" Bihar in "Sampling Theory" with 12 research papers published in various journals of national and international repute. For his continuous works he was awarded with presitigious "Dr. Radhakrishnan Award" on 4th Sept., 1992 by M.P. State U.G.C. and was declared as "The Man of Physical Sciences".

His papers have been referred in various text and reference books published by national and international publishers. He beard the responsibility of chairmanship in "Mathematics & Statistics Session" set up to evaluate the oral presentation of young scientists up to 28 years of age in scientific conference sponsored by MAPCOST, Bhopal and hosted by J.N.K.V.V., during Feb. 27-28, 1993 to give them "Young Scientist Award" at State level. His lecture in workshop on "Frontiers of Research in Finite Population Inferences" during (Nov. 1-9, 1993) financed by Govt. of India secured him second rank.

Dr. Ktishwaha is an appointed member of (i) Executive Board and (ii) Editorial Board of "Bulletin Sample Survey Research Group, India", "Academic Council of R.D. University, Jabalpur," to guide Ph.D. Thesis. He has published 23 research papers in the journals of National & International repute. Till now two classical books and one Objective Agricultural Statistics have been published recently by the author.

At present, he is working as a Senior Associate Professor (Statistics) in the Department of Mathematics and Statistics, Jawaharlal Nehru Krishi Vishwavidyalaya, Jabalpur, M.P., India.